Monographs in Mathematical Economics

Volume 1

More information about this series at http://www.springer.com/series/13278

Takako Fujiwara-Greve

Non-Cooperative Game Theory

Takako Fujiwara-Greve
Department of Economics
Keio University
Minato-ku, Tokyo
Japan

ISSN 2364-8279 ISSN 2364-8287 (electronic)
Monographs in Mathematical Economics
ISBN 978-4-431-56415-7 ISBN 978-4-431-55645-9 (eBook)
DOI 10.1007/978-4-431-55645-9

Springer Tokyo Heidelberg New York Dordrecht London

Softcover reprint of the hardcover 1st edition 2015

Printed on acid-free paper

Springer Japan KK is part of Springer Science+Business Media
(www.springer.com)

To Henrich, Jan, and Ryo

Preface

This book is based on my Japanese book *Hikyouryoku Game Riron* (Non-Cooperative Game Theory), published in 2011 by Chisen-Shokan, with some modifications for international readers. Both books grew out of my lecture notes used for teaching at Keio University, Waseda University, and the Norwegian Business School (BI) over more than a decade. Like my lectures, this book covers topics from basics to graduate-level ones. I also included many exercises that my students did in the class.

I constructed the chapters according to solution concepts, which are ways to predict outcomes, or, in other words, are theories on their own. Game theory is a collection of such theories, and users of game theory should choose an appropriate solution concept for each situation to make a good prediction. However, there is no complete agreement of what solution concept **should** be used for particular games, even among game theorists. For example, I advocate using *sequential equilibrium* for extensive form games with imperfect information, but some people might use *perfect Bayesian equilibrium*. Therefore, in this book, I do not make a correspondence between a class of games and a solution concept. Rather, I line up solution concepts and let the readers decide which one to apply when they face a game to analyze.

Although I made every effort to avoid incorrect or misleading expressions, no book is free of errors. Even my own opinion may change over time. Therefore, I set up a website for corrections and clarifications at http://web.econ.keio.ac.jp/staff/takakofg/gamebook.html.
(Please note that, in the long run, the URL may change. The life of a book is usually longer than the life of a web page.) I also regret that I could not cover some important topics in non-cooperative game theory, such as epistemic game theory and learning models.

For beginners, I recommend reading chapters and sections without a star and doing some exercises. Juniors and seniors at universities can read chapters and sections with a single star. For those who want to study game theory at a graduate level, chapters and sections with two stars are useful, and after that, readers should go on to the research papers in the references.

I thank Toru Maruyama for coordinating with Springer Japan to publish this book. Two anonymous referees provided valuable comments to improve the quality of the book. Many Japanese people who helped me in writing the Japanese version of this book are already acknowledged there, so here I list mainly non-Japanese individuals who helped me in my study and career. But first of all, I must thank Masahiro Okuno-Fujiwara, who wrote a recommendation letter that helped me get into the Stanford Graduate School of Business (GSB) Ph.D. program. Without his help, my whole career would not have developed like this. Now he is one of my most important co-authors as well. My teachers at Stanford GSB, in particular, David Kreps, John Roberts, Faruk Gul, and Robert Wilson, were helpful at many points of my study there. After GSB days, I am grateful to Ehud Kalai, Geir Asheim, Carsten Krabbe Nielsen, Christian Riis, Joel Watson, Nicholas Yannelis, Marcus Berliant, Michihiro Kandori, Akihiko Matsui, and Masaki Aoyagi for their support. Daisuke Oyama, who used the Japanese version of the book in his course, gave me comments that led to improvements of this book. I also thank Masayuki Yao, Hayato Shimura, and other students in my recent courses at Keio University who checked the draft carefully. Allison Koriyama proofread the entire manuscript and corrected the language thoroughly. All remaining errors are mine. Finally, but of course not the least, I thank my husband and a co-author, Henrich R. Greve for his constant support from the first term of the Ph.D. course at GSB.

Tokyo
March 2015

Takako Fujiwara-Greve

Contents

The star marks indicate difficulty: no star means introductory level, * means intermediate level, and ** means advanced level. Chapters with varying difficulties have sections with different number of stars. Otherwise all sections are of the same level as the chapter.

Chapter 1
Games in Game Theory

1.1 Games in Game Theory

There are many *games* in our lives: video games like Tetris, sports games, such as baseball and volleyball, and board games such as chess, shogi, and go, to name a few. Although these are all called "games", they have different characteristics. Video games, unless connected to other machines, are played by one person against a computer. There are no interactions between multiple decision-makers, thus they are more like puzzle solving. In sports games, multiple decision-makers interact during the game. However, the outcomes of sports games may not be completely determined by players' decisions. For example, even if a baseball pitcher decided to pitch in the strike zone, the actual throw may not become a strike. There may be many causes for the discrepancy between the decision and the realized action, but a particularly important one is a technical error. That is, the outcomes of sports games are often fundamentally affected by physical skills, and not just by decisions. By contrast, the outcomes of board games are solely determined by players' decisions.

Game theory aims "to show that there is a rigorous approach to" social sciences (Preface to First Edition of von Neumann and Morgenstern [4]). Therefore, non-social situations such as one person facing a computer do not fall within its domain of interest. Sports can be analyzed only partially for strategic decisions, and we must make reserved predictions about the final outcomes because they are heavily affected by players' skills and other environmental factors that cannot be taken into account beforehand. Board-game-like situations in which multiple decision-makers interact and their decisions determine social outcomes are the best fit for rigorous approach to social sciences. Note, however, that there are many such situations, from rock-paper-scissors to international negotiations. Game theory can be applied to a wide variety of social decision problems.

T. Fujiwara-Greve, *Non-Cooperative Game Theory*, Monographs in Mathematical Economics 1, DOI 10.1007/978-4-431-55645-9_1

1.2 Non-cooperative and Cooperative Games

Game theory distinguishes between situations in which a decision-maker acts independently from all other decision-makers and those in which multiple decision-makers can act as a group. This is because there is a significant difference in the resulting decisions and actions. *Non-cooperative game theory*, which this book is about, analyzes only the former case, in which decision-makers cannot make a "binding agreement" to enforce some action on one another. The latter case is dealt with by *cooperative game theory*. The distinction follows Nash's [3] statement: "This (cooperative game) theory is based on an analysis of the interrelationships of the various coalitions which can be formed by the players of the game. Our (non-cooperative game) theory, in contradistinction, is based on the **absence** of coalitions in that it is assumed that each participant acts independently, without collaboration or communication with any of the others."

It is, however, fruitless to debate which assumption is correct. Whether we can assume that decision-makers can make a binding agreement or not depends on the situation to be analyzed. The book by von Neumann and Morgenstern [4], which laid the foundation of game theory, develops extensive analyses of both non-cooperative and cooperative situations. Nonetheless, in microeconomics where the applications and extensions of game theory are most prevalent, it is plausible that the decision-makers such as firms and consumers cannot make binding agreements.

The aim of non-cooperative game theory is to find out **what happens in a society** (or, a strategy combination) when individuals make independent and strategic decisions, while cooperative game theory aims to determine **what people get** (or, a payoff imputation) when various coalitions are tried out to improve participants' collective welfare. Thus, one must choose which theory to employ depending on one's interest and the relevant social problem.

This book deals with non-cooperative games only. We refer readers to, for example, Aumann [1], Osborne and Rubinstein [5], and Owen [6] for readings of cooperative game theory.

1.3 Components of a Game

A *game* is completely determined by four components: decision-makers, options that they can choose, their objectives in choosing an option, and what they know about the structure of the interaction. *Game theory* is a general theory that tries to provide a logical method to predict outcomes for any game specified by these components.

Decision-makers in game theory are called *players*. They can be humans, organizations such as firms, or even animals who can be interpreted as acting agents.

In the basic framework of game theory, decisions and actions of a player are assumed to be the same. For example, consider the rock-paper-scissors game, where (usually) two people show a hand each simultaneously, with one of the shapes of rock (a balled-up fist), paper (an open hand), or scissors (two fingers out). In this game, "choosing the action rock" means that the person in fact shows a rock with her or his hand. In an extended setting, game theory can also analyze situations where there is a discrepancy between a decision to do something and the actual realization of the intended action, as long as there is a well-defined (e.g., statistical or probabilistic) relationship between the intention and possible realizations. For example, a firm may decide to develop a new product (and in fact starts an R&D activity), but it may fail to develop the product. Even in such cases, if it is possible to relate the R&D activity to an objective probability distribution of events such as success and failure, then we can introduce a player called *Nature* and assume that Nature chooses possible events with the relevant probabilities.

This *Nature* is a player who controls things that actual players in the game cannot control. The introduction of such a player to deal with uncontrollables is an ingenious idea. Thanks to this theoretical innovation, game theory can analyze many uncertain situations mathematically and logically.

After specifying the set of players, we need to determine the set of options that each player can choose from. There are two ways to think about options, depending on whether one can list all options before the game starts or not. For example, in a rock-paper-scissors game, each player can choose one hand out of the three, and thus we should formulate that the three hands are the sole options. By contrast, in games such as chess and go, players must decide on actions at many points of time and depending on the circumstances. For such games, options may be formulated in two ways: actions at each stage of the game and "action plans" to be used throughout the game.

Game theory thus distinguishes games with different number of rounds of decision-making. Games with essentially a single (per player) and simultaneous occasion of decision-making, e.g., rock-paper-scissors, are called *normal-form games* (*simultaneous move games*, *strategic-form games*). Simultaneity need not be literal, as long as the decision-makers do not know the choices of the others, when they choose their own decisions. In these games, the only options are the ways of playing the entire game, called *strategies*.

Games with multiple stages of decision-making, e.g., chess and go, are called *extensive-form games*. For extensive-form games, we need to describe step by step who decides what at which stage of the game. A *strategy* of an extensive-form game is an **action plan** throughout the game, contingent on the history of the game.[1] A decision at a given point in an extensive-form game is called an *action*, to be distinguished from a strategy.

Next, let us turn to player objectives. Each player evaluates all possible endings of a game (called *outcomes*) and assigns real numbers (called *payoffs*) to them, with the interpretation that the larger the number, the better. For a normal-form game, an outcome is determined when all players each choose a strategy. Similarly, for an extensive-form game, an outcome is determined when all players choose an action at

[1] This usage of the word *strategy* is consistent with military terminology. See, for example, Book 2, Chap. 1 of Clausewitz [2].

all of their decision points, which is the same as all players each choosing a strategy. Therefore the payoff of a player depends not only on her/his own strategy choice but also on all other players' strategy choices. The relationship is described as a *payoff function* from the set of all possible outcomes (which is essentially the set of all possible strategies of all players) to the set of real numbers.

A player's objective is to choose a strategy that gives as high a payoff as possible, which is sometimes called the *rationality* hypothesis in game theory. Note that the payoff numbers need not be the same as some observable numbers, such as the player's income, consumption amount, a firm's sales and so on. The payoff numbers are mostly ordinal measures (cf. the notion of the expected payoff in Sect. 3.6) to indicate a player's evaluation of outcomes. It is perfectly fine to consider a payoff function that takes into account others' welfare. In this sense, a payoff function in game theory is not a utility function in economic theory.

Because a player alone cannot determine an outcome of a game, nor own payoff value, an individual player is not solving a simple optimization problem. Economic theory, from Adam Smith to Arrow-Debreu General Equilibrium Theory, formulated each economic agent's objective as a single decision-maker's maximization problem, even in a social setting. By contrast, game theory explicitly includes the effect of others' decisions/actions into one's objective function. Therefore, it is not a set of single-person optimization problems but a **social optimization problem**, or, a convoluted problem of conflicting optimizations. It is a mathematical challenge worth the attention of the great mathematician von Neumann.

Lastly, the informational structure that specifies what players know and what they do not is an important (and in fact a mathematically complex) part of a game. When all players know the set of players, the set of strategies of each player, and the payoff functions of all players **very well**, the game is said to have *complete information*. Otherwise, the game has *incomplete information*. (The precise definition of "knowing very well" will be explained in Sect. 2.3.) Information crucially affects each player's decision making, and, if the game has incomplete information, we have a big problem of formulating "a game that players do not know well". This is not easy. In this book we start with complete information games. There, players know the game structure as well as we (analysts) do, so that our analysis can be completed safely.

For a rigorous analysis of extensive-form games, we need a more detailed classification of games in terms of their informational structure. Whether past players' decisions are known to later player(s) affects the latter's decision making. *Perfect information* games are those in which all players know all players' past decisions at any stage of the game. Otherwise, the game has *imperfect information*. For example, chess is a perfect information game, for players with good enough memory.

The informational structure where a player remembers her or his own decisions and things that (s)he knew is called *perfect recall*. Without perfect recall, a player may forget what (s)he did/knew and may not make a rational decision. Thus, this assumption is important. However, a game with perfect recall is not necessarily a game with perfect information. It is fine not to know some of the other players' decisions, but information does not decrease over time in a game with perfect recall.

References

1. Aumann R (1989) Lectures on game theory. Westview Press, Boulder
2. von Clausewitz C (1989) On War, Reprint edn. Princeton University Press, Princeton
3. Nash J (1951) Non-cooperative games. Ann Math 54:286–295
4. von Neumann J, Morgenstern O (1944) Theory of games and economic behavior. Princeton University Press, Princeton
5. Osborne M, Rubinstein A (1994) A course in game theory. MIT Press, Cambridge
6. Owen G (2013) Game theory, 4th edn. Emerald Group Publishing, Bingley

Chapter 2
Strategic Dominance

2.1 Prisoner's Dilemma

Let us start with perhaps the most famous example in Game Theory, the *Prisoner's Dilemma*.[1] This is a two-player normal-form (simultaneous move) game. Two suspects, A (a woman) and B (a man), are taken into custody. The district attorney is sure that they have committed a crime together but does not have enough evidence. They are interrogated in separate rooms and cannot communicate. Both A and B have the same options: defection from the accomplice to reveal the evidence or cooperation with the other to keep silent. What will they do?

To predict a strategic outcome of this situation, we need to understand the suspects' objectives. The suspects should want to minimize possible punishment or penalties. The DA tells the suspects separately as follows."If you defect from your accomplice and tell me where the evidence is, I will release you right away. But if the other suspect defects first, I will prosecute only you and with the evidence you will be sentenced for a long time."

A careful suspect might ask, "what happens if both of us defect?" In this case the plea bargain does not work, and both suspects get prosecuted, sharing the guilt. Yet another possibility is that no one defects, in which case we assume that the prosecutor still goes for the trial. But this trial is not as scary as the one with strong evidence after defection.

Let us formulate the payoffs for the suspects in terms of the lengths of sentence of the four possible outcomes. If player A defects and player B does not, A gets 0 years of sentence with the plea bargain and B gets 5 years. Similarly, if player B defects and player A does not, then A gets 5 years and B gets 0. Since a shorter sentence is better, we should assign a smaller number to 5 years of sentence than 0. As an example, let us define the payoff as -1 times the term of sentence. Then no prosecution means a payoff value of 0, and a 5 year sentence is a payoff of -5.

[1] Alternatively, it is written as "Prisoners' Dilemma". This book follows Luce and Raiffa [4]. Also notice that the players are not yet prisoners.

T. Fujiwara-Greve, *Non-Cooperative Game Theory*, Monographs in Mathematical Economics 1, DOI 10.1007/978-4-431-55645-9_2

Table 2.1 Player A's payoff matrix

A's strategy\B's strategy	Defect	Cooperate
Defect	−3	0
Cooperate	−5	−1

If both suspects defect, no plea bargain holds, and the prosecutor gets strong evidence of the crime. But the suspects share the guilt, so let's assume that both get a 3 year sentence, and the payoff is −3 each. If both cooperate with each other and keep silent, although they still will be prosecuted, the evidence is weak. Assume that, in this case, they both get a 1 year sentence, and a payoff of −1.

Now the game is specified. If you were player A (or B), what would you do to maximize your payoff (or minimize the term of sentence)?

2.2 Strict Dominance

Although the game is completely specified in the previous section by words, it is convenient to introduce a mathematical formulation for later generalizations. Denote the set of strategies of a player by adding a subscript for the name of the player to the letter S (which stands for Strategy). That is, player A's set of strategies is denoted as $S_A = \{C, D\}$ (where C stands for cooperation and D stands for defection), and player B's set of strategies is denoted as $S_B = \{C, D\}$ as well. Player A's payoff (or term of sentence) cannot be determined by her strategy alone. Rather, it is determined by a combination of both players' strategies. The set of all combinations of both players' strategies is the Cartesian product $S_A \times S_B$ of the two strategy sets.[2] Hence, player A's payoff is computed by a *payoff function* $u_A : S_A \times S_B \to \Re$. To be concrete, for the current Prisoner's Dilemma game, u_A is specified as follows:

$$u_A(D, D) = -3, u_A(C, D) = -5, u_A(D, C) = 0, u_A(C, C) = -1. \quad (2.1)$$

(Note that the first coordinate in the parentheses is player A's strategy, and the second coordinate is player B's.)

Let us consider player A's payoff maximization. Table 2.1 is a *matrix representation* of the payoff function (2.1), where the rows are her strategies and the columns are player B's strategies. It is easy to see from this table that, if B defects, it is better for A to also defect, because defection gives her a 3 year sentence while cooperation (with B) gives her 5 years. If player B cooperates with player A, still it is better for A to defect, because defection leads to release, but cooperation means a 1 year sentence.

---

[2]If there is a natural order among players in the description of the game, the product is usually taken that way. In this case the alphabetical order of the names suggests that A is the first coordinate. One can equivalently formulate the product and the game with player B as the first player.

In summary, strategy D gives a strictly greater payoff than that of strategy C, regardless of the opponent's strategy choice. Hence, the payoff-maximizing player A would not choose strategy C.

This logic can be generalized. Consider an arbitrary n-player normal-form game. Let $\{1, 2, \ldots, n\}$ be the set of players with the generic element i, S_i be the set of strategies of player i, and $u_i : S_1 \times S_2 \times \cdots \times S_n \to \Re$ be the payoff function of player i. For convenience we also introduce the following notation:

$$S := S_1 \times S_2 \times \cdots \times S_n,$$
$$S_{-i} := S_1 \times \cdots \times S_{i-1} \times S_{i+1} \times \cdots \times S_n.$$

S is the set of all *strategy combinations* (*strategy profiles*) of all players. S_{-i} is the set of strategy combinations of players other than i.

Definition 2.1 Given a player i, a strategy $s_i \in S_i$ is *strictly dominated*[3] by another strategy $s_i' \in S_i$ if, for any $s_{-i} \in S_{-i}$,

$$u_i(s_i, s_{-i}) < u_i(s_i', s_{-i}).$$

That is, for any possible strategy combination s_{-i} by the other players, s_i gives strictly less payoff value than s_i' does. Hence, at least s_i' is better than s_i in terms of payoff maximization. In general, we postulate that **a payoff-maximizing (rational) player would not use a strategy that is strictly dominated by some other strategy**.

For the above Prisoner's Dilemma game, player A would not use strategy C, and hence she must choose strategy D. This indirect way of predicting the choice of strategy D is important. In general, a player may have more than two strategies, and all we can say right now is that choosing a strategy that is strictly dominated by some other strategy is in contradiction with payoff maximization. With more than two strategies, it is often insufficient to "predict" that a player would choose a strategy that is not strictly dominated.[4]

Next, we formulate the payoff function of player B. Recall that the first coordinate of a strategy combination is player A's strategy. Hence, we should write:

$$u_B(D, D) = -3, u_B(C, D) = 0, u_B(D, C) = -5, u_B(C, C) = -1. \quad (2.2)$$

[3] For the extended definition of strict dominance under "mixed strategies", see Sect. 3.6.

[4] However, if there is a strategy that strictly dominates all other strategies (of the relevant player), then this strategy (called the *dominant strategy*) should be chosen by a rational player. In the Prisoner's Dilemma, the strategy that is not strictly dominated by some other strategy coincides with the dominant strategy, and thus it is easy to predict an outcome by the dominant strategy. In general, there are few games with dominant strategies. Hence we do not emphasize the prediction that a player chooses the dominant strategy. Note also that, if there is a strategy combination such that all players are using a dominant strategy, the combination is called a *dominant-strategy equilibrium*.

We can see that this payoff function is symmetric[5] (we can permutate the players to obtain the same function) to player A's payoff function (2.1). Therefore, by the same logic for player A, strategy C is strictly dominated by strategy D for player B, and we predict that B does not choose strategy C.

To summarize, our forecast for the Prisoner's Dilemma game is that both players defect, and the social outcome would be the strategy combination (D, D).

What is the dilemma here? In order to choose a strategy, players can use simple logic. They do not even need to know the other's payoff function. But the dilemma occurs when they realize the resulting payoff combination. With the rational outcome of (D, D), both players get 3 year sentences, but if they choose the outcome (C, C), then they could have reduced the sentence to 1 year each! That is, the strategy combination (D, D) is socially not desirable. This social valuation can be mathematically defined as follows.

Definition 2.2 A strategy combination $(s_1, s_2, \ldots, s_n) \in S$ is *efficient*[6] if there is no strategy combination $(s'_1, s'_2, \ldots, s'_n) \in S$ such that
(1) for any player $i \in \{1, 2, \ldots, n\}$, $u_i(s_1, s_2, \ldots, s_n) \leqq u_i(s'_1, s'_2, \ldots, s'_n)$, and
(2) for some player $j \in \{1, 2, \ldots, n\}$, $u_j(s_1, s_2, \ldots, s_n) < u_j(s'_1, s'_2, \ldots, s'_n)$.

Using this definition, the rational players' predicted strategy choice (D, D) in the Prisoner's Dilemma is not efficient, because there exists (C, C) which satisfies (1) and (2).

This is one of the most fundamental problems in game theory: **the outcome that results from payoff-maximizing players' decisions is not always socially efficient**. However, it is in a way plausible, because non-cooperative game theory assumes that each player independently chooses decisions and does not consider the effects on others. Making self-interested players act in consideration of others is a completely different problem from finding a strategic outcome of arbitrary games. Game theory's main target is the latter, and the former should be formulated as a problem of designing games or finding a class of games with predicted outcomes (*equilibria*) that possess particular characteristics. For most of this book, we describe the main problem of the game theory, to provide a coherent prediction to arbitrary games. The Prisoner's Dilemma is "solved" by constructing cooperative equilibria for the infinitely repeated version of it (Chap. 5), or by adding incomplete information to the finitely repeated version (Chap. 8).

[5] Its matrix representation is symmetrical to the one in Table 2.1.

[6] Strictly speaking, this definition is *Strong Efficiency*. There is also *Weak Efficiency*, which requires that there is no $(s'_1, s'_2, \ldots, s'_n) \in S$ such that $u_i(s_1, s_2, \ldots, s_n) < u_i(s'_1, s'_2, \ldots, s'_n)$ for all $i \in \{1, 2, \ldots, n\}$.

Table 2.2 Player A's new payoff matrix

A's strategy\B's strategy	Defect	Cooperate
Defect	−3	−1
Cooperate	−5	0

2.3 Common Knowledge of a Game

So far, each player has only needed to know her/his own payoff function in order to make a rational decision. However, such method does not work with even a small change in the game structure. For example, let us modify the above Prisoner's Dilemma a little. Keeping player B's payoff function intact, assume that player A wants to defect if B defects, but she wants to cooperate with him if he does so. That is, we switch the last two payoff values in (2.1):

$$u_A(D, D) = -3, u_A(C, D) = -5, u_A(D, C) = -1, u_A(C, C) = 0. \tag{2.3}$$

The new matrix representation is Table 2.2.

Now, no strategy of player A is strictly dominated by some other strategy. What should player A do? Suppose that A knows the structure of the game, specifically, that A knows B's payoff function (2.2) and that he is rational. Then A can predict what B will do. By the logic in Sect. 2.2, player A can predict that B would choose strategy D. Then, if A defects, she would get −3, while if she cooperates, she would receive −5. Therefore it is rational for A to choose strategy D.

In this way, if players know all others' payoff functions and their rationality, it may be possible to "read" others' strategic decisions and find out their own rational choices. In this book, until the end of Chap. 5, we restrict our attention to games with complete information, which means that all players not only know the structure of the game (the set of players, the set of strategies of each player, and the payoff function of each player) and rationality of all players, but also they know that "all players know the structure of the game and rationality of all players", and they know that they know that "all players know the structure of the game and rationality of all players", and so on, *ad infinitum*. This infinitely deep knowledge is called *common knowledge* and is essential for the next analysis.[7]

2.4 Iterative Elimination of Strictly Dominated Strategies

The point of the analysis of the modified Prisoner's Dilemma in Sect. 2.3 was that, when the game and rationality are common knowledge, it may be possible to **eliminate** strictly dominated strategies of some player(s) from consideration and to solve for an outcome. Since a player takes into account the other player's decision-making,

[7] For a more detailed explanation of common knowledge, see, for example, Aumann [1], Chap. 5 of Osborne and Rubinstein [7], and Perea [8].

Table 2.3 Modified Prisoner's Dilemma

A's strategy\B's strategy	Defect	Cooperate
Defect	−3, −3	−1, −5
Cooperate	−5, 0	0, −1

let us combine the two players' payoff matrices. For example, the modified Prisoner's Dilemma can be expressed by the (double) matrix representation of Table 2.3, where the first coordinate of the payoff combinations is player A's payoff (based on Eq. (2.3)) and the second coordinate is B's (based on Eq. (2.2)).

Table 2.3 in fact completely specifies a game. The set of players is shown at the top left cell, and player A's strategies are represented by rows (hence she is the *row player*), while player B's strategies are represented by columns (hence he is the *column player*). Payoff functions are defined by two matrices of first coordinates (for A) and second coordinates (for B). This double matrix representation of a game allows us to encompass the problem of both players' optimization in the same table and is quite convenient. Note, though, that the row player A can only choose the top or bottom row, and the column player B can only choose the left or right column. This means that, if players try to move from (D, D) to (C, C), they cannot make that happen unilaterally.

Let us repeat the logic of Sect. 2.3 using Table 2.3. Assume that the structure of the game, i.e., Table 2.3, as well as the rationality of both players are common knowledge. Player A does not have a strictly dominated strategy, but player B's strategy C is strictly dominated by strategy D, and thus rational player B would not choose strategy C. Moreover, player A knows this. Therefore, A can consider a reduced Table 2.4, eliminating strategy C from B's choices.

Focusing on Table 2.4, player A can compare the first coordinates of the rows to conclude that strategy C is strictly dominated by strategy D. Hence, the outcome becomes (D, D).

The above example does not require common knowledge (that is, the infinitely many layers of knowledge) of the game and rationality. It suffices that player A knows the game and rationality of player B. For more complex games, we need deeper knowledge. Consider the game represented by Table 2.5, where the two players are called 1 and 2, player 1 has three strategies called x, y, and z, and player 2 has three strategies called X, Y, and Z. (This game, with three strategies for each of the two players, has 3×3 payoff matrix for each player and belongs to the class called 3×3 *games*. In this sense, the Prisoner's Dilemma is a 2×2 *game*.)

To predict a strategic outcome for the game represented by Table 2.5, consider player 1, who compares the first coordinates of payoff pairs across rows. There is no

Table 2.4 Reduced matrix representation

A's strategy\B's strategy	Defect
Defect	−3, −3
Cooperate	−5, 0

Table 2.5 A 3 × 3 game

1\2	X	Y	Z
x	3, 5	2, 2	2, 3
y	2, 2	0, 4	4, 1
z	1, 1	1, 2	1, 5

strict dominance relationship between strategies x and y, because if player 2 chooses strategy X or Y, x is better than y, but if player 2 chooses strategy Z, y is better than x. However, strategy z is strictly dominated by strategy x. Therefore, the rational player 1 would not choose strategy z. Similarly, player 2 chooses among the columns X, Y, and Z by comparing the second coordinates of the payoff values. There is no strict dominance among the three strategies of player 2 in Table 2.5. If player 1 chooses strategy x, then strategy X is better for player 2 than the other two strategies. If player 1 chooses y, then Y is the best, and if player 1 uses z, then strategy Z is the best for player 2.

So far, the players can reason using only the knowledge of their own payoff functions. Next, player 2, knowing the payoff function and rationality of player 1, can predict that player 1 would not use strategy z. Hence player 2 can focus on a reduced Table 2.6 after eliminating strategy z.

In the reduced Table 2.6, strategy Z is strictly dominated by strategy X for player 2. However, for player 1, strategies x and y are not strictly dominated by another.

By the common knowledge of the game, player 1 also knows the payoff function and rationality of player 2. In addition, player 1 knows that player 2 knows the payoff function and rationality of player 1. Therefore, player 1 can deduce that player 2 would eliminate the possibility of strategy z by player 1 and focus on Table 2.6. From that, player 1 can predict that player 2 would not choose strategy Z. This reasoning gives us a further reduced game, Table 2.7.

In Table 2.7, strategy y is strictly dominated by strategy x. Thus, we can predict strategy x as player 1's rational choice. This can be also predicted by player 2, who will choose strategy X. (Note that in this last step, we use the assumption that player 2 knows that player 1 knows that player 2 knows player 1's payoff function.) Therefore, the predicted outcome of the game represented by Table 2.5 is unique and is (x, X).

Table 2.6 Reduced 3 × 3 game

1\2	X	Y	Z
x	3, 5	2, 2	2, 3
y	2, 2	0, 4	4, 1

Table 2.7 Further reduced 3 × 3 game

1\2	X	Y
x	3, 5	2, 2
y	2, 2	0, 4

Table 2.8 Weak domination

1\2	L	R
U	11, 0	10, 0
D	10, 0	10, 0

This is an example where we can iteratively eliminate strictly dominated strategies (for all players) to arrive at a prediction of the strategic outcome of a game. That is, the iterative elimination of strictly dominated strategies is an equilibrium concept. Two remarks are in order. First, iterative elimination of an arbitrary size game is possible only under the common knowledge of the game and rationality. Second, in general, the remaining outcome after iterative elimination may not be unique. See, for example, the game of Table 2.8.

A convenient property of the iterative elimination process of strictly dominated strategies is that the order of elimination does not matter, i.e., when there are multiple strategies that are strictly dominated by some other strategy, the resulting outcome(s) of the elimination process do not change by the order of deletion among them. This is because, even if some strictly dominated strategies are not eliminated at one round, they will be still strictly dominated in the later rounds and therefore will be eliminated in the end.[8] (The elimination process stops when there are no more strictly dominated strategies for any player.)

2.5 Weak Dominance

Recall that the strict dominance relationship between strategies requires that the strict inequality among payoffs must hold, for any strategy combination by other players. To extend this idea, one may think it irrational to choose a strategy that never has a greater payoff than some other strategy and in at least one case has a strictly lower payoff. Consider the game represented by Table 2.8.

In this game, no player has a strictly dominated strategy. If player 2 chooses strategy R, strategy D gives player 1 the same payoff as strategy U does. For player 2, both strategies give the same payoff for any strategy by player 1.

However, choosing strategy D is not so rational for player 1, because strategy U guarantees the same payoff as that of strategy D for any strategy by player 2, and in the case where player 2 chooses strategy L, it gives a greater payoff. Let us formalize this idea.

Definition 2.3 Given a player i, a strategy $s_i \in S_i$ is *weakly dominated* by another strategy $s_i' \in S_i$ if,
(1) for any $s_{-i} \in S_{-i}$, $u_i(s_i, s_{-i}) \leqq u_i(s_i', s_{-i})$, and
(2) for some $s_{-i}' \in S_{-i}$, $u_i(s_i, s_{-i}') < u_i(s_i', s_{-i}')$.

---

[8]For a formal proof, see Ritzberger [9], Theorem 5.1.

Table 2.9 Weak domination 2

1\2	L	R
U	11, 0	10, 10
D	10, 0	10, 0

Let us consider the iterative elimination of **weakly** dominated strategies. This is a stronger equilibrium concept than the iterative elimination of strictly dominated strategies, because the process may eliminate more strategies. For example, in the game represented by Table 2.8, strict dominance does not eliminate any strategy (i.e., the "equilibria" by iterative elimination of strictly dominated strategies is the set of all strategy combinations), while weak dominance can eliminate strategy D for player 1. Nonetheless, player 2's strategies do not have a weak dominance relationship, hence the process stops there with the "equilibria" of (U, L) and (U, R). Thus the prediction by iterative elimination of weakly dominated strategies may not be unique either.

Moreover, a fundamental weakness of the iterative elimination of weakly dominated strategies is that, unlike with strict domination, the order of elimination affects the resulting outcomes. Consider a modified game of Table 2.8, represented by Table 2.9.

Suppose that we eliminate player 2's weakly dominated strategy L first. Then the reduced game does not have even a weakly dominated strategy. Therefore the "prediction" by this process is $\{(U, R), (D, R)\}$. However, if we eliminate player 1's weakly dominated strategy D first, then strategy L for player 2 is strictly dominated by R in the reduced game and can be eliminated. Hence under this order of elimination, the resulting outcome is (U, R) only. Therefore, the iterative elimination of weakly dominated strategy is not a well-defined equilibrium concept.

In addition, if we consider Nash equilibrium (Chap. 3) as the standard equilibrium concept, iterative elimination of weakly dominated strategies may delete a strategy that is a part of a Nash equilibrium (see Problem 3.9). Therefore, it is safer to apply iterative elimination for strictly dominated strategies only. Iterative elimination of strictly dominated strategies does not eliminate a strategy that is part of a Nash equilibrium (see Sect. 3.5). However, if there are many equilibria in a game, it is better to focus on the ones that do not involve weakly dominated strategies. (This is advocated by Luce and Raiffa [4] and Kohlberg and Mertens [3].)

2.6 Maximin Criterion

Most games do not have dominance among any player's strategies. Can players deduce some strategy combination by some reasoning, at least based on their objective of payoff-maximization? Let us focus on two-player, *zero-sum* games.

Definition 2.4 A two-player, *zero-sum game* is a normal-form game $G = (\{1, 2\}, S_1, S_2, u_1, u_2)$ such that the sum of the two players' payoff values is zero for any strategy combination. That is, for any $(s_1, s_2) \in S$,

$$u_1(s_1, s_2) = -u_2(s_1, s_2).$$

For example, games in which one player wins and the other loses can be formulated as zero-sum games.

In a two-player, zero-sum game, maximizing one's own payoff is equivalent to minimizing the opponent's payoff. Therefore, when the structure of the game and player rationality are common knowledge, a player can reason that, if her chosen strategy is to be a part of a stable outcome, then it must be that the opponent's strategy against it is the one that minimizes her payoff, given her strategy. We can examine the payoff matrices of the players to find out which strategy combination is consistent with this reasoning by both players. For the game represented by Table 2.10, player 1 can deduce that if she chooses strategy x, then player 2 would choose strategy Y, which gives her payoff of -4. This "payoff that you receive from a strategy when your opponent minimizes your payoff, given the strategy" is called the *reservation payoff*. In a two-player, zero-sum game, each strategy is associated with a reservation payoff, as Table 2.11 shows for the game of Table 2.10. Note that the row player's reservation payoffs are shown in the rows of the rightmost column, while the column player's are shown at the bottom of the columns.

One way to proceed with this reasoning is that, because player 2 will try to minimize player 1's payoff, player 1 must choose a strategy with the highest reservation payoff. That is, player 1 should solve the following maximization problem:

$$\max_{s_1 \in S_1} \min_{s_2 \in S_2} u_1(s_1, s_2).$$

Table 2.10 A zero-sum game

1\2	X	Y
x	3, −3	−4, 4
y	−2, 2	−1, 1
z	2, −2	1, −1

Table 2.11 Reservation payoffs

1\2	X	Y	1's reservation payoff
x	3, −3	−4, 4	−4
y	−2, 2	−1, 1	−2
z	2, −2	1, −1	1
2's reservation payoff	−3	−1	

Table 2.12 A non-zero-sum game

1\2	A	B	1's reservation payoff
a	2, −2	0, 0	0
b	1, −1	3, 3	1
2's reservation payoff	−2	0	

Strategically choosing one's action by solving the above problem is called *maximin criterion* (Luce and Raiffa [4]), and the above value is called the *maxmin value* of player 1.

Similarly, player 2 solves

$$\max_{s_2 \in S_2} \min_{s_1 \in S_1} u_2(s_1, s_2).$$

By the definition of a zero-sum game, it holds that $u_2(s_1, s_2) = -u_1(s_1, s_2)$. Hence the latter problem is equivalent to

$$\min_{s_2 \in S_2} \max_{s_1 \in S_1} u_1(s_1, s_2).$$

This is the value that player 2 wants player 1 to receive. The value of this minimization is called the *minmax value* of player 1.[9]

For the example of Table 2.11, the strategy that maximizes player 1's reservation payoff is strategy z, and the one for player 2 is strategy Y. Moreover, when the two players play this strategy combination, their maximal reservation payoffs realize in this game. Therefore, we can predict that the outcome of this game is (z, Y). (In general, however, we need to expand the set of strategies to *mixed strategies*, defined in Sect. 3.6, in order to warrant the consistency of maxmin value and minmax value for zero-sum games. This is the cornerstone result called the Minimax Theorem by von Neumann [5] and von Neumann and Morgenstern [6]. For general Minimax Theorems, see Fan [2])

It is, however, easy to see that the maximin criterion does not work in general games. Consider the game represented by Table 2.12. The strategy that maximizes player 1's reservation payoff is b, while the one for player 2 is B. However, for the strategy combination (b, B), their payoffs are not the maximal reservation payoffs. In other words, it is not consistent for the players to choose a strategy that maximizes their own payoff under the assumption that "the opponent minimizes my payoff".

Nonetheless, it is still a valid reasoning to maximize your payoff under the assumption that **your opponent maximizes her/his own payoff**. This idea leads to the notion of Nash equilibrium later. In two-player, zero-sum games, your opponent's minimization of your payoff and her maximization of her own payoff are equivalent, but in general games, they are not equivalent. The latter concerns rationality of the opponent.

[9]The definitions of maxmin value and minmax value in this section are within the "pure" strategies. A general definition of the minmax value is given in Sect. 5.7.

Table 2.13 A three-player game

1 \ 2	X	Y
x	3, 5, 4	0, 2, 1
y	2, 2, 5	2, 4, 3

3: L

1 \ 2	X	Y
x	3, 5, 2	2, 2, 0
y	2, 2, 0	0, 4, 1

R

2.7 Matrix Representation of 3-Player Games

So far, we have shown examples of two-player games only. In this section, we describe how three-player games can be represented by matrices. For example, suppose that there are three players, 1, 2, and 3, and player 1's set of strategies is $\{x, y\}$, player 2's is $\{X, Y\}$, and player 3's is $\{L, R\}$. We can represent this three-player game by two tables, where in each table, the rows are player 1's strategies and the columns are player 2's strategies. Player 3 is assumed to choose the matrices (or the tables) corresponding to L or R. In this case, player 3 is called the *matrix player*.

The players' payoffs are usually listed in order of the row player's, then the column player's, and lastly the matrix player's. To compare player 3's payoffs among strategy combinations, we must jump from one table to the other. For example, when player 1 chooses strategy x and player 2 chooses Y, player 3 must compare the third coordinate's value 1 at the top-right corner of Table L, which corresponds to the strategy combination (x, Y, L), and the third coordinate's value 0 at the top-right of Table R, which corresponds to the strategy combination (x, Y, R). By comparing this way, our reader should be able to conclude that in the game represented by Table 2.13 (consisting of two tables), strategy R is strictly dominated by strategy L for player 3. In this way, three-person games can be nicely represented by matrices.

Games with four or more players cannot be represented easily with matrices, and thus these games are usually specified by words and equations.

Problems

2.1 Construct matrix representations for the following games.

(a) There are two sales clerks, Ms. A and Mr. B. They have the same abilities, and their sales performance depends only on their effort. They both have the same feasible strategies, to make an effort (strategy E), or not to make an effort (strategy N). They choose one of these strategies simultaneously. Their payoffs are the following "points", which are based on their relative performances.

If both players pick the same strategy, then the sales figures are the same, and therefore each player gets 1 point. If one player makes an effort while the other does not, then the player who makes an effort has a higher sales figure, so that (s)he gets 3 points and the other gets 0 points.

(b) Consider the same strategic situation, except for the payoff structure. If a player makes an effort, (s)he incurs a cost of 2.5 points. The sales points are awarded in the same way as in (a).

(c) Suppose that there are two electric appliance shops, P1 and P2. Every morning, both shops put prices on their sale items. Today's main sale item is a laptop computer. The wholesale price is \$500 per computer, which a shop must pay to the wholeseller if it sells the computer. For simplicity, assume that each shop charges either \$580 or \$550, and there are only two potential buyers. If the two shops charge the same price, each shop sells one unit, while if one shop charges a lower price than the other, it sells two units and the other shop sells nothing. A shop's payoff is its profit, namely, the total sales (charged price times the number of units sold) minus the total cost (wholesale price times the number of units sold).

2.2 Find the outcomes for the iterative elimination of strictly dominated strategies of the games in Problem 2.1(a), (b), and (c).

2.3 Consider the following two-player normal-form game. Analyze how the set of strategy combinations ("equilibria") that survive iterative elimination of weakly dominated strategies may be affected by the order of elimination.

P1\P2	Left	Center	Right
Up	2, 3	2, 2	1, 2
Middle	2, 1	2, 2	2, 2
Down	3, 1	2, 2	2, 2

2.4 Let us define the *second-price, sealed-bid auction* of n bidders (players). There is a good to be auctioned. The strategy of a player $i \in \{1, 2, \ldots, n\}$ is a *bid*, which is a non-negative real number. Players submit one bid each, simultaneously, in sealed envelopes. The player who submits the highest bid wins the auction (gets to buy the good) at the price of the second-highest bid. If there are multiple bidders who submitted the highest bid, one of them is selected as a winner with equal probability.

The payoff of a player $i \in \{1, 2, \ldots, n\}$ is defined as follows. Let a non-negative real number v_i be the *valuation* of the good for player i, which is the benefit i gets if i obtains the object. Let $\mathbf{b} = (b_1, b_2, \ldots, b_n)$ be the bid combination of all players. If i wins the auction, her payoff is $v_i - f(\mathbf{b})$ (where $f(\mathbf{b})$ is the second-highest number among $\{b_1, b_2, \ldots, b_n\}$). Otherwise her payoff is 0.

Prove that, for any player $i \in \{1, 2, \ldots, n\}$, the "honest" strategy such that $b_i^* = v_i$ (where player i bids her true valuation of the good) weakly dominates all other strategies.

2.5 Consider the following two-player normal-form game.

1\2	a	b	c
A	4, 1	5, 2	−2, 1
B	3, 3	4, 4	1, 2
C	2, 5	1, 1	−1, 6
D	5, 0	2, 4	0, 5

(a) Let K_0 be the fact that "player 1 does not take a strictly dominated strategy and she knows her own payoff function. Likewise, player 2 does not take a strictly dominated strategy and knows his own payoff function". What is the set of strategyf combinations deduced only from K_0? (You can use notation such as $\{A, B\} \times \{a, b\}$.)

(b) Let K_1 be the fact that "player 1 knows K_0 and player 2's payoff function. Player 2 also knows K_0 and player 1's payoff function". What is the set of strategy combinations deduced only from K_1?

(c) Let K_2 be the fact that "player 1 and player 2 both know K_0 and K_1". What is the set of strategy combinations deduced only from K_2?

(d) What is the set of strategy combinations after iterative elimination of strictly dominated strategies?

2.6 Consider the following three-player normal-form game

1 \ 2	L	R
U	2, 2, 2	3, 1, 1
D	1, 1, 3	2, 3, 4

3: A

1 \ 2	L	R
U	1, 2, 3	31, 1, 0
D	0, 3, 2	30, 30, 30

B

Player 1 chooses between rows U and D, player 2 chooses between columns L and R, and player 3 chooses between matrices A and B, simultaneously. The ith coordinate in each payoff combination is player i's payoff. Find the set of strategy combinations that survive iterative elimination of strictly dominated strategies.

2.7 Construct the matrix representation of the following game.

There is a duopoly market, in which firm 1 and firm 2 are the only producers. Each firm chooses a price in $\{2, 3, 4\}$ simultaneously. The payoff of a firm $i \in \{1, 2\}$ is its sales, which is its price multiplied by its demand. The demand of a firm $i \in \{1, 2\}$ is determined as follows. Let p_i be firm i's price and p_j be the opponent's. The demand of firm i is

$$D_i(p_i, p_j) = \begin{cases} (4.6 - p_i) & \text{if} p_i < p_j \\ \frac{1}{2}(4.6 - p_i) & \text{if} p_i = p_j \\ 0 & \text{if} p_i > p_j. \end{cases}$$

For example, if firm 1 chooses $p_1 = 3$ and firm 2 also chooses $p_2 = 3$, then each firm's sales (payoff) is $3 \cdot \frac{1}{2}(4.6 - 3) = 2.4$.

2.8 Continue to analyze the game in Problem 2.7.

(a) Find the set of strategy combinations that survive iterative elimination of strictly dominated strategies.

(b) Assume that firms can choose a price from $\{1, 2, 3, 4\}$. The payoff is determined in the same way as in Problem 2.7. Construct a matrix representation.

(c) Find the set of strategy combinations that survive iterative elimination of strictly dominated strategies for the game (b).

(d) Find the set of strategy combinations that survive iterative elimination of both strictly and weakly dominated strategies for the game (b).

References

1. Aumann R (1976) Agreeing to disagree. Ann Stat 4(6):1236–1239
2. Fan K (1953) Minimax theorems. Proc Nat Acad Sci 39:42–47
3. Kohlberg E, Mertens J-F (1986) On the strategic stability of equilibria. Econometrica 54(5):1003–1037
4. Luce D, Raiffa H (1957) Games and decisions. Wiley, New York
5. von Neumann J (1928) Zur Theorie der Gesellschaftsspiele. Mathematische Annalen 100:295–320 (English translation (1959): On the theory of games of strategy. In: Tucker A, Luce R (eds) Contributions to the theory of games. Princeton University Press, Princeton, pp 13–42
6. von Neumann J, Morgenstern O (1944) Theory of games and economic behavior. Princeton University Press, Princeton, NJ
7. Osborne M, Rubinstein A (1994) A course in game theory. MIT Press, Cambridge, MA
8. Perea A (2001) Rationality in extensive form games. Klouwer Academic Publishers, Dordrecht, the Netherlands
9. Ritzberger K (2002) Foundations of non-cooperative game theory. Oxford University Press, Oxford, UK

Chapter 3
Nash Equilibrium

3.1 Nash Equilibrium

Let us re-interpret the predicted strategy combination (D, D) in the Prisoner's Dilemma discussed in Sect. 2.2. If you believe that your opponent defects, then changing your strategy from Defect to Cooperate decreases your payoff. This property holds for both players. Therefore, the (D, D) combination is **strategically stable**, that is, each player maximizes her/his own payoff given the belief of the opponent's strategy choice, which is the actual strategy chosen by the opponent. This logic is different from comparing a single player's strategies, as in the elimination of strictly dominated strategies. Rather, it concerns the stability of a strategy **combination**. In non-cooperative games, multiple players cannot coordinate moving from one strategy combination to another. Hence, starting from a strategy combination, we investigate whether a player wants to deviate to a different strategy, **given** the strategies of the opponents. If no player wants to deviate unilaterally from a strategy combination, then the original combination is strategically stable. This stability notion is one of the fundamental contributions to game theory by Nash [13].

Using this stability notion, we can predict a subset of strategy combinations, even for games without strictly dominated strategies. Consider the following example.

There are two students, Ann and Bob. They are planning to go to a concert together after school today. Right now, they are in separate classrooms and cannot communicate. (They are good students who refrain from using their cell phones while in a class, or you can assume that one of them forgot her or his phone at home today.) Each student must independently decide whether to go to (stay at) Ann's classroom (strategy A) or to go to (stay at) Bob's classroom (strategy B). If one stays and the other goes to that room, they can meet immediately and get the payoff of 1 each. In addition, the player who did not move saved the effort and gets an additional payoff of 1. The worst case is that they cannot meet. There are two ways in which this can happen. Ann chooses strategy B and Bob chooses strategy A, or Ann chooses A and Bob chooses B. In these cases, both players get a payoff of 0. (This game

T. Fujiwara-Greve, *Non-Cooperative Game Theory*, Monographs in Mathematical Economics 1, DOI 10.1007/978-4-431-55645-9_3

Table 3.1 Meeting game

A\B	A	B
A	2, 1	0, 0
B	0, 0	1, 2

is a gender-free payoff version of the *Battle of the Sexes* in Luce and Raiffa [11].) Table 3.1 is the (double) matrix representation of this game.

In this game, no player has a strictly dominated strategy. Hence we cannot reduce the outcomes using the logic of iterative elimination of strictly dominated strategies. However, the strategy combinations where the two players can meet, (A, A) and (B, B), differ from those where the two cannot meet, (A, B) and (B, A), from the perspective of strategic stability. From the strategy combinations (A, A) or (B, B), no player wants to change strategies unilaterally, while from other strategy combinations, a player wants to change strategies to increase the payoff. Therefore, we should exclude (A, B) and (B, A) as stable outcomes.

In order to formalize the above logic, let us define the strategies that maximize a player's payoff, while fixing the combination of all other players' strategies.

Definition 3.1 For any player $i \in \{1, 2, \ldots, n\}$ and any strategy combination by i's opponents, $s_{-i} \in S_{-i}$, a *best response*[1] of player i to s_{-i} is a strategy $s_i^* \in S_i$ such that

$$u_i(s_i^*, s_{-i}) \geqq u_i(s_i, s_{-i})$$

for all $s_i \in S_i$.

In general, even though the strategy combination by the opponents is fixed, there can be multiple best responses for player i. Thus we define the set of best responses to s_{-i}:

$$BR_i(s_{-i}) = \{s_i^* \in S_i \mid u_i(s_i^*, s_{-i}) \geqq u_i(s_i, s_{-i}), \ \forall\, s_i \in S_i\}.$$

In any strategically stable strategy combination, all players should be playing a best response to the rest of the strategy combination. This is the fundamental equilibrium concept in the non-cooperative game theory.

Definition 3.2 A strategy combination $(s_1^*, s_2^*, \ldots, s_n^*) \in S$ is a *Nash equilibrium* if, for any player $i \in \{1, 2, \ldots, n\}$,

$$s_i^* \in BR_i(s_1^*, s_2^*, \ldots, s_{i-1}^*, s_{i+1}^*, \ldots, s_n^*).$$

Alternatively, $(s_1^*, s_2^*, \ldots, s_n^*) \in S$ satisfies the following simultaneous inequalities:

[1] In some literature it is called a *best reply*.

Table 3.2 Prisoner's dilemma

A\B	C	D
C	−3, −3	0, −5
D	−5, 0	−1, −1

$$u_1(s_1^*, s_{-1}^*) \geqq u_1(s_1, s_{-1}^*), \quad \forall s_1 \in S_1,$$
$$u_2(s_2^*, s_{-2}^*) \geqq u_2(s_2, s_{-2}^*), \quad \forall s_2 \in S_2,$$
$$\vdots$$
$$u_n(s_n^*, s_{-n}^*) \geqq u_n(s_n, s_{-n}^*), \quad \forall s_n \in S_n.$$

To apply this notion to the Meeting Game in Table 3.1, for both players the unique best response to the opponent's strategy A is A. Hence (A, A) is a Nash equilibrium. Similarly, strategy B is the unique best response to strategy B by the opponent. Therefore, this game has two Nash equilibria, (A, A) and (B, B). The other two strategy combinations are not Nash equilibria, and thus we have narrowed down the set of strategy combinations with the Nash equilibrium concept.

Let us go back to the Prisoner's Dilemma formulated in Sect. 2.2. The (doble) matrix representation is shown in Table 3.2.

Regardless of the opponent's strategy, each player has a unique best response, which is to defect. Therefore, there is a unique Nash equilibrium of (D, D). This is an example where the unique Nash equilibrium and the unique outcome of iterative elimination of strictly dominated strategies coincide. For general relationships between Nash equilibria and the outcomes that survive the iterative elimination of strictly dominated strategies, see Propositions 3.1 and 3.2 in Sect. 3.5.

Consider one more example to get used to the derivation of Nash equilibria. "Chicken" is a dangerous game among youths, dramatized in movies such as "Rebel without a Cause" and "Stand by Me." The players are two boys, A and B. They hate each other and want to determine who is "chicken". In "Rebel without a Cause", two boys drive cars side by side towards a cliff to see who jumps out first. In "Stand by Me", two men drive cars towards each other on a single-lane road, until one of them swivels off. Whoever escapes first is chicken and loses face.

To make the situation tractable, we assume that the boys have only two strategies, Go and Stop. They choose one of them simultaneously. If one chooses Go and the other chooses Stop, the player who chose Go wins. If both choose Go, that is the worst case and both boys may die. If both choose Stop, both boys have a payoff between winning and losing. An example of such payoff functions is Table 3.3. This form of 2×2 games is called a *Chicken Game*.

Table 3.3 Chicken game

A\B	Go	Stop
Go	−100, −100	10, −10
Stop	−10, 10	−1, −1

This game has four strategy combinations. Clearly (Go, Go) is not a Nash equilibrium. If the opponent chooses Go, it is better to choose Stop than to die. (Stop, Stop) is not a Nash equilibrium, either. If the opponent chooses Stop, you can choose Go to win (and not to die). Now, both (Go, Stop) and (Stop, Go) are Nash equilibria. At each of these strategy combinations, both players maximize their payoff given the other's strategy.

Notice that in all of the above games of the Meeting Game (Table 3.1), the Prisoner's Dilemma (Table 3.2) and the Chicken Game (Table 3.3), the players have the same set of strategies. However, while the Nash equilibria in the Meeting Game and the Prisoner's Dilemma consist of players choosing the same strategy (these are called *symmetric equilibria*), the Nash equilibria in the Chicken Game are combinations in which players choose different strategies (these are called *asymmetric equilibria*). That is, even though the set of strategies is the same, the equilibrium strategy may not be the same across all players.

There are many ways to interpret a Nash equilibrium. One is that it is a *self-enforcing agreement*.[2] Suppose that players meet before playing a game and agree on a strategy combination to be used. After that, when the game is actually played, each player faces the decision whether to follow the agreement or to choose a different strategy, assuming that all others follow the agreement. If no player can increase her/his payoff by changing strategies unilaterally, the original agreement is self-enforcing. This is precisely the definition of a Nash equilibrium.

Another interpretation is that a Nash equilibrium is the limit of an iterative reasoning of best responses. This interpretation is easy to understand for two-player games, in which each player has a unique best response to each strategy by the opponent. Consider player 1's reasoning process as follows. First, player 1 anticipates that player 2 (the only opponent) will choose some strategy s_2. Then it is rational for player 1 to choose the best response $\{s_1\} = BR_1(s_2)$. (From now on, if there is a unique best response, we omit the parentheses to write this as $s_1 = BR_1(s_2)$.) However, player 1 may think that player 2 will anticipate this and thus chooses the best response $s_2' = BR_2(s_1)$. In this case, player 1 should choose $s_1' = BR_1(s_2')$. But player 2 can also anticipate this and should play $s_2'' = BR_2(s_1')$. Then player 1 should choose $s_1'' = BR_1(s_2'')$, and so on. If this iterative reasoning of best responses converges, then the strategy choices must satisfy $s_1^* = BR_1(s_2^*)$ and $s_2^* = BR_2(s_1^*)$. This is a Nash equilibrium. (Player 2 can use the same reasoning, and the limit is also a Nash equilibrium.)

3.2 Cournot Game

The concept of an equilibrium as a limit of a reasoning process or an iterative best response process was already formulated by Cournot [7], well before Nash's work. This is known as the *Cournot equilibrium* in economics. Let us make a simple

[2] This is different from the binding agreement used in defining a cooperative game by Nash [13].

economic model to define a Cournot equilibrium. The game has infinitely many strategies, and thus it demonstrates how to find a Nash equilibrium for a game with infinitely many strategies as well.

Assume that in the market for a certain good, there are only two producers, called firm 1 and firm 2. They play a simultaneous-move game, in which each firm chooses a quantity of the good to supply to the market. The set of strategies (common to both firms) is the set of all non-negative real numbers. After the simultaneous choice of quantities, the game ends. The payoff of a firm is its profit, which is the revenue minus the cost of production. The revenue is the market price of the good times the relevant firm's supply quantity. The production cost of firm i ($i = 1, 2$) is a coefficient $c_i (> 0)$ times its supply quantity. The coefficient is the *marginal cost* such that, for an additional 1 unit of production, firm i incurs an additional cost of c_i. This formulation means that the marginal cost is the same starting from any quantity. Assume that the market-clearing price when firm 1 supplies q_1 units and firm 2 supplies q_2 units of the good is

$$A - q_1 - q_2.$$

(We also assume that $A > \max\{c_1, c_2\}$.) Therefore, firm 1's payoff function is a function of both firms' quantities as follows:

$$u_1(q_1, q_2) = (A - q_1 - q_2)q_1 - c_1 q_1.$$

Analogously, firm 2's payoff function is

$$u_2(q_1, q_2) = (A - q_1 - q_2)q_2 - c_2 q_2.$$

Cournot [7] postulated that the market outcome is a result of iterative best responses: each firm computes the best response to the opponent's strategy and the best response to the best response to the opponent's strategy and so on. If firm 2's strategy is q_2, firm 1's best response is the maximizer of $u_1(\cdot, q_2)$. To see that this function $u_1(\cdot, q_2)$ has a maximum with respect to q_1, let us rearrange $u_1(q_1, q_2)$ as follows:

$$u_1(q_1, q_2) = (A - q_1 - q_2)q_1 - c_1 q_1 = -q_1^2 + (A - q_2 - c_1)q_1.$$

This is a quadratic function of q_1 and its coefficient of the square term is negative. Therefore the function is concave, and the maximum is attained at q_1 where the derivative of $u_1(\cdot, q_2)$ is 0.[3] By differentiation, the derivative is

$$\frac{\partial u_1(q_1, q_2)}{\partial q_1} = -2q_1 + A - q_2 - c_1. \tag{3.1}$$

[3] To be precise, it is $q_1 = 0$ when $A - q_2 - c_1 < 0$. However, as we see shortly, firm 2's range of "rational" strategies is limited as well, and thus we can look at the situation with $A - q_2 - c_1 \geqq 0$ only.

(Note that the derivative (3.1) is a decreasing function of the opponent's strategy. The same property holds for firm 2's payoff function as well. This type of game is said to have *strategic substitutability*. See Bulow et al. [6].)

By solving $-2q_1 + A - q_2 - c_1 = 0$ for q_1, we obtain the (unique) best response[4] to q_2 as

$$q_1 = \frac{1}{2}(A - q_2 - c_1). \tag{3.2}$$

Similarly, when firm 1's strategy is q_1, firm 2's best response is

$$q_2 = \frac{1}{2}(A - q_1 - c_2).$$

Using these best response formulas, we can compute the chain of best responses for the two firms. For example, suppose that firm 1 thinks that firm 2 would choose $q_2^{(0)} = \frac{1}{4}(A - c_2)$. Firm 1's best response to this strategy is

$$q_1^{(0)} = \frac{1}{2}(A - q_2^{(0)} - c_1) = \frac{1}{8}(3\,A - 4\,c_1 + c_2).$$

Firm 2's best response to $q_1^{(0)}$ is

$$q_2^{(1)} = \frac{1}{2}(A - q_1^{(0)} - c_2) = \frac{1}{16}(5\,A + 4\,c_1 - 9\,c_2).$$

Continuing the chain of best responses, we have

$$\begin{aligned} q_1^{(1)} &= \frac{1}{2}(A - q_2^{(1)} - c_1) = \frac{1}{32}(11\,A - 20\,c_1 + 9\,c_2), \\ q_2^{(2)} &= \frac{1}{2}(A - q_1^{(1)} - c_2) = \frac{1}{64}(21\,A + 20\,c_1 - 41\,c_2), \\ q_1^{(2)} &= \frac{1}{2}(A - q_2^{(2)} - c_1) = \frac{1}{128}(43\,A - 84\,c_1 + 41\,c_2), \\ &\dots \end{aligned}$$

and $q_1^{(m)}$ and $q_2^{(m)}$ converge to

$$q_1^* = \frac{1}{3}A - \frac{2}{3}c_1 + \frac{1}{3}c_2, \quad q_2^* = \frac{1}{3}A + \frac{1}{3}c_1 - \frac{2}{3}c_2.$$

Cournot defined this limit as the equilibrium. In fact, at the limit, both $q_1^* = \frac{1}{2}(A - q_2^* - c_1)$ and $q_2^* = \frac{1}{2}(A - q_1^* - c_2)$ hold simultaneously. That is, q_1^* is the best response to q_2^* and vice versa. In our terminology, (q_1^*, q_2^*) is a Nash equilibrium.

---

[4] Again, when $A - q_2 - c_1 < 0$, we should write $\max\{\frac{1}{2}(A - q_2 - c_1), 0\}$ as the best response, but we have argued that this is not a relevant case.

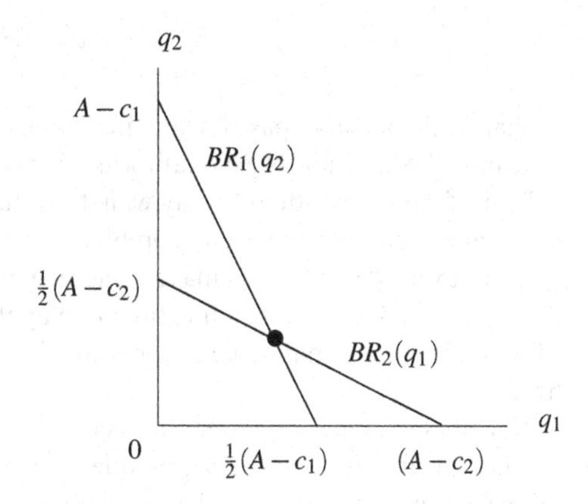

Fig. 3.1 Cournot-Nash equilibrium

We can interpret the above argument as an example of firm 1's reasoning process. Firm 2 can also reason iteratively, starting from an arbitrary initial belief of firm 1's strategy. Again, the process converges to (q_1^*, q_2^*). Hence, in this game, firms can reach the Nash equilibrium by playing the best response over the infinite horizon, as well as by reasoning iteratively starting from any belief of the opponent's strategy. (This statement does not hold for general games. Interested readers can read Sect. 3.8 for a general analysis of what players can reach with the iterative reasoning of best responses.)

The best responses of the two firms are linear functions in the opponent's strategy as illustrated in Fig. 3.1. The Cournot-Nash equilibrium is the point at which both players choose the best response to each other, and hence the two lines intersect.

Finally, let us consider the efficiency of the Cournot-Nash equilibrium. At the equilibrium, the market-clearing price is

$$\frac{1}{3}(A + c_1 + c_2),$$

and the equilibrium payoff of each firm is

$$u_1(q_1^*, q_2^*) = \frac{1}{9}(A - 2c_1 + c_2)^2, \quad u_2(q_1^*, q_2^*) = \frac{1}{9}(A + c_1 - 2c_2)^2.$$

For simplicity, assume that the marginal costs of the firms are the same, $c_1 = c_2 = c$, to make the game symmetric. Then, for each $i = 1, 2$, firm i's equilibrium quantity is $q_i^* = \frac{1}{3}(A - c)$, and the equilibrium payoff is

$$u_i(q_1^*, q_2^*) = \frac{1}{9}(A - c)^2.$$

However, if the firms reduce their production to $q_1^o = q_2^o = \frac{1}{4}(A - c)$, then each firm's payoff becomes

$$\frac{1}{8}(A-c)^2,$$

so that both players' payoffs are increased. Therefore, like with the Prisoner's Dilemma, the Cournot-Nash equilibrium is not efficient.

Firms facing this situation may want to collude to adjust production.[5] However, it is difficult to enforce a strategy combination which is not a Nash equilibrium. Suppose, for example, that firms make an agreement to choose the efficient combination $q_1^o = q_2^o = \frac{1}{4}(A-c)$. When it is time to play the game, assuming that the opponent follows q_j^o, the optimal strategy for firm i is not q_i^o, and thus it is rational to break the agreement.

Nonetheless, real firms sometimes collude to restrict production, raising the price and their profits. It seems to be possible to enforce non-Nash equilibrium actions on rational firms. We show how that is possible in Sect. 5.6.

3.3 Bertrand Game

In the model by Cournot, firms choose production quantities. Alternatively, we can think of the strategic choice of prices. A price competition model is introduced by Bertrand [4]. Although the model may be otherwise similar to the Cournot model, the structure of the best responses is quite different from that of the quantity choice game, and, moreover, it matters whether the products of the firms are viewed as the same by the consumers or not. Let us assume that there are only two firms (1 and 2) in the market. Their strategies are the prices for their own product, and the set of strategies is again all non-negative real numbers.

3.3.1 When Products Are Differentiated

In this subsection, we assume that the two firms' products are not the same for the consumers. Some consumers prefer one product over the other so much that not all consumers buy from the cheaper firm. For example, when firm 1's price is p_1 and firm 2's price is p_2, the demand for firm 1's product is

$$D_1(p_1, p_2) = A - p_1 + p_2,$$

and the demand for firm 2's product is

$$D_2(p_1, p_2) = A - p_2 + p_1.$$

---

[5]The definition of efficiency in game theory is different from the one in microeconomics where consumers' welfare is taken into account.

These formulations represent the idea that even if a firm's price is higher than the opponent's, its demand may not become 0.

As in the Cournot game, we assume that firm i's marginal cost is constant and is denoted as $c_i > 0$. The total production cost for a firm producing x units is $c_i \cdot x$. We also assume that $A > \max\{c_1, c_2\}$. Firm i's payoff function (profit) is

$$\begin{aligned} u_i(p_i, p_j) &= (p_i - c_i)D_i(p_i, p_j) \\ &= -p_i^2 + (A + c_i + p_j)p_i - Ac_i - c_i p_j, \end{aligned}$$

where j is the opponent. This is a quadratic function of p_i and the coefficient of the square term is negative. Hence it is maximized at p_i that makes the partial derivative 0. The partial derivative with respect to p_i[6] is

$$\frac{\partial u_i(p_1, p_2)}{\partial p_i} = -2p_i + A + c_i + p_j.$$

The best response which makes the above 0 is

$$p_i = \frac{1}{2}(A + c_i + p_j).$$

A Bertrand-Nash equilibrium (p_1^*, p_2^*) of this game is a solution to the system of simultaneous equations:

$$\begin{cases} p_1 = \frac{1}{2}(A + c_1 + p_2) \\ p_2 = \frac{1}{2}(A + c_2 + p_1). \end{cases}$$

By computation, there is a unique solution such that

$$p_1^* = \frac{1}{3}(3A + 2c_1 + c_2), \quad p_2^* = \frac{1}{3}(3A + c_1 + 2c_2).$$

In particular, when we consider the symmetric cost case of $c_1 = c_2 = c$, then $(p_1^*, p_2^*) = (A + c, A + c)$ so that both firms charge a price higher than the marginal cost c.

3.3.2 When Products Are Perfect Substitutes

When the two firms' products are viewed as the same good (such products are called *perfect substitutes* in economics), all consumers buy the cheaper one. Hence the demand for firm i's product is a discontinuous function as follows:

---

[6]The partial derivative of player i's payoff function is increasing in the opponent's strategy p_j. This type of game is said to have *strategic complementarity*.

$$D_i(p_1, p_2) = \begin{cases} A - p_i & \text{if } p_i < p_j \\ \frac{1}{2}(A - p_i) & \text{if } p_i = p_j \\ 0 & \text{if } p_i > p_j. \end{cases}$$

This formulation implies that, when the two firms charge the same price, they share the market demand equally. Let us also assume that $A > \max\{c_1, c_2\}$ again. Now that the payoff function is discontinuous, it is not possible for us to find a best response to p_j by differentiation. However, we can find a Nash equilibrium by reasoning. For simplicity, assume that $c_1 = c_2 = c$.

First, we analyze whether a strategy combination in which the firms' prices differ can be a Nash equilibrium. Without loss of generality, suppose that $p_1 < p_2$. We further distinguish where the marginal cost c lies in relation to p_1 and p_2. If $c < p_1 < p_2$, then firm 2, getting 0 payoff, can change its strategy to p_2' such that $c < p_2' < \min\{A, p_1\}$ and earn a positive payoff. (Since any non-negative real number can be chosen, such pricing is feasible.) Therefore, at least one player is not playing a best response to the other's strategy, which means that a strategy combination such that $c < p_1 < p_2$ is not a Nash equilibrium.

If $c = p_1 (< p_2)$, then firm 1 monopolizes the market but earns 0 payoff. There exists a slightly higher price p_1' such that $c < p_1' < \min\{A, p_2\}$, so that firm 1 can still monopolize the market and earn a positive payoff. Hence a strategy combination such that $c = p_1 < p_2$ is not a Nash equilibrium. For the same reason, a combination such that $p_1 < c < p_2$ is not a Nash equilibrium, either. How about the case where $p_1 < p_2 \leqq c$? In this case, firm 1 is getting a negative payoff, and it can increase its payoff to 0 by charging $p_1' > p_2$. Therefore, such a strategy combination is not a Nash equilibrium. In summary, there is no Nash equilibrium in which the two firms charge different prices.

Next, suppose that the two firms choose the same price p. There are three cases, $p > c$, $p = c$, and $p < c$. When $p > c$, each firm gets the demand of $\frac{1}{2}(A - p)$ and a payoff of

$$u_i(p, p) = (p - c)\frac{1}{2}(A - p).$$

However, there exists a small $\varepsilon > 0$ such that by reducing the price to $p - \varepsilon$, a firm can earn

$$u_i(p - \varepsilon, p) = (p - \varepsilon - c)(A - p + \varepsilon) > (p - c)\frac{1}{2}(A - p).$$

Therefore, such a strategy combination is not a Nash equilibrium. When $p < c$, both firms have negative payoff, so that one can increase its payoff to 0 by charging a higher price than the opponent's. That is, such a strategy combination is not a Nash equilibrium.

Finally, consider the strategy combination such that $p_1 = p_2 = c$. Both firms earn 0 payoff, but this strategy combination is a Nash equilibrium. To see this, suppose that a firm changes its price. If it raises the price, there is no demand and thus the payoff is still 0. If it lowers its price below c, the demand increases, but the firm's payoff is negative. Therefore no firm can increase its payoff by changing the price, given that the other charges $p_j = c$.

This example shows that, even if the payoff functions are discontinuous on a continuum set of strategies, there may be a Nash equilibrium. An important implication for economics is that the equilibrium prices equal the marginal cost, and both firms earn 0 profit. This is the same outcome as in a perfectly competitive market (which usually is assumed to happen when there are so many firms that they cannot influence the market price), and is a quite different outcome from the Nash equilibrium under product differentiation. This is because the market is more competitive when the products are perfect substitutes, so that consumers only look at prices. (Note, however, that the marginal cost pricing equilibrium depends on the assumption that real number pricing is feasible. Even if the products are perfect substitutes, if the prices must be chosen from a discrete set, the equilibrium price may not go down to the marginal cost. See Problem 3.2(b).)

The Bertrand-Nash equilibrium is not necessarily efficient (in the sense of Definition 2.2). For example, in the perfect substitute case, there exists a strategy combination in which both firms coordinate to charge the same price which is higher than the marginal cost, to earn a positive profit.

3.4 Location Choice Game

We introduce a *location choice game* (also called *Hotelling's model*), which is quite useful not only in economics but also in political science. This game is also an infinite-strategy game.[7] On a line segment, there is a continuum of "positions". The line segment can be interpreted as the spectrum of consumer preferences over the qualities of some product, voter preferences over the policies on an agenda, and so on. Two players, player 1 and 2 (interpreted as sellers or politicians), simultaneously choose their positions s_1 and s_2 in the segment, and the game ends.

The payoff of a player depends on the locations of the two players and the distribution of consumers (voters) on the line segment. For simplicity, let us assume that the consumers are uniformly distributed on the segment $[0, 1]$. Each consumer gives a payoff of 1 to the nearest player. An interpretation is that a consumer purchases the good from the nearby seller and gives a profit of 1 to the seller. (In the voting model, a voter gives one ballot to the nearby politician.) If the two players are located at the same distance from a consumer/voter, she gives a payoff of 1 to one of them with a probability of $1/2$.

Specifically, the payoff of player 1 for each strategy combination $(s_1, s_2) \in [0, 1]^2$ is as follows. When $s_1 < s_2$, all consumers/voters located to the left of the mid-point $\frac{1}{2}(s_1 + s_2)$ of s_1 and s_2 (nearer to s_1 than to s_2) give a payoff of 1 to player 1. Because the consumers/voters are distributed according to the uniform distribution, the total mass is also $\frac{1}{2}(s_1 + s_2)$, which is player 1's payoff. See Fig. 3.2.

[7] Versions of finite-strategy location games exist. See, for example, Watson [20], Chap. 8.

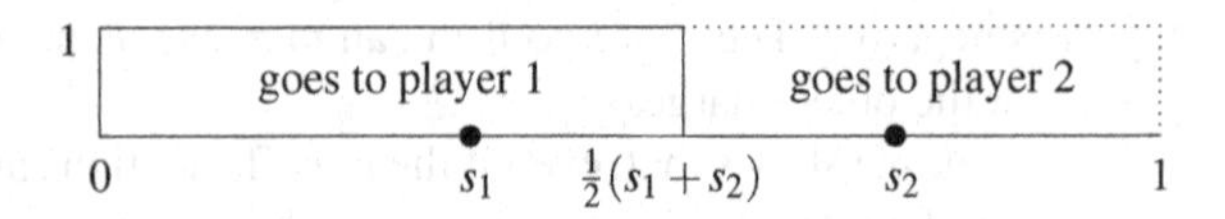

Fig. 3.2 Location choice game

When $s_1 > s_2$, by the opposite logic, player 1's payoff is the measure of the consumers at or to the right of the mid-point of s_1 and s_2, that is $1 - \frac{1}{2}(s_1 + s_2)$. Finally, when $s_1 = s_2$, all consumers have the same distance to both players. Hence half of them will go to player 1, resulting in the payoff of $\frac{1}{2}$. Player 2's payoff function is analogously derived.

Let us show that the unique Nash equilibrium of this Location Choice Game is $(s_1^*, s_2^*) = (\frac{1}{2}, \frac{1}{2})$. First, we show that this combination is a Nash equilibrium. When the opponent locates at $s_j^* = \frac{1}{2}$, the choice of $s_i < \frac{1}{2}$ gives the measure of the region $\left[0, \frac{1}{2}(s_i + \frac{1}{2})\right]$, and $s_i > \frac{1}{2}$ gives the measure of the region $\left[\frac{1}{2}(s_i + \frac{1}{2}), 1\right]$. The former is $\frac{1}{2}(s_i + \frac{1}{2}) < \frac{1}{2}$ (because $s_i < \frac{1}{2}$), and the latter is $1 - \frac{1}{2}(s_i + \frac{1}{2}) < \frac{1}{2}$ as well (because $s_i > \frac{1}{2}$). By locating at $\frac{1}{2}$, the player gets $\frac{1}{2}$. Therefore, the unique best response to $s_j^* = \frac{1}{2}$ is $s_i^* = \frac{1}{2}$, that is, $(s_1^*, s_2^*) = (\frac{1}{2}, \frac{1}{2})$ is a Nash equilibrium.

Next, we show that there is no other Nash equilibrium. Take any strategy combination such that $s_1 \neq s_2$. Then at least one player can improve the payoff by locating nearer to the opponent. Hence such combination is not a Nash equilibrium. Consider any strategy combination such that $s_1 = s_2 = s$ but $s \neq \frac{1}{2}$. Both players get the payoff of $\frac{1}{2}$ under this strategy combination. However, because $s \neq \frac{1}{2}$, one of the regions $[0, s]$ or $[s, 1]$ has a measure greater than $\frac{1}{2}$. A player can move to that region a little to increase her/his payoff beyond $\frac{1}{2}$. Therefore, such a strategy combination is not a Nash equilibrium either.

The above logic can be applied to general distributions of consumers/voters. With the uniform distribution, the mean and the median[8] are both $\frac{1}{2}$. The critical point is the median. If there is a unique median in the distribution of consumers/voters, then both players locating at the median is the unique Nash equilibrium of the location choice game. This result is called the Median Voter Theorem in political science. A good reference is Osborne [15], Sect. 3.3.

3.5 Strategic Dominance and Nash Equilibrium*

In the Prisoner's Dilemma, the unique Nash equilibrium coincides with the unique outcome that survives the iterative elimination of strictly dominated strategies. In general, however, it can happen that the iterative elimination of strictly dominated strategy does not have any bite, while the Nash equilibrium concept selects among

[8] A *median* of a cumulative distribution F is a real number m such that $F(X \leqq m) = F(X \geqq m) = \frac{1}{2}$.

possible outcomes, as in the Meeting Game. Let us clarify the relationship between the equilibrium concept by iterative elimination of strictly dominated strategies and Nash equilibrium.

Proposition 3.1 *For any game* $G = (\{1, 2, \ldots, n\}, S_1, S_2, \ldots, S_n, u_1, u_2, \ldots, u_n)$, *if a strategy combination* $s^* = (s_1^*, s_2^*, \ldots, s_n^*) \in S$ *is a Nash equilibrium, then it is not eliminated during the process of iterative elimination of strictly dominated strategies.*

Proof Suppose that a player's strategy in the combination s^* is eliminated during the process of iterative elimination of strictly dominated strategies. There may be multiple such strategies, so let the first one that is eliminated in the process be s_i^*.

During the elimination process, there must exist $s_i' \in S_i$ that strictly dominates s_i^* in some reduced game. That is, for any $s_{-i} \in S_{-i}$ which are not eliminated at the time when s_i^* is eliminated,

$$u_i(s_i', s_{-i}) > u_i(s_i^*, s_{-i}).$$

By assumption, all other s_j^*'s survive when s_i^* is eliminated. Hence,

$$u_i(s_i', s_{-i}^*) > u_i(s_i^*, s_{-i}^*).$$

This contradicts the assumption that s^* is a Nash equilibrium. □

Thus, by eliminating strictly dominated strategies, we do not lose Nash equilibria. One might wish for the elimination process to get rid of non-Nash equilibrium strategies. This is not always possible, but a sufficient condition is known.

Proposition 3.2 *For any game* $G = (\{1, 2, \ldots, n\}, S_1, S_2, \ldots, S_n, u_1, u_2, \ldots, u_n)$ *such that* $|S_i| < \infty$ *for all* $i = 1, 2, \ldots, n$, *if a single strategy combination* $s^* = (s_1^*, s_2^*, \ldots, s_n^*) \in S$ *survived the process of iterative elimination of strictly dominated strategies, then* s^* *is the unique Nash equilibrium of* G.

Proof We prove that s^* is a Nash equilibrium and there is no other Nash equilibrium. First, suppose that s^* is not a Nash equilibrium. Then there exists a player i and $s_i' \in S_i \setminus \{s_i^*\}$ such that

$$u_i(s_i', s_{-i}^*) > u_i(s_i^*, s_{-i}^*).$$

Since this s_i' must have been eliminated, there exists $s_i'' \in S_i \setminus \{s_i'\}$ such that for any $s_{-i} \in S_{-i}$ that survived until the time that s_i' is eliminated,

$$u_i(s_i'', s_{-i}) > u_i(s_i', s_{-i}).$$

In particular,

$$u_i(s_i'', s_{-i}^*) > u_i(s_i', s_{-i}^*)$$

holds. If $s_i'' = s_i^*$, then we have a contradiction.

If $s_i'' \neq s_i^*$, then s_i'' is also eliminated, and thus there exists $s_i''' \in S_i \setminus \{s_i''\}$ such that for any $s_{-i} \in S_{-i}$ that survived until the time that s_i'' is eliminated,

$$u_i(s_i''', s_{-i}) > u_i(s_i'', s_{-i}).$$

In particular,

$$u_i(s_i''', s_{-i}^*) > u_i(s_i'', s_{-i}^*)$$

also holds. If $s_i''' = s_1^*$, then it is a contradiction. This argument will not continue forever since there are only finitely many strategies. Hence, s^* is a Nash equilibrium.

Next, if there is a Nash equilibrium other than s^*, then Proposition 3.1 implies that it should also survive the iterative elimination process. Hence s^* is the unique Nash equilibrium. □

3.6 Existence of Nash Equilibrium and Mixed Strategies

For the "equilibria" by iterative elimination of strictly dominated strategies, existence is guaranteed. There must be at least one strategy per player that survives the elimination process. However, Nash equilibrium may not exist. Consider the following example.

The players are child 1 and 2. They have a penny each. The sides (Heads or Tails) are the possible strategies each child may choose. The two players show one side of a coin simultaneously, and the game ends. The payoff of a player depends on the combination of the sides. If the sides are matched, child 1 wins and receives the coin from child 2. If the sides do not match, child 2 wins and receives the coin from child 1. This game is called *Matching Pennies*, and its matrix representation is Table 3.4.

This game does not have a Nash equilibrium. For any strategy combination, there is a player (the loser) who can increase her/his payoff by changing the sides of the coin.

Usually, in such a situation, it is important to fool the opponent so that (s)he cannot read your strategy. For example, tennis players try to hide their intention for the direction of a service, whether towards the opponent's forehand or backhand. If the receiver cannot guess which side will be attacked, staying in the middle is a good idea. This makes the server unable to read which side the receiver is guessing, and thus the server's reaction is also to not focus on one side.

Table 3.4 Matching pennies

1\2	Heads	Tails
Heads	1, −1	−1, 1
Tails	−1, 1	1, −1

This idea can be formulated as probabilistic choices of strategies. Let us extend our concept of strategies that we have considered so far as actions or decisions to be made for sure (with probability 1), and call them *pure strategies*. We now allow players to choose these pure strategies probabilistically so that their opponent(s) cannot be sure of a pure strategy, and call such an action or decision a *mixed strategy*. When there is a finite number of pure strategies,[9] the definition of a mixed strategy is given as follows.

Definition 3.3 For a player i, a *mixed strategy* is a probability distribution on S_i, or a function σ_i from S_i to $\Re$ such that

$$0 \leqq \sigma_i(s_i) \leqq 1, \quad \forall s_i \in S_i, \quad \text{and} \quad \sum_{s_i \in S_i} \sigma_i(s_i) = 1.$$

The real number $\sigma_i(s_i)$ is the probability that the mixed strategy σ_i assigns to the pure strategy s_i. In general, the set of probability distributions on a set X is written as $\Delta(X)$, and thus we write the set of mixed strategies of a player i as $\Delta(S_i)$. A *completely mixed strategy* or a *strict mixed strategy* is a mixed strategy that puts a positive probability on all of one's pure strategies. A pure strategy can be interpreted as a degenerate mixed strategy, which chooses the pure strategy with probability 1, and therefore belongs to the set of mixed strategies.

Let us extend the game to allow mixed strategies for all players. When players use mixed strategies, pure-strategy combinations (outcomes of the game) will be generated probabilistically so that players need to evaluate **probability distributions** over outcomes. Standard game theory assumes that each player's objective in this case is to maximize the expected value of some payoff function (called the *von Neumann Morgenstern utility*), and the payoff function is defined over the set of pure-strategy combinations as before. Let $(\sigma_1, \ldots, \sigma_n)$ be a combination of mixed strategies. We assume that, for each player i, there is $u_i : S \to \Re$ such that player i evaluates $(\sigma_1, \ldots, \sigma_n)$ by the expected value of u_i, which can be written (when there are finitely many pure strategies) as

$$Eu_i(\sigma_1, \ldots, \sigma_n) = \sum_{(s_1, \ldots, s_n) \in S} \sigma_1(s_1) \cdots \sigma_n(s_n) u_i(s_1, \ldots, s_n),$$

where the product $\sigma_1(s_1) \cdots \sigma_n(s_n)$ is the probability that a pure-strategy combination $(s_1, \ldots, s_n)$ occurs. The function Eu_i is called *the expected payoff* function.

Strictly speaking, assuming payoff-maximizing players and extending their choices to mixed strategies do not imply that players maximize their expected payoff. In his book [14], von Neumann proved that the expected payoff-maximizing behavior is equivalent to having a preference relation for the set of probability distributions (lotteries) over the set of outcomes, which satisfies axioms called Independence and

[9] To be precise, we can use the summation as long as pure strategies are countably many. When there are uncountably many pure strategies, we need integration.

Continuity. (For details, see, for example, Chap. 8 of Rubinstein [19], Kreps [10], and references therein.)

Even if we assume Independence and Continuity of the preference relation for the set of lotteries over outcomes, the expected payoff function $Eu_i : \Delta(S_1) \times \cdots \times \Delta(S_n) \to \Re$ is unique only up to affine transformations of the underlying payoff function u_i. That is, an arbitrary monotone transformation of u_i may alter equilibria under expected payoff maximization, which does not happen when players use only pure strategies.

Noting these precautions, we consider the mixed-strategy extension $G' = (\{1, \ldots, n\}, \Delta(S_1), \ldots, \Delta(S_n), Eu_1, \ldots, Eu_n)$ of the normal-form game $G = (\{1, \ldots, n\}, S_1, \ldots, S_n, u_1, \ldots, u_n)$. Let us also extend the set of best responses to

$$BR_i(\sigma_{-i}) = \{\sigma_i \in \Delta(S_i) \mid Eu_i(\sigma_i, \sigma_{-i}) \geqq Eu_i(x, \sigma_{-i}) \ \forall x \in \Delta(S_i)\}.$$

(See also Problem 3.1.)

A *Nash equilibrium* $(\sigma_1^*, \sigma_2^*, \ldots, \sigma_n^*)$ in the set of mixed strategy combinations is defined by

$$\sigma_i^* \in BR_i(\sigma_1^*, \sigma_2^*, \ldots, \sigma_{i-1}^*, \sigma_{i+1}^*, \ldots, \sigma_n^*), \quad \forall i \in \{1, 2, \ldots, n\}.$$

Let us find a mixed-strategy Nash equilibrium for the Matching Pennies game of Table 3.4. For each player, the pure strategies are Heads and Tails. The set of mixed strategies is

$$\Delta(S_1) = \Delta(S_2) = \{\sigma \mid 0 \leqq \sigma(x) \leqq 1, \ \forall x = H, T, \ \sigma(H) + \sigma(T) = 1\}.$$

When a player has only two pure strategies, a mixed strategy can be identified by the probability of one of the pure strategies. Let us use the probability of Heads as representing a mixed strategy for each player.

When q is a mixed strategy of child 2, the expected payoff for child 1 using a mixed strategy p is computed as follows.

$$\begin{aligned} Eu_1(p, q) = & \ pqu_1(H, H) + p(1-q)u_1(H, T) + (1-p)qu_1(T, H) \\ & + (1-p)(1-q)u_1(T, T) \\ = & \ pq - p(1-q) - (1-p)q + (1-p)(1-q) \\ = & \ p(4q - 2) + 1 - 2q. \end{aligned}$$

If $q > \frac{1}{2}$, then this is a strictly increasing function of p, so that maximizing p is equivalent to maximizing the expected payoff. Hence the unique best response is $p = 1$, that is, to play the pure strategy Heads. By contrast, if $q < \frac{1}{2}$, then this is a decreasing function of p, so that the unique best response is $p = 0$ or the pure strategy Tails. Finally, if $q = \frac{1}{2}$, then the expected payoff is constant for any p. This means that all mixed strategies are best responses. Mathematically, we can write

$$BR_1(q) = \begin{cases} 1 & \text{if } q > \frac{1}{2} \\ \Delta(S_1) & \text{if } q = \frac{1}{2} \\ 0 & \text{if } q < \frac{1}{2}. \end{cases}$$

Notice that the best response is a correspondence such that it takes a set value at $q = \frac{1}{2}$.

Analogously, when child 1's mixed strategy is p, the expected payoff of child 2 using a mixed strategy q is

$$\begin{aligned} Eu_2(p, q) = &\ pqu_2(H, H) + p(1-q)u_2(H, T) + (1-p)qu_2(T, H) \\ &+ (1-p)(1-q)u_2(T, T) \\ = &\ -pq + p(1-q) + (1-p)q - (1-p)(1-q) \\ = &\ q(2-4p) - 1 + 2p. \end{aligned}$$

Hence, if $p < \frac{1}{2}$, then the expected payoff is strictly increasing in q so that $q = 1$ is the unique best response; if $p > \frac{1}{2}$, then the best response is $q = 0$ only, and if $p = \frac{1}{2}$, all mixed strategies are best responses. We can summarize this as follows.

$$BR_2(p) = \begin{cases} 0 & \text{if } p > \frac{1}{2} \\ \Delta(S_2) & \text{if } p = \frac{1}{2} \\ 1 & \text{if } p < \frac{1}{2}. \end{cases}$$

In a Nash equilibrium strategy combination, each strategy is a best response to the other. Take a mixed strategy p for child 1 such that $p > \frac{1}{2}$. Child 2's unique best response is $q = 0$, but the unique best response to $q = 0$ for child 1 is $p = 0$, not in the range of $p > \frac{1}{2}$. Therefore, there is no Nash equilibrium in mixed strategies such that $p > \frac{1}{2}$. Similarly, there is no Nash equilibrium such that $p < \frac{1}{2}$, nor $q \neq \frac{1}{2}$. Finally, consider the combination $p = \frac{1}{2}$ and $q = \frac{1}{2}$. Because any mixed strategy is a best response to the opponent's strategy, $\frac{1}{2}$ is included. Therefore, the Matching Pennies game has a unique Nash equilibrium in mixed strategies such that both players play the mixed strategy "choose Heads with probability $\frac{1}{2}$ and choose Tails with probability $\frac{1}{2}$".

The best response correspondences of the Matching Pennies game are depicted in Fig. 3.3. The square area of $[0, 1] \times [0, 1]$ is the set of all mixed strategy combinations. Child 1 wants to match the sides of the coins, hence the best response correspondence is upward sloping. Child 2 wants to show the opposite side, which makes the correspondence declining in the probability of Heads by child 1. Like with the Cournot game, the Nash equilibrium is the intersection of the best response correspondences.

The following Proposition is useful to find a Nash equilibrium, when it involves a non-degenerate mixed strategy, as in the Matching Pennies game.

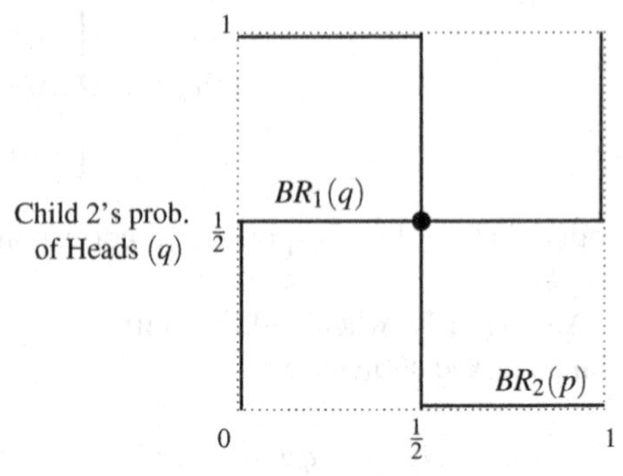

Fig. 3.3 Best response correspondence of matching pennies game

Table 3.5 Rock-Paper-Scissors game

1\2	R	P	S
R	0, 0	−1, 1	1, −1
P	1, −1	0, 0	−1, 1
S	−1, 1	1, −1	0, 0

Proposition 3.3 *For any Nash equilibrium* $(\sigma_1^*, \ldots, \sigma_n^*)$*, all pure strategies in the support*[10] *of* σ_i^* *give the same expected payoff, given that the opponents' strategies are* σ_{-i}^*.

Proof Fix a Nash equilibrium $(\sigma_1^*, \ldots, \sigma_n^*)$, and consider a non-degenerate mixed strategy σ_i^* in it. Suppose, to the contrary, that $s_i \in supp(\sigma_i^*)$ has greater expected payoff against σ_{-i}^* than $s_i' \in supp(\sigma_i^*)$ does. Then by increasing the probability of s_i, player i can increase the total expected payoff, so that σ_i^* is not a best response to σ_{-i}^*. □

Note also that a strategy that is not strictly dominated by any pure strategy may be strictly dominated by a mixed strategy. See Sect. 3.8.

When a player has more than two pure strategies, finding a mixed strategy Nash equilibrium graphically by drawing the best response correspondences is difficult. However, we can use Proposition 3.3 to eliminate non-Nash strategy combinations, as well as to make a guess about the support of a Nash equilibrium mixed strategy.

For example, consider the *rock-paper-scissors game* in Table 3.5. Each of the two players has three pure strategies, R, P, and S. The players show one hand each, simultaneously, and the game ends.

First, let us show that this game has no Nash equilibrium in pure strategies. This is rather obvious, because if the opponent plays a pure strategy, there is a unique best response to win, but in that case the opponent would not choose the original pure strategy.

Second, we check whether there is a Nash equilibrium in which a player uses only two pure strategies with a positive probability. For example, if there is a Nash

[10] The *support* of a probability distribution σ_i on S_i is the set $supp(\sigma_i) := \{s_i \in S_i \mid \sigma_i(s_i) > 0\}$.

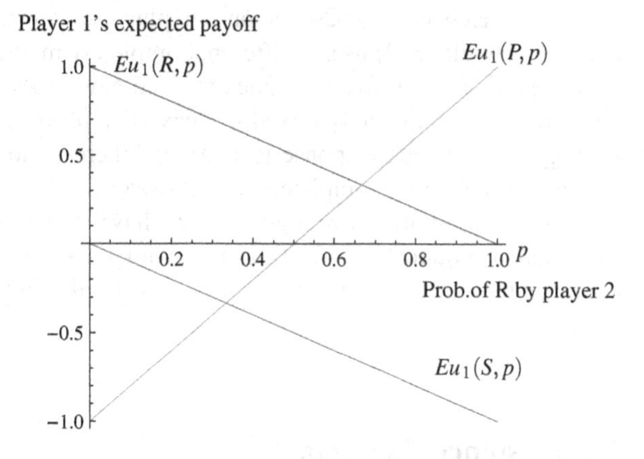

Fig. 3.4 Player 1's expected payoff from each pure strategy

equilibrium in which player 2 assigns a positive probability only to R and S, then by Proposition 3.3, (a) player 2 receives the same expected payoff from R and S, and because it is a Nash equilibrium strategy, (b) the pure strategy P would not give a greater expected payoff than R or S to player 2. In addition, player 1 must be playing a best response to player 2's mixed strategy. Let p be the probability that player 2 chooses R and $1-p$ be the probability that he chooses S, and we compute player 1's best response to the mixed strategy. The expected payoffs for player 1 from each pure strategy R, P, and S are

$$\begin{aligned} Eu_1(R,p) &= p\cdot 0 + (1-p)1 = 1-p \\ Eu_1(P,p) &= p\cdot 1 + (1-p)(-1) = -1+2p \\ Eu_1(S,p) &= p(-1) + (1-p)0 = -p. \end{aligned}$$

Clearly, player 1's pure strategy S is strictly dominated by R if player 2 plays only R or S with a positive probability. The expected payoffs of the three pure strategies as functions of p are depicted in Fig. 3.4. It shows that unless $p = \frac{2}{3}$, where $Eu_1(R,p) = Eu_1(P,p)$, player 1 has a unique pure-strategy best response to p. When player 1 uses a pure strategy, however, player 2 does not have the same expected payoff from R and S, contradicting (a). When $p = \frac{2}{3}$, player 1 assigns a positive probability only to R and P, but then the pure strategy R is strictly dominated by P for player 2. This contradicts (b). In summary, there is no mixed-strategy Nash equilibrium such that player 2 puts a positive probability only on R and S. By analogous arguments, there is no mixed-strategy Nash equilibrium in which a player assigns a positive probability to only two pure strategies.

Finally, we show that there is a unique Nash equilibrium in which a player uses a completely mixed strategy. This is in fact a combination where each player assigns an equal probability to all pure strategies, written as $(\frac{1}{3}, \frac{1}{3}, \frac{1}{3})$ (the probabilities are usually listed in the order of pure strategies in the matrix representation). If the opponent is using this mixed strategy, any pure strategy gives the same expected payoff of 0. Hence it is a best response to play the same mixed strategy. Moreover, if a player (say player 2) uses a different completely mixed strategy, then not all pure strategies of player 1 give the same expected payoff. By the above logic, if player 1 does not play a completely mixed strategy, then player 2's initial completely mixed strategy is not a best response to it. Hence there is no other Nash equilibrium in which a player uses a completely mixed strategy.

To summarize the above argument, we have shown that the rock-paper-scissors game has a unique Nash equilibrium in mixed strategies, which is a symmetric equilibrium such that both players assign $\frac{1}{3}$ to all pure strategies, i.e., $((\frac{1}{3}, \frac{1}{3}, \frac{1}{3}), (\frac{1}{3}, \frac{1}{3}, \frac{1}{3}))$.

3.7 Existence Theorem**

Nash [12] proved that any finite game (where the number of players and the number of their pure strategies are both finite) possesses a Nash equilibrium in mixed strategies. Therefore, predicting a game's outcome as a Nash equilibrium is not vacuous. Recall that an "equilibrium" is a stable outcome of a game. One interpretation of a Nash equilibrium is that, starting from the strategy combination, when all players try to maximize their own (expected) payoff given the others' strategies, they reach the original strategy combination. That is, a Nash equilibrium strategy combination is a *fixed point*[11] of (expected) payoff-maximizing behavior. Thus the existence of a Nash equilibrium is warranted by a Fixed Point Theorem.

Theorem 3.1 *(Existence Theorem for Nash Equilibria) For any normal-form game* $G = (\{1, 2, \ldots, n\}, S_1, S_2, \ldots, S_n, u_1, u_2, \ldots, u_n)$ *such that* n *is finite and* S_i *is finite for each* $i \in \{1, 2, \ldots, n\}$*, there exists a Nash equilibrium in mixed strategies in the extended game* $G' = (\{1, 2, \ldots, n\}, \Delta(S_1), \Delta(S_2), \ldots, \Delta(S_n), Eu_1, Eu_2, \ldots, Eu_n)$.

Proof Let $\Sigma := \Delta(S_1) \times \cdots \times \Delta(S_n)$ be the set of mixed strategy combinations. For each $i \in \{1, 2, \ldots, n\}$ and each mixed strategy combination $\sigma \in \Sigma$, let $\sigma_{-i} := (\sigma_1, \ldots, \sigma_{i-1}, \sigma_{i+1}, \ldots, \sigma_n)$ be the combination of players other than i in σ. For each $\sigma \in \Sigma$, define

$$BR(\sigma) := BR_1(\sigma_{-1}) \times BR_2(\sigma_{-2}) \times \cdots \times BR_n(\sigma_{-n}).$$

This is a correspondence from Σ to Σ.

If $\sigma^* = (\sigma_1^*, \sigma_2^*, \ldots, \sigma_n^*)$ is a Nash equilibrium, by definition,

$$(\sigma_1^*, \sigma_2^*, \ldots, \sigma_n^*) \in BR_1(\sigma_{-1}^*) \times BR_2(\sigma_{-2}^*) \times \cdots \times BR_n(\sigma_{-n}^*),$$

[11] This subsection uses some concepts from Euclidean topology. The Appendix Section A.1 contains basic definitions and results (without proofs).

so that σ^* is a fixed point of the correspondence BR.

We use Kakutani's Fixed Point Theorem as follows.

Kakutani's Fixed Point Theorem. Let $X \subset \Re^k$ be a non-empty, compact, and convex set. If a correspondence $F : X \to\to X$ is non-empty, convex, and closed-valued and upper hemi-continuous, then there exists $x^* \in X$ such that $x^* \in F(x^*)$.

In order to apply Kakutani's theorem, we need to show that the set of mixed strategy combinations Σ is compact and convex, and the correspondence BR satisfies the sufficient properties. We ask readers to prove that Σ is compact and convex in Problem 3.10.

Let us show that for any $\sigma \in \Sigma$, the value $BR(\sigma)$ is non-empty, i.e., for any $i \in \{1, 2, \ldots, n\}$ and any $\sigma \in \Sigma$, $BR_i(\sigma_{-i})$ is non-empty. By the definition for best response, $BR_i(\sigma_{-i})$ is the set of maximizers of a continuous function Eu_i over the compact set $\Sigma_{-i} := \Delta(S_1) \times \cdots \times \Delta(S_{i-1}) \times \Delta(S_{i+1}) \times \cdots \times \Delta(S_n)$. Hence it is non-empty by Bolzano-Weierstrass' Theorem (Proposition A.4 in Appendix).

Next, we prove that, for any $i \in \{1, 2, \ldots, n\}$ and any $\sigma \in \Sigma$, $BR_i(\sigma_{-i})$ is convex. Take arbitrary $\sigma_i, \sigma_i' \in BR_i(\sigma_{-i})$, and any real number $\alpha \in [0, 1]$. Consider the convex combination $p_i := \alpha\sigma_i + (1-\alpha)\sigma_i'$. It suffices to prove that $p_i \in BR_i(\sigma_{-i})$. Without loss of generality, let $i = 1$. By the definition of the expected payoff,

$$\begin{aligned}
Eu_1(p_1, \sigma_{-1}) &= \sum_{(s_1,\ldots,s_n)\in S} p_1(s_1)\sigma_2(s_2)\cdots\sigma_n(s_n)u_1(s_1, \ldots, s_n) \\
&= \sum_{(s_1,\ldots,s_n)\in S} (\alpha\sigma_1 + (1-\alpha)\sigma_1')(s_1)\sigma_2(s_2)\cdots\sigma_n(s_n)u_1(s_1, \ldots, s_n) \\
&= \alpha \sum_{(s_1,\ldots,s_n)\in S} \sigma_1(s_1)\sigma_2(s_2)\cdots\sigma_n(s_n)u_1(s_1, \ldots, s_n) \\
&\quad + (1-\alpha) \sum_{(s_1,\ldots,s_n)\in S} \sigma_1'(s_1)\sigma_2(s_2)\cdots\sigma_n(s_n)u_1(s_1, \ldots, s_n) \\
&= \alpha Eu_1(\sigma_1, \sigma_{-1}) + (1-\alpha)Eu_1(\sigma_1', \sigma_{-1}). \qquad (3.3)
\end{aligned}$$

Since $\sigma_1, \sigma_1' \in BR_1(\sigma_{-1})$, for any $x \in \Delta(S_1)$,

$$Eu_1(\sigma_1, \sigma_{-1}) \geqq Eu_1(x, \sigma_{-1}),$$

and

$$Eu_1(\sigma_1', \sigma_{-1}) \geqq Eu_1(x, \sigma_{-1})$$

hold. By (3.3), for any $x \in \Delta(S_1)$, we have

$$Eu_1(p_1, \sigma_{-1}) = \alpha Eu_1(\sigma_1, \sigma_{-1}) + (1-\alpha)Eu_1(\sigma_1', \sigma_{-1}) \geqq Eu_1(x, \sigma_{-1}),$$

that is, p_1 is also a maximizer. Therefore $p_1 \in BR_1(\sigma_{-1})$, and thus $BR_i(\sigma_{-i})$ is convex for any i. Because a finite product of convex sets is also convex (Proposition A.5 in Appendix), $BR(\sigma)$ is convex for any $\sigma \in \Sigma$.

The correspondence BR is not only closed-valued but also compact-valued. That is, for any $i \in \{1, 2, \ldots, n\}$ and any $\sigma \in \Sigma$, $BR_i(\sigma_{-i})$ is closed and bounded. Since the range Σ of BR is bounded, we only need to show that $BR_i(\sigma_{-i})$ is closed. Let $i = 1$ again, and take any convergent sequence $\{\sigma_1^{(1)}, \sigma_1^{(2)}, \ldots\}$ in $BR_1(\sigma_{-1})$. Denote $\sigma_1 := \lim_{k\to\infty} \sigma_1^{(k)}$. Since each $\sigma_1^{(k)}$ belongs to $BR_1(\sigma_{-1})$, for any $x \in \Delta(S_1)$, it holds that

$$Eu_1(\sigma_1^{(k)}, \sigma_{-1}) \geqq Eu_1(x, \sigma_{-1}).$$

By the continuity of the expected payoff, the weak inequality holds at the limit as well:

$$Eu_1(\sigma_1, \sigma_{-1}) \geqq Eu_1(x, \sigma_{-1}).$$

Therefore $BR_1(\sigma_{-1})$ is closed.

Lastly, we show that $BR(\sigma)$ is upper hemi-continuous. Since Σ is compact and BR is shown to be compact-valued, we prove that the graph of BR is closed, which is equivalent to upper hemi-continuity in this case (Proposition A.6 in Appendix). The graph of the correspondence BR is a set,

$$gr(BR) = \{(\sigma, \sigma') \in \Sigma \times \Sigma \mid \sigma' \in BR(\sigma)\}.$$

Take an arbitrary convergent sequence $\{(\sigma^{(1)}, \sigma'^{(1)}), (\sigma^{(2)}, \sigma'^{(2)}), \ldots\}$ in $gr(BR)$ and let $(\sigma, \sigma') := \lim_{k\to\infty}(\sigma^{(k)}, \sigma'^{(k)})$. Because Σ is compact, $(\sigma, \sigma') \in \Sigma \times \Sigma$. For any $k = 1, 2, \ldots$ and any $i \in \{1, 2, \ldots, n\}$,

$$\sigma_i'^{(k)} \in BR_i(\sigma_{-i}^{(k)}),$$

so that for any $x \in \Delta(S_i)$,

$$Eu_i(\sigma_i'^{(k)}, \sigma_{-i}^{(k)}) \geqq Eu_i(x, \sigma_{-i}^{(k)}).$$

Again, using the continuity of Eu_i, we have

$$Eu_i(\sigma_i', \sigma_{-i}) \geqq Eu_i(x, \sigma_{-1}).$$

This implies that for any i, $\sigma_i' \in BR_i(\sigma_{-i})$, or $\sigma' \in BR(\sigma)$. Hence the graph of the correspondence BR is closed.

We have proved that the correspondence BR from Σ to itself satisfies all the sufficient conditions of Kakutani's Fixed Point Theorem. Therefore, there exists a fixed point (a Nash equilibrium in mixed strategies) in Σ. □

For existence of Nash equilibria in games with uncountable number of strategies (such as Cournot game), see Sect. 1.3.3. of Fudenberg and Tirole [8] and references therein.

3.8 Rationalizability**

The stability logic of a Nash equilibrium can be restated as "if you believe that all others follow a Nash equilibrium strategy combination, then you are also willing to follow the strategy prescribed by the Nash equilibrium". It is, however, unclear why all players come to hold the same belief of a particular Nash equilibrium strategy combination. One such an example is the Cournot game in Sect. 3.2. The players can reach the Nash equilibrium through a sequence of best response reasoning, based on common knowledge of the game and rationality of both players, starting from an arbitrary belief about the opponent's action.

However, in general games, even if players have common knowledge of the game's structure and everyone's rationality, they may not be able to play a Nash equilibrium. To see this, consider the Meeting Game (or the Battle of the Sexes) in Table 3.1. If Ann believes that Bob will play strategy A, then she plays A. If, at the same time, Bob believes that Ann will play strategy B, he chooses B. As a result, they play (A, B). This shows that, if players differ in their beliefs, they cannot reach a Nash equilibrium. Furthermore, even if we consider a sequence of best response to best response, and so on, the play cycles between (A, B) and (B, A) if the players start from different beliefs.[12]

What can players play from the common knowledge of the structure of the game and everyone's rationality? For simplicity, we consider a two-player game $G = (\{1, 2\}, S_1, S_2, u_1, u_2)$. We can think of the following reasoning process.

(0) For each $i \in \{1, 2\}$, player i thinks that the opponent can choose any strategy from $\Delta(S_j)$ $(j \neq i)$. Define $\Sigma_j^0 = \Delta(S_j)$.

(1) Based on (0), player i reasons that player j has a belief $p_i \in \Sigma_i^0$ about i's choice. Player i may also think that player j would choose a best response to p_i. Hence the set of possible mixed strategies chosen by player j is the set of mixed strategies which maximize $Eu_j(\cdot, p_i)$ for **some** $p_i \in \Sigma_i^0$:

$$\Sigma_j^1 = \{p_j \in \Sigma_j^0 \mid \exists p_i \in \Sigma_i^0;\ \ Eu_j(p_j, p_i) \geqq Eu_j(q, p_i)\ \forall q \in \Sigma_j^0\}.$$

(2) The opponent also makes the same reasoning. Hence player i can guess that player j thinks that player i would choose some $p_i \in \Delta(\Sigma_i^1)$. (Because player j

[12]Note that some learning processes such as the fictitious play admit convergence of belief distributions on the opponents' strategies to a Nash equilibrium distribution in a class of games including 2×2 games. However, in this section we are not seeking to make the players' reasoning converge to a Nash equilibrium, and thus details are omitted. Interested readers are referred to classic papers by Brown [5] and Robinson [18], and recent papers by Hofbauer and Sandholm [9] and Berger [2].

Table 3.6 An example

1\2	L	R
U	3, 1	0, 0
M	1, 0	1, 2
D	0, 1	3, 0

may not be sure of a particular element in Σ_i^1, it is possible that player j believes in a probability distribution over Σ_i^1.) Player i again needs to consider only best responses by player j to some $p_i \in \Delta(\Sigma_i^1)$. Hence,

$$\Sigma_j^2 = \{p_j \in \Sigma_j^1 \mid \exists p_i \in \Delta(\Sigma_i^1);\ \ Eu_j(p_j, p_i) \geqq Eu_j(q, p_i),\ \forall q \in \Sigma_j^1\}$$

is the set of possible choices by the opponent j.

The set of strategies remaining after continuing this reasoning process *ad infinitum* are those that players can play from only the common knowledge of the game and mutual rationality. Mathematically, it is defined as follows (by Bernheim [3] and Pearce [17]).[13]

Definition 3.4 For each $i = 1, 2$, let $\Sigma_i^0 = \Delta(S_i)$. For each $m \geqq 1$ and each $i = 1, 2$, define

$$\Sigma_i^m := \{p_i \in \Sigma_i^0 \mid \exists p_j \in \Delta(\Sigma_j^{m-1});\ \ Eu_i(p_i, p_j) \geqq Eu_i(q, p_j)\ \forall q \in \Sigma_i^{m-1}\}.$$

Then the set of *rationalizable strategies* of player i is

$$R_i := \cap_{m=0}^{\infty} \Sigma_i^m.$$

Let us find the set of rationalizable strategies of the game in Table 3.6.

First, player 1's pure strategies U and D belong to Σ_1^1. This is because each one is a best response to $p_2 = L$ and R respectively. However, M is not a best response to any mixed strategy by player 2. Hence $M \notin \Sigma_1^1$. Both pure strategies of player 2 belong to Σ_2^1.

Next, we can see that R for player 2 does not belong to Σ_2^2. This is because R is not a best response to any belief within $\Delta(\{U, D\})$. For player 1, $\Delta(\Sigma_2^1) = \Delta(\{L, R\})$ implies that both U and D continue to belong to Σ_1^2.

Finally, since player 2 has only L to choose after the second step, only U belongs to Σ_1^3. The process stops here, and the unique rationalizable strategy for player 1 is U and the unique rationalizable strategy of player 2 is L.

One can see that the above reasoning process resembles the process of iterative elimination of dominated strategies. Notice, however, that the game in Table 3.6 does not have a strictly dominated strategy for any player in the sense of Definition 2.1. In the process of rationalizability, we consider mixed strategies, and thus we should

[13]Good references are Osborne [15] Chap. 12 and Osborne and Rubinstein [16] Chap. 4.

extend the definition of dominance in mixed strategies. (We also describe domination by some strategy, instead of a binary relation.)

Definition 3.5 Given a player i, a mixed strategy $\sigma_i \in \Delta(S_i)$ is *strictly dominated* if there exists $\sigma_i' \in \Delta(S_i)$ such that for any $s_{-i} \in S_{-i}$,

$$Eu_i(\sigma_i, s_{-i}) < Eu_i(\sigma_i', s_{-i}).$$

In the game of Table 3.6, M is strictly dominated, for example, by $\frac{1}{2}U+\frac{1}{2}D$. As we explain below, for two-player games, if a strategy is strictly dominated in the sense of Definition 3.5, then it is never a best response to any belief of the opponent's strategy. Hence, the strategies remaining after eliminating strictly dominated strategies in mixed strategies correspond to the rationalizable strategies.

Lemma 3.1 *Take any two-player normal-form game $G = (\{1, 2\}, S_1, S_2, u_1, u_2)$ with finite pure strategies of all players. For any $i \in \{1, 2\}$, player i's strategy $\sigma_i \in \Delta(S_i)$ is strictly dominated if and only if σ_i is not a best response to any $p_j \in \Delta(S_j)(j \neq i)$.*

Proof We follow the proof in Pearce [17]. Fix $i \in \{1, 2\}$. Assume that $\sigma_i \in \Delta(S_i)$ is strictly dominated. Then there exists $\sigma_i' \in \Delta(S_i)$ such that for any $s_j \in S_j$, $Eu_i(\sigma_i, s_j) < Eu_i(\sigma_i', s_j)$ holds. Hence for any $p_j \in \Delta(S_j)$,

$$\sum_{s_j \in S_j} p_j(s_j)Eu_i(\sigma_i, s_j) < \sum_{s_j \in S_j} p_j(s_j)Eu_i(\sigma_i', s_j) \iff Eu_i(\sigma_i, p_j) < Eu_i(\sigma_i', p_j),$$

that is, σ_j is not a best response to p_j.

To prove the converse, assume that $\sigma_i \in \Delta(S_i)$ is never a best response to any belief $p_j \in \Delta(S_j)$. Then for each $p_j \in \Delta(S_j)$, there exists a better strategy $\hat{\sigma}_i(p_j) \in \Delta(S_i)$ than σ_i such that $Eu_i(\hat{\sigma}_i(p_j), p_j) > Eu_i(\sigma_i, p_j)$.

Construct a new zero-sum game $\overline{G} = (\{1, 2\}, S_1, S_2, \overline{u}_1, \overline{u}_2)$ by defining $\overline{u}_i(s_i, s_j) = u_i(s_i, s_j) - Eu_i(\sigma_i, s_j)$ and $\overline{u}_j(s_i, s_j) = -\overline{u}_i(s_i, s_j)$. By the Existence Theorem for Nash Equilibria 3.1, the zero-sum game $\overline{G}$ has a Nash equilibrium (σ_i^*, σ_j^*) in mixed strategies. By the definition of Nash equilibrium, for any $\sigma_j \in \Delta(S_j)$,

$$\begin{aligned} E\overline{u}_j(\sigma_i^*, \sigma_j) &\leqq E\overline{u}_j(\sigma_i^*, \sigma_j^*) \\ \iff E\overline{u}_i(\sigma_i^*, \sigma_j) &\geqq E\overline{u}_i(\sigma_i^*, \sigma_j^*) \\ &\geqq E\overline{u}_i(\hat{\sigma}_i(\sigma_j^*), \sigma_j^*) \quad \text{(Nash equilibrium)} \\ &> E\overline{u}_i(\sigma_i, \sigma_j^*) = 0. \quad \text{(Definition of } \hat{\sigma}_i \text{ and } \overline{u}_i.) \end{aligned}$$

That is, there exists σ_i^* such that for any $\sigma_j \in \Delta(S_j)$, we have $E\overline{u}_i(\sigma_i^*, \sigma_j) > 0$, or equivalently,

$$Eu_i(\sigma_i^*, \sigma_j) > Eu_i(\sigma_i, \sigma_j).$$

Hence σ_i is strictly dominated. □

The process of rationalizability eliminates strategies that are never a best response to any belief. By Lemma 3.1, this is equivalent to eliminating strictly dominated strategies. Therefore, the following equivalence holds.

Proposition 3.4 *Take any two-player, normal-form game* $G = (\{1, 2\}, S_1, S_2, u_1, u_2)$ *with finitely many pure strategies of all players. For any* $i \in \{1, 2\}$, *player* i*'s strategy* $\sigma_i \in \Delta(S_i)$ *is rationalizable if and only if it survives the iterative elimination of strictly dominated strategies.*

How about the relationship to a Nash equilibrium? A Nash equilibrium is a strategy combination such that (i) all players choose an expected payoff-maximizing strategy given a belief of the other players' strategies, and (ii) the belief corresponds to the Nash equilibrium strategies of other players. Therefore, for two-player games, each player's belief is a strategy that the rational opponent can take, and this is so for any step of iterative best response reasoning. That is, any strategy with a positive probability in a Nash equilibrium is rationalizable.

Proposition 3.5 *For any two-player game* $G = (\{1, 2\}, S_1, S_2, u_1, u_2)$, *if* $(\sigma_1^*, \sigma_2^*) \in \Delta(S_1) \times \Delta(S_2)$ *is a Nash equilibrium, then each pure strategy in the support of* σ_i^* *is rationalizable for any* $i \in \{1, 2\}$.

However, the set of rationalizable strategy **combinations** is in general much larger than the set of Nash equilibria. For example, consider the Meeting Game in Table 3.1. The set of Nash equilibria in mixed strategies consists of three strategy combinations. By contrast, because the set of rationalizable strategies for each player is all mixed strategies, the set of rationalizable strategy combinations corresponds to the set of all mixed strategy combinations.

There is more than one way to extend the notion of rationalizability to games with three or more players. This is because the range of beliefs differ depending on whether players assume that their opponents' choices are independent or they assume that their opponents may correlate their choices using some common randomization device. (For correlation of strategies, see also Sect. 4.8.)

For example, from player 1's point of view, if the other players $j = 2, 3, \ldots, n$ randomize pure strategies only independently, then the set of possible strategy combinations by the opponents is $\Delta(S_2) \times \Delta(S_3) \times \cdots \times \Delta(S_n)$. However, it is the set of all probability distributions over $S_2 \times S_3 \times \cdots \times S_n$ if the opponents can correlate their strategies. The latter set $\Delta(S_2 \times S_3 \times \cdots \times S_n)$ is a much larger set than $\Delta(S_2) \times \Delta(S_3) \times \cdots \times \Delta(S_n)$.

We cite the example by Osborne and Rubinstein [16] to illustrate in a three-person game how the set of rationalizable strategies changes depending on whether a player believes in only independent randomizations by the opponents or allows for the possibility of correlations.

The game in Table 3.7 is a three-player game, in which player 1 chooses between rows U and D, player 2 chooses between columns L and R, and player 3 chooses among the matrices M_1, M_2, M_3, and M_4. All players have the same payoff, as shown in the table.

Table 3.7 Three-Player common interest game

1\2	L	R
U	8	0
D	0	0

3: M_1

1\2	L	R
U	4	0
D	0	4

3: M_2

1\2	L	R
U	0	0
D	0	8

3: M_3

1\2	L	R
U	3	3
D	3	3

3: M_4

Each of the pure strategies of players 1 and 2 is a best response to some belief. Strategy M_2 of player 3 is a best response if players 1 and 2 correlate their strategies to play (U, L) with probability $\frac{1}{2}$ and (D, R) with $\frac{1}{2}$. However, it is never a best response if players 1 and 2 play only independent mixed strategies. Let p be the probability of U by player 1 and q be the probability of L by player 2. The expected payoff of each pure strategy of player 3 is

$$\begin{aligned} Eu_3(p, q, M_1) &= 8pq \\ Eu_3(p, q, M_2) &= 4pq + 4(1-p)(1-q) \\ Eu_3(p, q, M_3) &= 8(1-p)(1-q) \\ Eu_3(p, q, M_4) &= 3. \end{aligned}$$

For any combination of (p, q), M_2 gives a lower expected payoff than the maximum of the expected payoffs of the other three pure strategies.

Chapter 4 of Osborne and Rubinstein [16] shows an equivalence result of n-person, finite pure strategy games such that, when players believe in correlated strategy combinations of their opponents, the set of rationalizable strategies coincides with the set of strategies that survive iterative elimination of strictly dominated strategies.

Because the set of rationalizable strategy combinations is larger than the set of Nash equilibria in general, we cannot say that players will play a Nash equilibrium, even if all players have common knowledge of the game and rationality of all players and if they can reason infinitely many times. Then, what amount of knowledge is sufficient to play a Nash equilibrium or another equilibrium for arbitrary games? This question has only partial answers so far (see, for example, Chap. 5 of Osborne and Rubinstein [16] and Aumann and Brandenberger [1]).

Problems

3.1 We have defined the set of best responses in mixed strategies as follows:

$$BR_i(\sigma_{-i}) = \{\sigma_i \in \Delta(S_i) \mid Eu_i(\sigma_i, \sigma_{-i}) \geqq Eu_i(x, \sigma_{-i}) \ \forall x \in \Delta(S_i)\}.$$

Alternatively, we can compare the expected payoffs with those of pure strategies only:

$$\overline{BR}_i(\sigma_{-i}) = \{\sigma_i \in \Delta(S_i) \mid Eu_i(\sigma_i, \sigma_{-i}) \geqq Eu_i(x, \sigma_{-i}) \ \forall x \in S_i\}.$$

Prove that, for any $\sigma_{-i} \in \Delta(S_1) \times \cdots \times \Delta(S_{i-1}) \times \Delta(S_{i+1}) \times \cdots \times \Delta(S_n)$,

$$BR_i(\sigma_{-i}) = \overline{BR}_i(\sigma_{-i}).$$

(You can assume that the set of pure strategies is finite for all players.)

3.2 Find all Nash equilibria in pure strategies of the following games.
(a) Consider a market in which only two firms, 1 and 2, operate. The two firms choose production quantities simultaneously, and the game ends. When firm 1 chooses quantity q_1 and firm 2 chooses q_2, the market price is

$$P(q_1, q_2) = a - (q_1 + q_2).$$

The firms incur production costs proportional to their quantity such that firm 1's total cost is $TC_1(q_1) = c \cdot q_1$ and firm 2's total cost is $TC_2(q_2) = c \cdot q_2$, where $0 < c < a$. The payoff of a firm is the profit. Firm 1 has a capacity constraint of production so that its set of pure strategies (quantities) is $S_1 = [0, \frac{a-c}{4}]$. Firm 2 has no capacity constraint and chooses a pure strategy from $S_2 = [0, \infty)$.

(b) Consider a market in which only two firms, 1 and 2, operate. The two firms choose prices p_1 and p_2 simultaneously, and the game ends. As in Sect. 3.3.2, their products are perfect substitutes, and all consumers buy from the cheaper firm. Specifically, when the price combination is (p_1, p_2), the demand that firm $i = 1, 2$ gets is

$$D_i(p_1, p_2) = \begin{cases} 5 - p_i & \text{if } p_i < p_j \\ \frac{1}{2}(5 - p_i) & \text{if } p_i = p_j \\ 0 & \text{if } p_i > p_j. \end{cases}$$

Unlike the model in Sect. 3.3.2, the two firms choose integer prices from $S_1 = S_2 = \{1, 2, 3, 4, 5\}$. Their production costs are the same, and if firm i produces q_i units, the total cost is $TC_i(q_i) = q_i$. For each $i = 1, 2$, firm i's payoff is its profit $u_i(p_i, p_j) = (p_i - 1) \cdot D_i(p_i, p_j)$.

3.3 Find all Nash equilibria in mixed strategies of the following two-player games.

(a)

P1\P2	L	R
U	0, 1	3, 3
M	5, 2	0, 0
D	1, 8	1, 7

(Hint: This is easy. Draw a graph of the expected payoffs of the pure strategies of player 1.)
(b)

P1\P2	L	R
U	0, 1	3, 3
M	5, 2	0, 0
D	2, 8	2, 7

(Hint: This is difficult. There is a mixed strategy equilibrium involving D.)

(c)

P1\P1	L	R
U	0, 3	3, 2
M	5, 0	0, 4
D	1, 1	1, 1

(d) The armies of country 1 and country 2 are marching from the south and the north respectively towards a valley. Let the generals of country 1 and 2 be players. As they are approaching from the opposite directions, they cannot know each other's decision. Each of the generals has two pure strategies, Advance and Retreat, and chooses one simultaneously, and the game ends. If both generals choose to advance, they meet in the valley and 1000 people get killed in each army. Thus the payoff is -1000 for each. If one advances and the other retreats, the one that advanced occupies the valley and gets the payoff of 100, while the one who retreated gets the payoff of -10. If both retreat, their payoff is 0 each.

3.4 (Weibull [21]) Consider the normalized symmetric 2×2 game in Table 3.8.
(a) Find all Nash equilibria (in pure and mixed strategies) when $a_1 < 0$ and $a_2 > 0$.
(b) Find all Nash equilibria (in pure and mixed strategies) when $a_1 > 0$ and $a_2 > 0$.
(c) Find all Nash equilibria (in pure and mixed strategies) when $a_1 < 0$ and $a_2 < 0$.
(d) Find all Nash equilibria (in pure and mixed strategies) when $a_1 > 0$ and $a_2 < 0$.

3.5 Six businessmen, called P1, P2, P3, P4, P5, and P6, are considering whether to invest in golf (pure strategy G) or in wine tasting (pure strategy W). When the businessmen meet one another at a party, they would like to discuss a shared hobby.

First, the businessmen choose G or W simultaneously. After that, three pairs are randomly formed. The probability of being matched with any one of the five others is the same. After a matching, payoffs are given according to Table 3.9, depending only on the pair's strategy.

Table 3.8 Normalized symmetric 2×2 game

P1\P2	1	2
1	a_1, a_1	0, 0
2	0, 0	a_2, a_2

Table 3.9 Golf-Wine game

Pi\Pj	G	W
G	2, 2	1, 0
W	0, 1	3, 3

To find the pure Nash equilibria of this game, it is convenient to focus on the players' numbers instead of individual names. Is there a Nash equilibrium in pure strategies with the following property? If yes, prove that it is a Nash equilibrium. If no, show who wants to change strategies.
(a) All players choose the same pure strategy.
(b) Only one player chooses G and the other five choose W.
(c) Two players choose G and the other four choose W.
(d) Three players choose G and the other three choose W.

3.6 Firm X has been paying a fixed salary to employees. There are two employees called Ms. A and Mr. B. Each employee chooses a level of effort as a (pure) strategy. Specifically, their set of pure strategies is $S_A = S_B = \{1, 2, 3\}$, and choosing effort level k means that the employee gets disutility of k. The firm benefits from the employees' effort, and if Ms. A chooses k and Mr. B chooses k', then the firm's sales is $5(k + k')$.

The players are the two employees, and they choose an effort level simultaneously. Their payoff is the salary minus effort disutility.

(a) Suppose that the fixed salary is 4 for both employees, regardless of their strategy choices. Write down the matrix representation of the two-player normal-form game and find all Nash equilibria in pure strategies. Compute the profit of the firm at each Nash equilibrium (sales minus the salary payment to the employees).

(b) Firm X decides to change the payment system to an incentive-based pay system. The firm compares the effort levels of the employees, and the one with the higher level is paid 10, while the one with the lower level is paid 1. If both employees choose the same effort level, they get 4 each as before.

Write down the matrix representation of the new game and find all Nash equilibria in pure strategies. Compute the profit of the firm at each Nash equilibrium.

(c) Discuss the economic implications from this analysis.

3.7 Consider two diary farms B and C. They send their cows to the commons. If farm B sends b cows and farm C sends c cows to the commons, the milk production (in kiloliters, $k\ell$) per a cow is

$$f(b, c) = 300 - (b + c).$$

This is the same for all cows of both farms. The function shows that as the number of cows increases, the grass that each cow can eat decreases, so that milk production decreases. Moreover, even if one farm's number of cows is constant, if the other farm increases its number of cows, all cows' production is affected. The cost of milk production is $90 \cdot z$ if a farm keeps z cows, and this is the same for both farms.

Let the price of milk per 1 $k\ell$ be 1. When farm B sends b cows and farm C sends c cows to the commons, farm B's payoff (profit) is

$$\Pi_B(b, c) = \{300 - (b + c)\}b - 90\,b,$$

and farm C's payoff is

$$\Pi_C(b, c) = \{300 - (b + c)\}c - 90\,c.$$

(a) Given farm C's number of cows c, compute the number b^* which maximizes farm B's payoff, as a function of c.

(b) When farms B and C choose b and c simultaneously, find the (pure strategy) Nash equilibrium combination.

(c) From the "social welfare" viewpoint of the two farms, the efficient number of cows is the one that maximizes the sum of the payoffs of the two farms. Let $x = b+c$ and find the efficient x.

(d) Compare the sum of the cows in the Nash equilibrium of (b) and the efficient number of total cows in (c). (This is a problem called the *Tragedy of Commons*.)

3.8 Prove that a strictly dominated strategy in the sense of Definition 3.5 does not have a positive probability in any Nash equilibrium.

3.9 Make an example of a normal-form game which has a Nash equilibrium that assigns a positive probability to a strategy that is weakly dominated by some mixed strategy.

3.10 For a normal-form game with a finite number of players and a finite number of pure strategies of all players, let Σ be the set of all mixed strategy combinations. Prove that Σ is compact and convex. (This is a part of the Proof of Theorem 3.1.)

3.11 Find all rationalizable strategy combinations of the following game.

P1\P2	L	R
U	0, 1	3, 3
M	5, 2	0, 0
D	1, 8	1, 7

3.12 A beach is divided into three zones, as illustrated below. From the left, they are called Zone 1, Zone 2, and Zone 3 respectively.

1	2	3

Each zone has 300 beachgoers, and they buy one ice cream per person from the nearest ice cream shop. Answer the following questions (a)–(d).

(a) Consider a simultaneous move game of two ice cream shops A and B. The players are the owners of the two shops, and each player's payoff is the number of ice creams sold. A pure strategy is choosing in which zone to locate. Thus the sets of pure strategies are $S_A = S_B = \{1, 2, 3\}$. The sales are computed by the fact that people buy from the nearest shop. For example, if both shops locate in Zone 1, then for all beachgoers the two shops are considered to be the same distance from them. Hence each shop sells 450 ice creams. If A locates in Zone 1 and B locates in Zone 2, then all people in Zone 1 buy from A, and people in Zones 2 and 3 buy from B. Therefore A's payoff is 300 and B's payoff is 600.

Continue to compute this way and make a matrix representation of the two-person normal-form game of players A and B, where A is the row player and B is the column player.

(b) Based on (a), find all pure Nash equilibria.

(c) Consider a simultaneous move game of three ice cream shops, A, B, and C. Again, the owners are the players, and their pure strategies are choosing in which zone to locate: $S_A = S_B = S_C = \{1, 2, 3\}$. Their payoff is the number of ice creams they sell. For example, if A and B locate in Zone 1 and C locates in Zone 3, then the beachgoers in Zone 1 are split between A and B. For the people in Zone 2, all shops are the same distance from them, so each shop gets 100 each. All people in Zone 3 will buy from C. Therefore, A's payoff is $150 + 100 = 250$, B's is also 250, and C's payoff is $100 + 300 = 400$.

Continue to compute this way and make a matrix representation of the three-person normal-form game of players A, B, and C, where A is the row player, B is the column player, and C is the matrix player.

(d) Based on (c), find all pure Nash equilibria of the three-person game.

References

1. Aumann R, Brandenberger A (1995) Epistemic conditions for Nash equilibrium. Econometrica 63(5):1161–1180
2. Berger U (2005) Fictitious play in $2 \times n$ games. J Econ Theory 120(2):139–154
3. Bernheim D (1984) Rationalizable strategic behavior. Econometrica 52(4):1007–1028
4. Bertrand J (1883) Théorie des Richesses: revue de Théories mathématiques de la richesse sociale par Léon Walras et Recerches sur les principes mathématiques de la théorie des richesses par Augustin Cournot. Journal des Savants
5. Brown G (1951) Iterative solution of games by fictitious play. In: Koopmans TC (ed) Activity analysis of production and allocations, Wiley, New York, pp 374–376
6. Bulow J, Geanakoplos J, Klemperer P (1985) Multiemarket oligopoly: strategic substitutes and complements. J Polit Econ 93(3):488–511
7. Cournot A (1838) Recherches sur les Principes Mathèmatiques de la Théorie des Richesses (Researches into the Mathematical Principles of the Theory of Wealth), Paris. (English translation, 1897)
8. Fudenberg D, Tirole J (1991) Game theory. MIT Press, Cambridge
9. Hofbauer J, Sandholm W (2002) On the global convergence of stochastic fictitious play. Econometrica 70(6):2265–2294

10. Kreps D (1988) Notes on the theory of choice. Westview Press, Boulder
11. Luce D, Raiffa H (1957) Games and Decisions. Wiley, New York
12. Nash J (1950) Equilibrium points in n-person games. In: Proceedings of the National Academy of Sciences of the United States of America 36(1):48–49
13. Nash J (1951) Non-cooperative games. Ann Math 54(2):286–295
14. von Neumann J, Morgenstern O (1944) Theory of games and economic behavior. Princeton University Press, Princeton
15. Osborne M (2003) An introduction to game theory. Oxford University Press, Oxford
16. Osborne M, Rubinstein A (1994) A course in game theory. MIT Press, Cambridge
17. Pearce D (1984) Rationalizable strategic behavior and the problem of perfection. Econometrica 52(4):1029–1050
18. Robinson J (1951) An iterative method of solving a game. Ann Math 54(2):296–301
19. Rubinstein A (2012) Lecture notes in microeconomic theory: the economic agent, 2nd ed. (See also his website for updates.) Princeton University Press, Princeton
20. Watson J (2007) Strategy: an introduction to game theory, 2nd edn. Norton, New York
21. Weibull J (1995) Evolutionary game theory. MIT Press, Cambridge

Chapter 4
Backward Induction

4.1 Extensive Form Games

So far, we have focused on situations in which players essentially make decisions only once and simultaneously for the entire game. However, there are games in which sequential decision-making should be explicitly considered. For example, suppose that in the Meeting Game described in Sect. 3.1, Ann's class ends earlier than Bob's, and Ann can choose whether to stay at her classroom or to go to Bob's classroom before Bob decides. When his class ends, Bob can see which strategy Ann has taken, and then he can choose between Ann's room and his room. In this case, we should explicitly deal with the different strategic situations for Ann and Bob. While Ann has the same situation as in the simultaneous-move game (choose rooms without knowing Bob's choice), Bob's strategic situation is not the same as before. He can choose which room to go to, depending on what Ann did. Moreover, this is known to Ann, and hence Ann should consider what Bob would do, depending on what she does. The **contingent decision-making** by Bob and **reasoning of others' reactions** by Ann are new aspects that appear when sequential decisions are involved.

As before, the strategies are to be lined up before the game is actually played. Before playing the game, Bob must consider the two possibilities for Ann's choice. Hence, a *strategy* of a sequential decision-making game is a complete plan of what to do in every possible situation that a player may encounter during the game. In this sequential decision Meeting Game, a strategy for Bob is not the action of going to Ann's room or staying, but a pair of actions: what to do if Ann chose A and what to do if she chose B.

Games with sequential decision-making are called *extensive-form games*, and their expressions and analyses should be different from those of normal-form (strategic-form) games. We explain this by formulating a normal-form game from the sequential decision Meeting Game (the resulting normal-form game is called the *induced normal form* of the original game) and comparing with a different expression and analysis to conclude how a rational equilibrium should be derived for extensive-form games.

T. Fujiwara-Greve, *Non-Cooperative Game Theory*, Monographs in Mathematical Economics 1, DOI 10.1007/978-4-431-55645-9_4

Table 4.1 Induced normal-form of the sequential decision meeting game

A\B	(A, A)	(A, B)	(B, A)	(B, B)
A	2, 1	2, 1	0, 0	0, 0
B	0, 0	1, 2	0, 0	1, 2

First, we formally construct the set of *strategies* of the two players. Ann is the first mover, and it is clear that her actions at her unique decision point are her strategies: $S_a = \{A, B\}$. By contrast, Bob's strategies are pairs of actions for each of Ann's possible choices. With a slight abuse of the notation, let us denote Bob's *actions* to go to Ann's room or to stay at his room as A and B respectively. A strategy for Bob is an action plan, written as $s_B = (x, y) \in \{A, B\}^2$ meaning that Bob chooses action $x \in \{A, B\}$ (resp. $y \in \{A, B\}$) when he learns that Ann chose action A (resp. B). Then his set of strategies consists of four possible plans; $S_b = \{(A, A), (A, B), (B, A), (B, B)\}$. Using these strategy sets, we can construct a 2×4 matrix representation of the sequential decision Meeting Game as shown in Table 4.1.

The payoffs are computed as follows. If Ann chooses A, then Bob's strategies (A, A) and (A, B) both result in the same outcome, that the two players meet at Ann's classroom. Thus Ann's payoff is 2 and Bob's is 1. Similarly, both (B, A) and (B, B) give the payoff combination of $(0, 0)$. If Ann chooses B, (A, B) and (B, B) give the same payoff combination $(1, 2)$, while (A, A) and (B, A) yield $(0, 0)$.

We can see that the induced normal-form game of an extensive-form game has more strategies than the simultaneous-move case, and many strategy combinations give the same payoff vector. This is because some parts of the action plans by later players do not matter for the cases that are excluded by the earlier movers' choices. Consequently, the number of Nash equilibria generally increases when decisions are sequentially made instead of simultaneously made. Among them, some are considered to be irrational.

For example, consider Bob's strategy (B, B), which means that whatever Ann does, he stays at his room. Ann's best response to (B, B) is B, because if she chooses A, Bob will not come. Among Bob's best responses to Ann's strategy B is (B, B) (although not a unique best response). Hence the combination $(B, (B, B))$ is a Nash equilibrium. However, this strategy combination is irrational **during the game**. If Ann somehow chose strategy A, would Bob continue to follow his plan of (B, B)? If Bob is rational, he should change his action to A to meet her and get a positive payoff.

As this example illustrates, some Nash equilibria of the induced normal-form game of an extensive-form game may not reflect rational decision-making during the game, taking into account that players are sequentially making decisions. Nash equilibria are based on payoff comparison **before the game is played**, and the comparison ignores the cases that are not reached if the original action plans are adhered to.

However, in order to explicitly analyze rational equilibria under sequential decisions, we should consider all possible strategic situations that players may potentially face during the game. For this purpose, a tree diagram (called the *game tree*) is

Fig. 4.1 A tree diagram

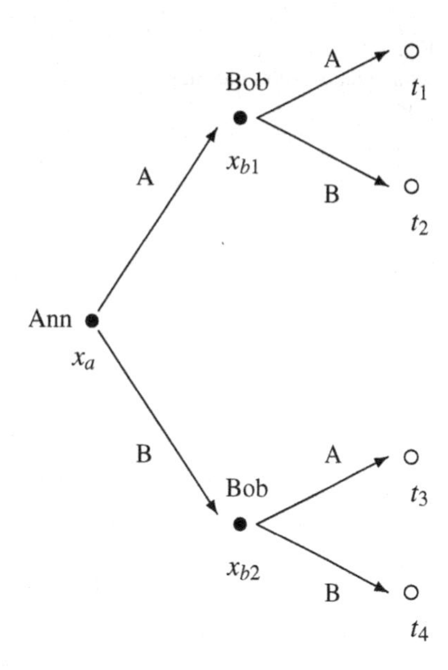

adopted in analyzing extensive-form games, and an equilibrium notion that incorporates rational choice during a game (or to exclude irrational choices at unreachable points in the game) is advocated by Selten [17, 18]. For example, Fig. 4.1 is the game tree of the sequential decision Meeting Game.

A tree diagram consists of nodes and edges. There are two classes of nodes; *decision nodes* (or *moves*), which are illustrated as black dots in Fig. 4.1, and *terminal nodes*, which are white dots. Decision nodes are the points where a player makes choices, and terminal nodes are the outcomes of the game. Therefore, all decision nodes must belong to some player. We denote by X_i the set of decision nodes which belong to player i and by Z the set of all terminal nodes. In the figures, we designate the name of the player who has the decision node, and, because the same player may have multiple decision nodes, we often name each decision node. In Fig. 4.1, $X_a = \{x_a\}$ is the set containing Ann's unique decision node, and $X_b = \{x_{b1}, x_{b2}\}$ is the set of Bob's decision nodes. The nodes t_1 to t_4 are terminal nodes.

Nodes are connected by edges, or directed arrows. Arrows correspond to *actions*, and they start from a decision node and end in either another decision node or a terminal node. The direction indicates that an action belongs to the player at the origin decision node. (Recall that an "action" does not mean a literal activity but a decision of a player. Hence, staying in your room is also an action.) For example, the two arrows originating from x_a are the two feasible actions for Ann, A and B, respectively from the top.

Decision nodes are further classified into two groups, the *origin* (or the root) and the others. The origin is the unique decision node which starts the game. Hence, there are only arrows starting from it and no arrows leading to it. Ann is the first-mover in this game, and thus her decision node x_a is the origin. The other decision nodes

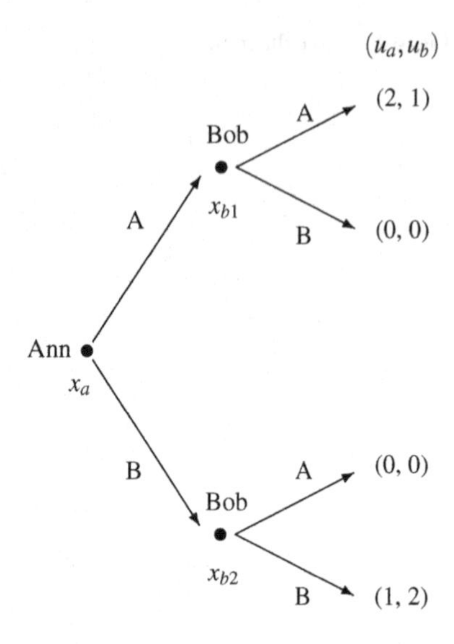

Fig. 4.2 Tree for the sequential decision meeting game

express strategic choice situations which take into account what has happened so far (called the *path* or the *history* of the game up to that point). For example, the choices by Bob at his decision node x_{b1} are his possible actions after the history that Ann chose action A, while the choices at x_{b2} are his possible actions after the history that Ann chose B. Choosing action A at x_{b1} and choosing A at x_{b2} are based on different histories, and thus must be formulated as different decisions. Moreover, in this example, Bob can observe Ann's choice, so he can choose different actions at the two decision nodes x_{b1} and x_{b2}.

Each terminal node represents a possible game ending. For example, t_1 shows the outcome in which Ann chooses A and Bob chooses A. Each outcome is evaluated by the players, and thus each terminal node must be accompanied by a payoff vector. With payoff vectors at the terminal nodes, Fig. 4.2 is a complete description of the sequential decision Meeting Game. In Fig. 4.2, Ann's payoff is the first coordinate, and Bob's is the second coordinate.[1]

A nice feature of using a game tree to analyze an extensive-form game is that we can explicitly illustrate the information structure, that is, what players will know and what they will not know during the game. This is done by drawing *information sets* for each player in the figure. A player's information sets must constitute a partition[2] of her/his decision nodes, and thus the collection of all information sets of a player is called his/her *information partition*. A player knows "an information set", meaning that (s)he can distinguish the decision nodes belonging to different information sets, but cannot distinguish the decision nodes within the same information set.

---

[1]The payoffs are usually ordered according to the decision order in the extensive-form game.

[2]A partition of a (finite) set Y is a collection of subsets of Y, $\{Y_1, Y_2, \ldots, Y_K\}$ such that for any k, k', $Y_k \cap Y_{k'} = \emptyset$ and $\cup_{k=1}^{K} Y_k = Y$.

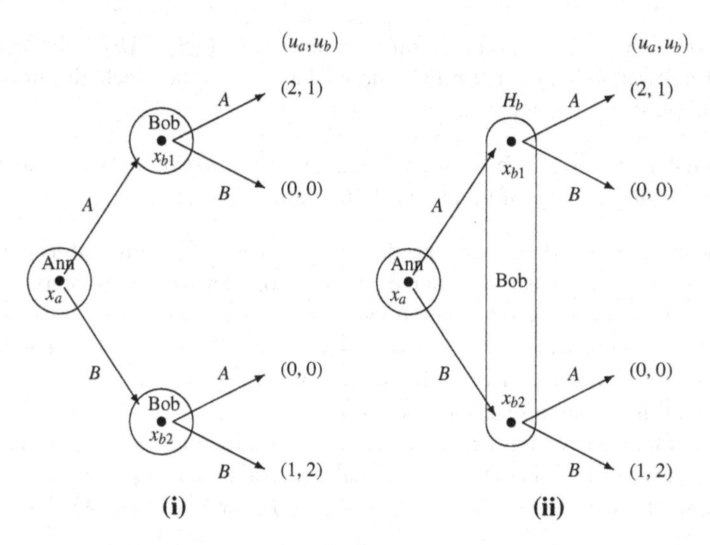

Fig. 4.3 Different information structures. **i** Perfect information game. **ii** Imperfect information game

Let us consider two possible information structures for the second-mover Bob, in the sequential decision Meeting Game. One is as we explained so far, that Bob will know which action Ann has taken, when it is his turn to choose. Then he can distinguish the two decision nodes x_{b1} and x_{b2} so that his information partition is $\{\{x_{b1}\}, \{x_{b2}\}\}$. Because he can distinguish information sets $\{x_{b1}\}$ and $\{x_{b2}\}$, he can choose actions separately in these sets. This information structure is illustrated in Fig. 4.3i by two circles around the two decision nodes. This is a *perfect information* game, in which all players know what has happened in the past perfectly.

The other possibility is that Bob cannot know Ann's actions. In this case, his two decision nodes are not distinguishable and thus belong to one information set. Bob's information partition becomes a singleton, consisting of the entire set of decision nodes $\{\{x_{b1}, x_{b2}\}\}$. This is illustrated in Fig. 4.3ii by an oval[3] $H_b = \{x_{b1}, x_{b2}\}$ as the unique information set for Bob, and the two decision nodes are contained in H_b. (Since the two decision nodes are not distinguishable, his two actions **must** have the same name at both of them here.)

For later use, let us make a notation. (Until Sect. 4.5, we focus on extensive-form games with finite number of players and their possible actions. From then on, the following notation is implicitly extended.) Fix an extensive-form game. For each player i, recall that X_i is the set of her/his decision nodes. Let $\mathcal{H}_i$ be her/his information partition, which is a partition of X_i uniquely determined by the underlying game. When X_i is finite, we can write $\mathcal{H}_i = \{H_{i1}, H_{i2}, \ldots, H_{iK}\}$.[4] Because it is a partition, any pair of elements in $\mathcal{H}_i$ do not have an intersection, and $\cup_{k=1}^{K} H_{ik} = X_i$. Each

[3]Circles and ovals are not the only way to illustrate information sets. In some literature, decision nodes in the same information sets are connected by lines.

[4]Notation can vary in the literature. In this book, the capital H is used to indicate a set of histories.

element H_{ik} $(k = 1, 2, \ldots, K)$ is an information set of player i. Using this notation, a perfect information game is formally defined as a game in which all information sets for all players are singletons.

Definition 4.1 An extensive-form game has *perfect information* if, for any player $i \in \{1, 2, \ldots, n\}$ and any information set $H_i \in \mathcal{H}_i$, $|H_i| = 1$.

To incorporate the definition that decision nodes in the same information set are indistinguishable, exactly the same set of actions (arrows) must start from the decision nodes in the same information set. To formalize, for each player i, let $A_i(x)$ be the set of feasible actions at a decision node $x \in X_i$. Then it must be that, for any $x, x' \in H_i \in \mathcal{H}_i$, $A_i(x) = A_i(x')$. (Hence we can write the set of feasible actions as a function of information sets, such as $A_i(H_i)$.)

For general extensive-form games, we need more structure in the game tree. A list of requirements follows below. For a formal definition using a precedence relation of nodes, see for example the textbook by Kreps [10] or Kreps and Wilson [11].

- The origin of the game tree must be unique and a singleton information set. This means that the start of the game must be perfectly recognized.
- Each decision node belongs to only one player.
- A sequence of actions from the origin to a decision node or a terminal node, called a *path* or a *history*, must be unique.
  Then arrows do not make a cycle. Rather, the game tree is a branching process, which represents how a game expands sequentially and does not go back in the history.

A sequence of arrows and decision nodes starting from the origin and ending in a terminal node is called a *play* of the game. Under the above assumptions, each terminal node corresponds to a unique play of the game, and thus can be identified by a play. That is, no name is needed for terminal nodes. (In addition, if no confusion occurs, a play is specified by a sequence of actions only.) The *length* of an extensive-form game is the maximal number of decision nodes that are contained in some play of the game. A *finite extensive-form game* is an extensive-form game that has a finite number of arrows, and hence its length, a finite number of players, and finite feasible actions at every decision node of all players.

We can formalize the notion of perfect recall by using paths and information sets. For each player i and any decision node x of player i, let h be the (unique) path from the origin to x, and $\overline{X}_i(h)$ be the sequence of information sets and actions that belong to player i on the path h, in the order as they occur in h.

Definition 4.2 An extensive-form game has *perfect recall* if, for any player $i \in \{1, 2, \ldots, n\}$ and any information set $H_i \in \mathcal{H}_i$ of i, if the paths to decision nodes $x, x' \in H_i$ are h and h' respectively, then $\overline{X}_i(h) = \overline{X}_i(h')$.

Clearly, the two games in Fig. 4.3 have perfect recall. An example of a game without perfect recall is the Absent-minded Driver Game by Piccione and Rubinstein [14]. A driver is about to get on a highway at night to go home in the suburbs. He

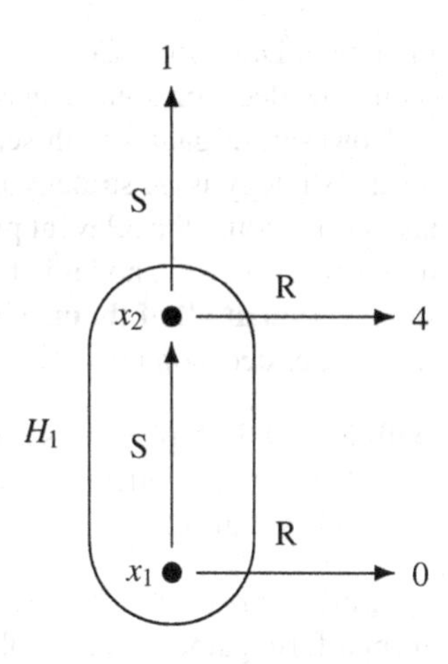

Fig. 4.4 Absent-minded driver game

must turn right at the second intersection, but he is tired, it is a dark and stormy night, and the landscape at every intersection looks the same. Thus, he knows that he will get confused at each intersection and not remember whether it was the first or the second. His decision nodes are the two intersections x_1 and x_2. At each decision node, he can choose either to go straight (action S) or to turn right (action R). If he turns right at the first intersection, he is in a dangerous area and gets a payoff of 0, and if he goes straight at the second intersection, he cannot find the way back home and must stay at a motel, which gives a payoff of 1. If he turns right at the second intersection, he arrives home and gets the highest payoff of 4.

However, he cannot distinguish the two intersections, and hence they belong to the same information set. This game is depicted as a game tree in Fig. 4.4. The decision node x_1 has the empty set as the (unique) path h_1 to it, and hence $\overline{X}(h_1)$ is also the empty set. By contrast, the path h_2 that leads to x_2 contains this player's information set H_1 and his action S at x_1. Hence $\overline{X}(h_2) = (H_1, S) \neq \overline{X}(h_1)$. This means that the extensive-form game of Fig. 4.4 does not have perfect recall.[5]

4.2 Strategies in an Extensive-Form Game

Any game can be formulated as both an extensive-form game and a normal-form game. For example, a simultaneous-move game like the Prisoner's Dilemma can be expressed as an extensive-form game such that the (artificially determined) second-mover player does not know the first-mover's actions, as in Fig. 4.3ii. (Just change the names of the players and actions, and adjust the payoff vectors according to the

[5]A careful reader may notice that the origin is not a singleton information set, but we can fix this problem by adding an artificial origin and an action, such as the moment he leaves a parking lot. We can also add a second player, his wife, and her actions after he gets home to make a multi-person game.

Prisoner's Dilemma.) Conversely, as we explained at the beginning of Sect. 4.1, a sequential decision game can be converted to the induced normal-form game.

However, in games with sequential decision-making, we must be careful about what a strategy is. A strategy is a contingent plan of actions, and the choices must take into account the relevant player's information. In particular, a player cannot distinguish the decision nodes in the same information set, so that (s)he must choose **the same action at all of them**. Thus, a strategy must choose an action per information set, not per decision node.

Definition 4.3 A player i's *pure strategy* in an extensive-form game is a function such that, for each information set H_i, a feasible action at that information set $a \in A_i(H_i)$ is assigned.

A *mixed strategy* in an extensive-form game is defined in the same way as for normal-form games. It is a probability distribution over the set of all pure strategies. In addition, for extensive-form games, we have another probabilistic notion of a strategy. It is a plan in which a player randomizes actions at each information set.

Definition 4.4 A player i's *behavioral strategy*[6] in an extensive-form game is a function such that, for each information set H_i, a probability distribution is assigned over the set $A_i(H_i)$ of feasible actions at that information set, independently from those on other information sets.

At first sight, you might think that there is not much difference between a mixed strategy and a behavioral strategy. However, they are fundamentally different notions. A mixed strategy determines a probability distribution **before the game is played**, over the pure strategies. A behavioral strategy chooses actions probabilistically, **during the game**, at each information set and independently.

To get an intuition, let us define the mixed strategies and behavioral strategies for Bob in the perfect information version of the sequential decision Meeting Game (Fig. 4.3i). Recall that the set of pure strategies for Bob is $S_b = \{(A, A), (A, B), (B, A), (B, B)\}$, where the first coordinate is the action when he is at the top information set (i.e., when Ann chooses action A), and the second coordinate is the action when he is at the bottom information set. A mixed strategy is a probability distribution (p, q, r, s) over the four pure strategies (in the order appearing in S_b, for example) such that $0 \leqq p, q, r, s$ and $p+q+r+s = 1$. Because of the last constraint, it effectively determines three non-negative real numbers between 0 and 1.

By contrast, a behavioral strategy for Bob is a pair of independent probability distributions, one over the actions A and B at the information set $H_{b1} = \{x_{b1}\}$, and the other over the actions A and B at the information set $H_{b2} = \{x_{b2}\}$. Essentially, he needs to determine the probabilities of choosing A at H_{b1} and A at H_{b2}, or two real numbers between 0 and 1.

[6] In some literature, e.g. Fudenberg and Tirole [7], it is called a behavior strategy.

Now it should be clear that the objects to which probabilities are assigned are completely different between a mixed strategy and a behavioral strategy. In addition, a behavioral strategy includes the requirement that a player randomizes independently across information sets.

In reality, we are often playing a behavioral strategy. For example, when we choose routes to a destination probabilistically, we usually do not line up all possible routes and randomize among them. Rather, we randomize at each intersection. A pure strategy can be viewed as a degenerate behavioral strategy as well.

To derive an equilibrium, it is sufficient to consider pure strategies and mixed strategies. However, behavioral strategies are more intuitive and are used in a very powerful mathematical method to solve a long-run optimization problem, known as *dynamic programming* (see Appendix A.2). Therefore, we use behavioral strategies to analyze extensive-form games. To be sure, we show that looking for equilibria among behavioral strategies has no loss of generality as compared to looking at mixed strategies.

Definition 4.5 For each player i, her/his two (mixed or behavior) strategies are *outcome equivalent* if, given a pure strategy combination by all other players, they induce the same probability distribution over the set of terminal nodes.

Proposition 4.1 *In any finite extensive-form game with perfect recall, for any player i and any behavioral strategy of player i, there is an outcome-equivalent mixed strategy of i.*

Proof Fix an arbitrary player i and her/his behavioral strategy b_i. Recall that b_i is a function such that, for each information set $H_i \in \mathcal{H}_i$, it assigns a probability $b_i(H_i)(a)$ to each feasible action $a \in A_i(H_i)$ such that $\sum_{a \in A_i(H_i)} b_i(H_i)(a) = 1$. We construct a mixed strategy that is outcome equivalent to b_i as follows. For each pure strategy s_i of player i and each information set H_i, let $s_i(H_i)$ be the pure action that s_i prescribes. Let the probability of s_i be the product of the probability that b_i assigns to each $s_i(H_i)$:

$$\sigma_i(s_i; b_i) = \underset{H_i \in \mathcal{H}_i}{\times} b_i(H_i)(s_i(H_i)).$$

The mixed strategy $\sigma_i(\cdot; b_i)$ is now shown to be outcome equivalent to b_i.

Because the game has perfect recall, for any history h to any decision node in the same information set, the sequence $\overline{X}_i(h)$ of information sets and actions that belong to player i is the same. Given a pure-strategy combination by all other players, the probability of a play (and the corresponding terminal node) is either 1 times the product of the probabilities of the actions by player i in $\overline{X}_i(h)$ or 0 times that. This probability is the same between b_i and the induced mixed strategy. □

Let us illustrate the idea of the above proof with a game, shown in Fig. 4.5. The second mover of this game has a unique information set, and thus his set of behavioral strategies and mixed strategies are the same. Hence we focus on player 1, who moves

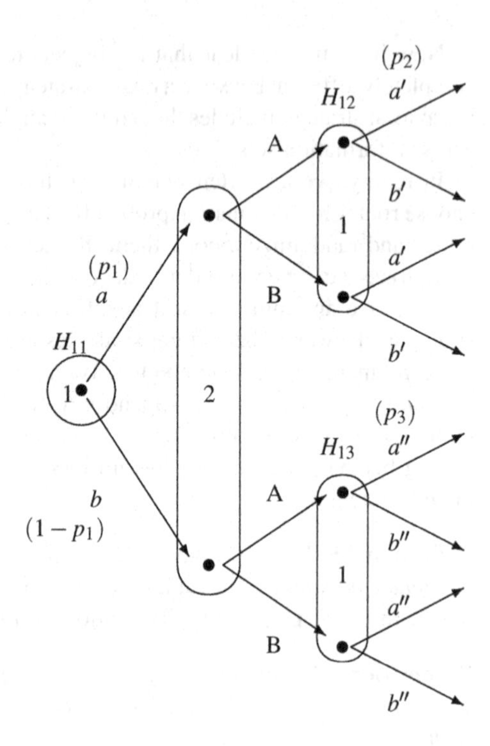

Fig. 4.5 Probabilities for a behavioral strategy

at the beginning and at the end of the game. Player 1 does not forget what she did at her first information set H_{11}, but does not know what action player 2 takes afterwards, as she chooses actions in information sets H_{12} and H_{13}. This is a game with perfect recall.

A behavioral strategy by player 1 assigns (independent) probabilities p_1, p_2, and p_3 to actions a, a', and a'' respectively in each information set. They are shown in parentheses in the figure. (Recall that, because player 1 cannot distinguish decision nodes in the same information set, she must choose the same action for the two decision nodes in each of H_{12} and H_{13}.) Given a pure strategy of player 2, the play (a, A, a') or the terminal node corresponding to it occurs with probability $p_1 \cdot p_2$ times 1 or 0 (depending on player 2's pure strategy), the play (a, A, b') or its terminal node occurs with probability $p_1(1 - p_2)$ times 1 or 0, and so on.

Let us construct a mixed strategy that is outcome equivalent. The number of pure strategies for player 1 is $2^3 = 8$ because she has two actions per information set. Specifically, the set of pure strategies of player 1 is $S_1 = \{(a, a', a''), (a, a', b''), (a, b', a''), (a, b', b''), (b, a', a''), (b, a', b''), (b, b', a''), (b, b', b'')\}$. In the Proof of Proposition 4.1, the probability of a pure strategy is defined as the product of the probability (assigned by the above behavioral strategy) of each action prescribed by the pure strategy. For example, the pure strategy (a, a', a'') has probability $p_1 \cdot p_2 \cdot p_3$ and (a, a', b'') has probability $p_1 \cdot p_2(1 - p_3)$. Then, the probability that the play (a, A, a') occurs is 1 or 0 times the probability that either the pure strategy (a, a', a'')

or (a, a', b'') occurs, which is $p_1 \cdot p_2 \cdot p_3 + p_1 \cdot p_2(1 - p_3) = p_1 \cdot p_2$. (We applied the fact that the probabilities at each information set are independent.) In this way, we can check that the probabilities of all terminal nodes are the same for the behavioral strategy and the mixed strategy.

If the game does not have prefect recall, however, there can be a behavioral strategy such that no mixed strategy can attain the same probability distribution over the terminal nodes. Reconsider the Absent-minded Driver Game in Fig. 4.4. Consider a behavioral strategy such that the player chooses S with probability p at any intersection. The terminal node determined by the play R has probability $1 - p$, the one by the play SR has probability $p(1 - p)$, and the one by the play SS occurs with probability p^2. This probability distribution cannot be generated by a mixed strategy. There is a unique information set, and the player has two pure strategies, "always R" and "always S". Hence the reachable terminal nodes by the pure strategies are the one with the play R and the one after SS. No probability distribution over the pure strategies yields the above probability distribution over the terminal nodes.[7]

The converse of Proposition 4.1 also holds. That is, in games with perfect recall, any mixed strategy of a player has an outcome-equivalent behavioral strategy. We do not write down the proof of this statement but explain the idea with Fig. 4.5 again. Consider a mixed strategy by player 1 such that the pure strategy (a, a', b'') has probability p and (b, a', a'') has probability $1 - p$. The terminal nodes that can have a positive probability are the ones corresponding to plays including (a, a') and those including (b, a'') only. The former class of terminal nodes have the probability p in total and the latter have $(1 - p)$ in total. Therefore, the behavioral strategy such that

$$b_1(H_{11})(a) = p, \;\; b_1(H_{12})(a') = 1, \;\; b_1(H_{13})(a'') = 1$$

is outcome-equivalent.

The outcome-equivalence of behavioral strategies and mixed strategies is proved by Kuhn [12] in full generality.

Theorem 4.1 (Kuhn's Theorem) *For any finite extensive-form game, any mixed strategy has an outcome-equivalent behavioral strategy, and any behavioral strategy has an outcome-equivalent mixed strategy, if and only if the game has perfect recall.*

Kuhn's theorem is proved for finite extensive-form games, and Aumann [2] extended the result to infinite-length extensive-form games such as infinitely repeated games (see Sect. 5.5). Thus, under perfect recall, we can restrict our attention to behavioral strategies to find equilibria.

---

[7] Fudenberg and Tirole [7] gives an "opposite" example (their Fig. 3.13) such that there is a probability distribution over the terminal nodes which is attainable by a mixed strategy but not by any behavioral strategy, under imperfect recall.

4.3 Backward Induction

After understanding the formulation of extensive-form games by game trees and the concept of strategies, we are ready to discuss what strategy combination should be defined as an equilibrium. We have argued using an example in Sect. 4.1 that some of the Nash equilibria of the induced normal-form game do not reflect rational behavior during the game. How do we fix this problem? Let us consider a very simple one-person sequential decision-making situation.

There is a tree, currently bearing a small fruit. If the fruit is picked now, it gives a payoff of 3. If it is not picked now, the decision can be delayed until one month later, and in the meantime the fruit will grow. Hence if the fruit is picked one month later, it gives a payoff of 5. There is also the choice not to pick it at either time point, which gives 0 payoff. This decision problem has two steps, whether to pick the fruit now or not and, if it is not picked now, whether to pick it or not one month later. The player wants to maximize the payoff in this problem. A tree diagram of this problem is Fig. 4.6. (Since it is a one-person game, or just a dynamic optimization problem, the name of the player at each information set is omitted.)

We have an obvious solution to this problem, even without seeing the tree diagram. That is the strategy (Not Pick, Pick), which means that the player does not pick the fruit now and picks it one month later. (From now on, if there is a clear order for information sets, we write the actions of a pure strategy in that order.) To think of this logic more carefully, the player needs to decide what to do one month later in order to decide what to do now. Implicitly, we have decided that the player must pick the fruit when it is one month later (by comparing 0 and 5 at that time), and because (s)he can get 5 later, (s)he should not pick the fruit now, which gives only 3. If the player was not going to pick the fruit one month from now, (s)he should better pick it now, but that future action is not rational one month later. This way of thinking is called *backward induction*. The player first solves the last decision problem, then goes back to the previous decision problem **given the optimal action in the future**. This way of solving is repeated until we reach the first decision node. Backward induction has been mathematically proven to be the correct method of long-run optimization by *dynamic programming*. (See Appendix A.2.)

This idea of backward induction can be applied to derive a solution to multi-person decision-making, under the assumption that the game has perfect and complete

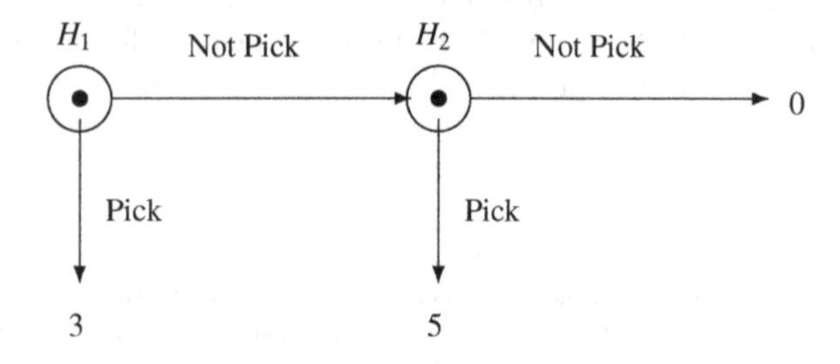

Fig. 4.6 A one-person sequential decision problem

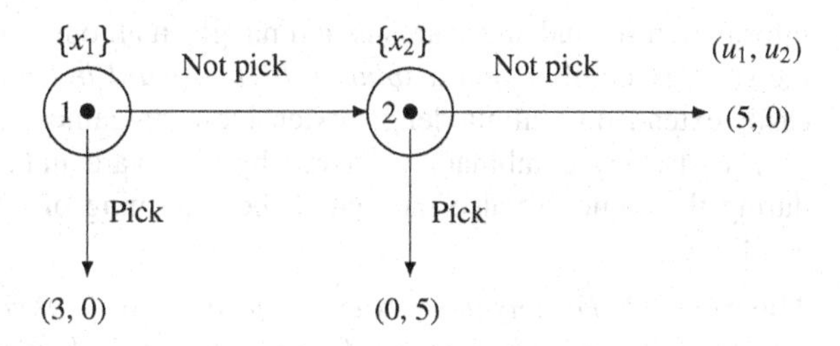

Fig. 4.7 A two person harvest game

information and the rationality of all players is also common knowledge. For example, consider a variant of the fruit picking problem. At the beginning, player 1 is under the tree, deciding whether to pick the fruit or not. If she does not pick the fruit, then one month later, player 2 gets the opportunity to pick the fruit or not. At each point, the player who picked the fruit gets the relevant payoff (3 at the beginning and 5 one month later). Assume that if player 2 does not pick the fruit one month later, the fruit is given to player 1. The game tree is depicted in Fig. 4.7. Player 1 has a single decision node x_1 (and a singleton information set $\{x_1\}$), and player 2 has a single decision node x_2 (and a singleton information set $\{x_2\}$). The first coordinate of each payoff vector is player 1's payoff, and the second coordinate is player 2's payoff. Both players know this game and that the opponent is rational.

Because each player has a unique information set, their set of pure strategies coincides with their set of actions. In order to maximize her payoff, player 1 must think backwards, that is, she solves player 2's decision problem to predict what will happen if she does not pick the fruit now. Player 1 knows player 2's payoff function and that he is rational. Therefore, player 1 can conclude that when given the opportunity, player 2 will pick the fruit. Given this optimal action by player 2 in the future, player 1's optimal action at the beginning of the game is to pick the fruit. Against the rational opponent, player 1 cannot hope to get the payoff of 5. The "backward induction solution" to this two-person game is a strategy combination (Pick, Pick), where both players pick the fruit when it is her/his move. Note also that, to reach to this solution of the game, player 1 only needs to know the game and rationality of player 2, but does not need to know that player 2 knows that player 1 is rational, and so on.

Suppose that the game is extended further, so that if player 2 does not pick the fruit one month later, then a month after that, player 1 gets to choose between picking or not. In this case, player 2 must also consider what player 1 would do a month later in order to decide what to do. Moreover, player 1 needs to know that player 2 knows her payoff function and rationality to choose her actions at the beginning of the game.

Let us generalize the above logic. For any complete and perfect information, finite extensive-form game, there is an optimal action for each player of the last information sets in the extensive-form game. Given the optimal actions of the last information sets, there is an optimal action for each player of the second to the last information sets. Repeat this until we reach the origin of the game tree. Then we have identified the strategies of all players such that at each information set, the action plan at that information set and afterwards is optimal given all other players' strategies. We call the strategy combination a *solution by backward induction*.[8] (This notion can be easily extended to infinite length extensive-form games. See Sect. 4.7.)

The strategy combination derived by backward induction is not only rational during the game but also rational at the beginning of the game, i.e., it is a Nash equilibrium.

Theorem 4.2 *For any finite extensive-form game Γ of complete and perfect information, the solution by backward induction is a Nash equilibrium of Γ.*

Proof We use mathematical induction with respect to the length ℓ of the extensive-form game. Clearly, when $\ell = 1$, the optimal action by the unique player is a Nash equilibrium by definition.

Assume that the theorem holds when the length of Γ is not more than ℓ. Consider Γ with length $\ell + 1$. We divide the game into two parts, the choice at the origin of the game and the rest of the game. Since the feasible actions at the origin are finite, denote them as $a = 1, 2, \ldots, k$. Each action $a = 1, 2, \ldots, k$ is followed by an extensive-form game with at most length ℓ, which is denoted by Γ^a. Let b^a be the solution by backward induction of Γ^a. Let i be the player who chooses an action at the origin.[9] If any player other than player i can increase her/his payoff by deviating from the strategy derived by backward induction, it is a contradiction to the mathematical induction assumption that each b^a is a Nash equilibrium in the restricted game of Γ^a. Therefore, if there is a player who can improve the payoff, it must be player i.

For each $a = 1, 2, \ldots, k$, let b^a_{-i} be the action plans of players other than i in b^a. Then the combination $(b^1_{-i}, b^2_{-i}, \ldots, b^k_{-i})$ is a behavioral strategy combination of the entire game Γ of players other than i. An arbitrary behavioral strategy b_i of player i can be decomposed as the probabilities $p(a)$ for each $a = 1, 2, \ldots, k$ at the origin and behavioral strategy $b_i \mid_a$ for game Γ^a for all $a = 1, 2, \ldots, k$. Then player i's expected payoff of Γ can be written as:

$$Eu_i(p, b_i \mid_1, \ldots b_i \mid_k, b^1_{-i}, \ldots b^k_{-i}) = \sum_{a=1}^{k} p(a) Eu_i(a, b_i \mid_a, b^a_{-i}),$$

where $Eu_i(a, b_i \mid_a, b^a_{-i})$ is the expected payoff when player i chooses action a at the origin and follows $b_i \mid_a$, while other players follow b^a_{-i} in game Γ^a. By the mathematical induction assumption, after each action a at the origin, the expected payoff $Eu_i(a, b_i \mid_a, b^a_{-i})$ is maximized by b^a_i, given all other players' behavioral

[8]In many textbooks, this equilibrium concept is only implicit, and they use the concept of (subgame) perfect equilibrium (see Chap. 5) for all extensive-form games with complete information. However, we (maybe influenced by Gibbons [8]) think that, to understand the subgame perfect equilibrium concept fully, it is important to understand backward induction on its own.

[9]In general, we should allow for the possibility that Nature (to be introduced in Sect. 4.8) moves at the origin. In this case, the theorem holds as well, in the sense that no player can improve their payoff by choosing a different action than the one in b^a in each Γ^a.

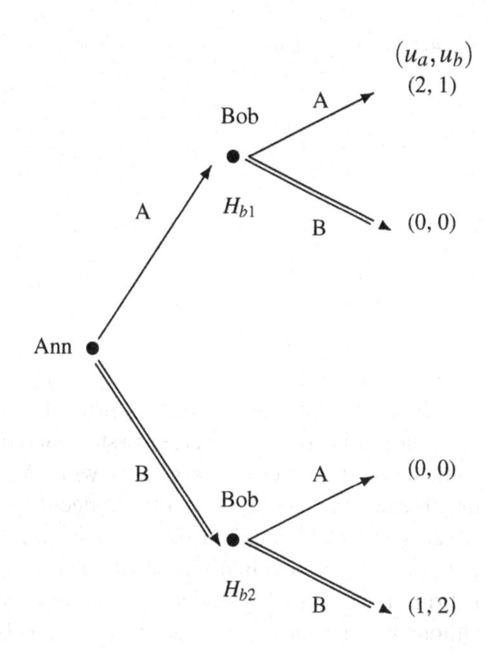

Fig. 4.8 A Nash equilibrium which is not a solution by backward induction

strategies. Moreover, the solution by backward induction puts a positive probability on only actions that maximize $Eu_i(a, b_i^a, b_{-i}^a)$ at the origin. Therefore, by changing behavioral strategies either at the origin or at later stages, player i cannot increase the expected payoff. □

The proof of Theorem 4.2 is essentially finite-length dynamic programming. The converse of Theorem 4.2 does not hold. That is, there can be a Nash equilibrium which is not optimal at some point of the extensive-form game. An example is the $(B, (B, B))$ Nash equilibrium of the sequential decision Meeting Game in Sect. 4.1. This strategy combination is illustrated by double arrows in Fig. 4.8. (To simplify figures, from now on we will omit the circles for singleton information sets.)

The strategy combination $(B, (B, B))$ means that Ann chooses the pure strategy B because Bob is choosing a strategy where he always stays at his classroom (action B in both information sets). However, if Ann chooses strategy A, it is not rational for him to choose action B.

4.4 Chain Store Paradox

Let us introduce a very famous problem posed by Selten [19], called the *chain store paradox*. In this game, there is a unique solution by backward induction, but it is rather unrealistic. We first consider a simple entry game in a market and then extend the game to finitely many markets with a chain store.

Fig. 4.9 An entry game

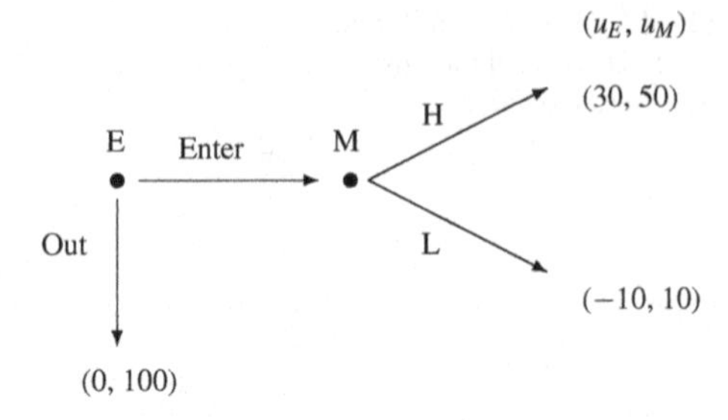

An entry game is defined as follows. There are two players, a potential entrant firm E and the current monopolist firm M. Firm E chooses to act first, whether to enter the market (action Enter) or to stay out (action Out). If firm E does not enter, the game ends, and E gets a payoff of 0, while M gets the monopolist payoff of 100. If firm E enters the market, then firm M decides whether to charge a high price (action H) as it has been doing, or to charge a low price (action L) and the game ends. (Firm E's pricing is not explicitly modeled, and is assumed to be the same as firm M's.) If firm M chooses a high price, the two firms share the high profit from the market, although firm E incurs some entry cost. Thus E's payoff is 30 and M's payoff is 50. If firm M chooses a low price, there is little profit from the market for each firm, but firm E must pay the entry cost. Thus E gets -10 and M gets 10.

The game tree of the entry game is depicted in Fig. 4.9. The payoffs are listed in the order of the decision making; the first coordinate is E's payoff, and the second is M's. Because each player makes decisions only once and M makes decisions only when E enters the market, each player has a unique and singleton information set, and the set of actions coincides with the set of strategies.

The solution by backward induction for the entry game is obvious: after entry, firm M compares 50 with 10 and chooses to charge a high price. Given this, firm E enters. Therefore, the solution should be (Enter, H). If we convert the game into the induced normal form, there is another Nash equilibrium (Out, L). See Problem 4.1. However, this equilibrium uses an **empty threat** that M chooses L if entry occurs, which is not rational at that information set. Therefore, the solution by backward induction is the only rational equilibrium for the extensive-form game.

Now, suppose that firm M is a chain store, operating in N markets as the local monopolist. In each market, there is a potential entrant firm. In market 1, E1 is the name of the potential entrant, and if E1 enters, M chooses one of the prices and an outcome for that market is determined. If E1 does not enter, that is also an outcome for that market. After the outcome for market 1 is observed, in market 2, a firm called E2 will decide whether to enter the market 2 or not. If entry occurs in market 2, then M chooses one of the prices, and this continues for N markets. Thus the players are E1, E2, ..., EN and M. They all observe the outcomes for all past markets. Each potential entrant's payoff is the same as that of firm E in Fig. 4.9 because they move only once. The chain store M gets payoffs in N markets, and thus the total payoff is its objective.

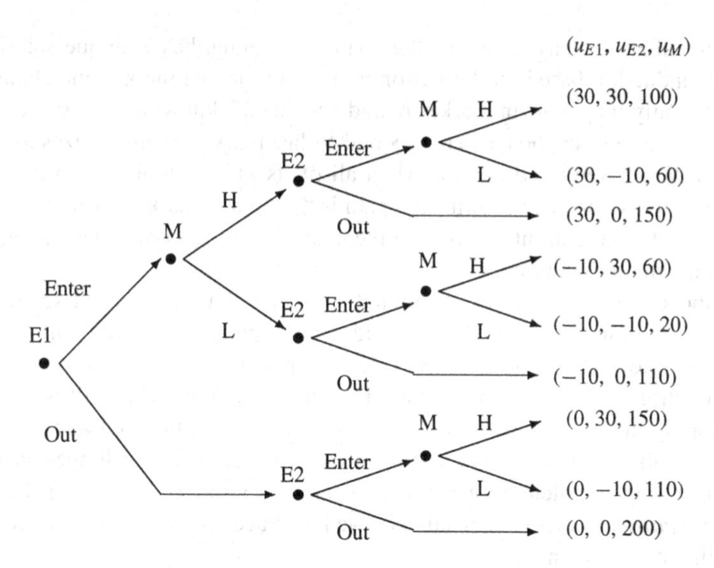

Fig. 4.10 Two-market chainstore game

To solve this large game, we consider the case of $N = 2$ and then generalize the logic. The two-market chain store game is depicted in Fig. 4.10.

The last information sets in Fig. 4.10 all belong to firm M. In each of these, M's optimal action is H, given the total payoff comparison. This is no coincidence. Each information set has different histories, but M cannot change the payoff in the first market. Therefore, the action choice in the second market (the last information sets) should maximize the payoff from the second market only. In effect, M is facing the one-shot game of Fig. 4.9.

Let us go back to the second-to-last information sets, all of which belong to the potential entrant E2 in market 2. Given firm M's myopic optimization in the last information sets, E2 can reason that M will choose action H if it enters, at each information set. Therefore, the optimal action in each information set is Enter.

The previous level information set belongs to firm M in the first market. At this information set, M is effectively choosing between two future sequences of actions: (Enter, H) after H in the first market or (Enter, H) after L in the first market. Out or L will not be chosen in the second market by the above reasoning. (This is a special feature of this game. We cannot generalize this property for general extensive-form games.) Therefore, in this information set, M should maximize its payoff in market 1, because the payoff from market 2 is the same after any action by M now. This implies that M's optimal action in market 1 is H.

Finally, consider firm E1's information set. It can now conclude that M will choose H in this market, and therefore the optimal action is Enter. In sum, the unique solution by backward induction for the $N = 2$ case is simply the repetition of the one-market equilibrium.

To generalize, for any $N < \infty$, the chain store game has a unique solution by backward induction based on the reasoning that, in the last market, the chain store maximizes only its payoff in market N and chooses H; knowing this, EN enters at all information sets, and hence in the second-to-last market, M maximizes its payoff in market $N - 1$ only and chooses H in all of its information sets so that EN-1 also enters, and so on. The equilibrium path is that in any market (even if N is very large), the potential entrant enters, and the chain store accommodates the entry by responding with a high price.

The above theoretical result of rational strategies is very clear but seems to be at odds with common sense. Why should a large chain store allow all entrants? Isn't it possible that M chooses low prices in first few markets to threaten later potential entrants? Then the loss incurred in the first few markets will be offset by the monopoly in later markets. This story is plausible, but the backward induction analysis excludes it. Note that the feasibility of backward induction hinges on a lot of things, including complete information and common knowledge of rationality. This *chain store paradox* can be resolved once we introduce incomplete information. We explain the resolution in Sect. 8.5.

4.5 Stackelberg Game

Cournot [6] considered a market model in which two firms simultaneously choose production quantities. By contrast, von Stackelberg [21] considered a model in which one firm (the leader) chooses a quantity first, and the other firm (the follower) chooses a quantity after observing the leader's choice. His solution uses backward induction. In this section we explain Stackelberg's model and its equilibrium by modifying the decision timing of the Cournot game in Sect. 3.2.

There are two firms, 1 and 2. Each firm can choose production quantities from the set of non-negative real numbers. Their payoff is the profit, which is the difference between the revenue and cost. The production cost of firm i ($i = 1, 2$) is its production quantity multiplied by a positive coefficient c_i. When q_1 is the firm 1's quantity and q_2 is the firm 2's, then the market price that sells all units is assumed to be

$$A - q_1 - q_2,$$

where A is greater than both $2c_1 - c_2$ and $3c_2 - 2c_1$. Thus, firm 1's payoff is a function of the quantity combination (q_1, q_2) of the two firms such that

$$u_1(q_1, q_2) = (A - q_1 - q_2)q_1 - c_1 q_1.$$

Similarly, firm 2's payoff function is

$$u_2(q_1, q_2) = (A - q_1 - q_2)q_2 - c_2 q_2.$$

Suppose that firm 1 chooses a quantity $q_1 \in [0, \infty)$ first, and, after observing q_1, firm 2 chooses $q_2 \in [0, \infty)$. This is a perfect-information extensive-form game. (We assume complete information until Chap. 6, and common knowledge of rationality until Chap. 9.) To derive a solution by backward induction, consider firm 2's decision problem first. In Sect. 3.2 we derived the best response $BR_2(q_1) = \frac{1}{2}(A - q_1 - c_2)$ of firm 2 to firm 1's quantity q_1.[10]

Now we go back to the leader's optimization problem. Firm 1 can compute the above best response of firm 2 and reason that its payoff is a function of only q_1 such that

$$\begin{aligned} u_1(q_1, BR_2(q_1)) &= \{A - q_1 - BR_2(q_1)\}q_1 - c_1 q_1 \\ &= \{A - q_1 - \frac{1}{2}(A - q_1 - c_2)\}q_1 - c_1 q_1. \end{aligned}$$

This is a concave function of q_1, and the slope at $q_1 = 0$ is positive by the assumption. Therefore, when the derivative is 0, the maximum is attained. By computation, the maximizer is $q_1^* = \frac{1}{2}(A - 2c_1 + c_2)$. When firm 1 chooses this quantity, firm 2's best response is $q_2^* = BR_2(q_1^*) = \frac{1}{4}(A + 2c_1 - 3c_2)$. The combination of production quantities is called the Stackelberg equilibrium in microeconomics, but in game theory, it is **the equilibrium path**. The solution by backward induction is a pair of strategies $(q_1^*, BR_2(\cdot))$. Note that firm 2's strategy is the best response function, which chooses an optimal production quantity at each information set corresponding to the observation of q_1.

Compared to the Nash equilibrium $(\frac{1}{3}(A - 2c_1 + c_2), \frac{1}{3}(A + c_1 - 2c_2))$ in the simultaneous-move case, on the equilibrium path, the leader produces more. This is because the leader can become aggressive, knowing that the follower will adjust its quantity so that the market price does not fall too much.

4.6 Ultimatum Game

Let us give another interesting example of an extensive-form game, called the *Ultimatum Game*. Suppose that there is a "pie" of size 1. Two players share this pie by the following process. First, player 1 proposes $x_1 \in [0, 1]$, which is interpreted as the share she gets. After observing the proposal, player 2 can choose either Yes or No and the game ends. That is, player 2 cannot offer a counter proposal, and player 1 is giving him an "ultimatum". If player 2 chooses Yes, then the pie is shared according to the proposal by player 1, and the payoff vector is $(x_1, 1 - x_1)$. (The first coordinate is player 1's payoff, which is the size of the pie she consumes. Player 2's payoff in the second coordinate is similar.) If player 2 chooses No, then both players get a payoff of 0.

---

[10] As we have noted in Sect. 3.2, if q_1 is large, it is possible that $\frac{1}{2}(A - q_1 - c_2)$ is negative. But such a large q_1 will be shown to be not rational, and thus we ignore those cases.

To solve this game by backward induction, let us find an optimal action for player 2 for each possible proposal $x_1 \in [0, 1]$ by player 1. When a proposal $x_1 < 1$ is made, then the share player 2 gets, $1 - x_1$, is positive if he agrees, while refusing gives him 0. Therefore, in this case there is a unique optimal action Yes. When $x_1 = 1$, both actions Yes and No (and all mixed actions) are optimal. To focus on the pure-strategy equilibrium, we have two pure strategies for player 2 that are optimal at each information set:

$$BR_2(x_1) = \text{Yes} \ \ \forall x_1 \in [0, 1];$$

$$BR_2'(x_1) = \begin{cases} \text{Yes if } x_1 < 1 \\ \text{No } \ \text{if } x_1 = 1. \end{cases}$$

If player 2 adopts the strategy BR_2, then there is a unique optimal strategy for player 1 at the beginning of the game, which is to demand $x_1 = 1$. By contrast, if player 2 uses BR_2', then there is no optimal strategy for player 1. As long as $x_1 < 1$, player 1 can increase her payoff by raising the proposal x_1, but when x_1 hits 1, her payoff drops to 0.[11]

Therefore, the Ultimatum Game has a unique solution by backward induction, in which player 1 proposes $x_1 = 1$ and player 2 uses BR_2. The equilibrium outcome is that player 1 takes the entire pie and the resulting payoff combination is $(1, 0)$. Although Yes and No give the same payoff for the rational player 2 in this case, this outcome is rather extreme. The driving force of the outcome is the game structure that only player 1 has the right to propose the shares and that player 2 cannot receive a positive payoff if he says No.

4.7 Alternating Offer Bargaining Game

Bargaining problems are important in cooperative games, but in non-cooperative game theory we also have a framework called bargaining games. This is almost equivalent to the following Ståhl-Rubinstein *alternating offer bargaining game*: see Rubinstein [16] and the book by Osborne and Rubinstein [13]. The model is like an extension of the Ultimatum Game in such a way that the two players make proposals alternatingly. Although the structure is simple, the game potentially allows infinitely alternating proposals and counter proposals.

Before analyzing the alternating offer bargaining game, we first extend the backward induction to an n-player perfect-information extensive-form game of arbitrary length. The restriction to perfect information is to simplify the definition to solve this particular game. The generalization of backward induction to general games is called *subgame perfect equilibrium*, which will be extensively analyzed in Chap. 5.

[11] When the set of feasible proposals by player 1 is not a continuum but a finite set, there is an optimal strategy. See Problem 5.1(b) for a related game.

Given a complete and perfect-information extensive-form game Γ, for each player $j = 1, 2, \ldots, n$ and each of her/his decision nodes x_j, we define the *subtree* $\Gamma(x_j)$ as the set of decision nodes including x_j and all "later" decision nodes of all players. A decision node y is later than another decision node x (or y is *after* x), if x is on some path from the origin of the game to y.

For each subtree $\Gamma(x_j)$, let $I(\Gamma(x_j)) \subset \{1, 2, \ldots, n\}$ be the set of players who have a decision node in $\Gamma(x_j)$. For any behavioral strategy b_k of player $k \in I(\Gamma(x_j))$, let $b_k \mid_{\Gamma(x_j)}$ be the restricted strategy of b_k on the subtree $\Gamma(x_j)$. That is, $b_k \mid_{\Gamma(x_j)}$ is a function that assigns each decision node of player k in the subtree the same action as b_k assigns; for each $x_k \in \Gamma(x_j)$, $b_k \mid_{\Gamma(x_j)} (x_k) = b_k(x_k)$.

Finally, a behavioral strategy combination $(b_1, \ldots, b_N)$ of Γ is defined as a solution by backward induction if, for each player $i \in \{1, 2, \ldots, n\}$ of Γ and any decision node x_i of player i, the restricted behavioral strategy $b_i \mid_{\Gamma(x_i)}$ on $\Gamma(x_i)$ maximizes i's payoff, given that any other player $k \in I(\Gamma(x_i)) \setminus \{i\}$ follows $b_k \mid_{\Gamma(x_i)}$ in the subtree $\Gamma(x_i)$.

When the extensive-form game is finite, the above definition is equivalent to requiring that the players at the last decision nodes choose actions which maximize their payoffs, the players at the second-to-last decision nodes choose actions taking this into account, and so on.

Now we specify the alternating offer bargaining game. There are two players, player 1 and player 2. There is a "pie" of size 1, to be shared by the two players. In the first period, player 1 chooses a proposal $x_1 \in [0, 1]$, which means that she gets x_1 of the pie and player 2 gets $1 - x_1$. After observing the proposal, player 2 chooses between Yes and No. If player 2 chooses Yes, the game ends, and they share the pie according to player 1's proposal to receive the payoff vector $(x_1, 1 - x_1)$. Up to this point, the game is the same as the Ultimatum Game.

If player 2 chooses No, however, the game continues to the second period. In the second period, it is player 2's turn to make a proposal $x_2 \in [0, 1]$, which means that player 1 gets x_2 of the pie and player 2 gets the rest. After observing the current proposal, player 1 decides whether to say Yes or No. If she says Yes, the game ends there. However (and importantly), since some time has passed, the total payoff of the two players is discounted by a *discount factor* $\delta \in (0, 1)$. That is, the size of the pie is now δ and if the players agree to share it by the ratio $(x_2, 1 - x_2)$, their actual payoff vector is $(\delta x_2, \delta(1 - x_2))$.

If player 1 chooses No in the second period, the game continues to the third period, which is player 1's turn to make a proposal $x_3 \in [0, 1]$. The game continues this way such that player 1 (resp. player 2) makes a proposal in odd (resp. even) periods. As time passes, the total payoff (the size of the pie) keeps shrinking by the rate δ. In period t, a proposal $x_t \in [0, 1]$ means that player 1 gets the share x_t of the pie in that period, which is discounted to be of the size δ^{t-1}. Therefore if an agreement is made, the payoff vector is $(\delta^{t-1}x_t, \delta^{t-1}(1 - x_t))$. If the players keep rejecting the other's proposal, the game potentially continues ad infinitum. The discounting of the future agreement payoff makes the optimization problem tractable, and it is in accordance with the saying "a bird in the hand is worth two in the bush".

Despite the potential infinite length, this game has a quite striking unique solution by backward induction. The original proof by Rubinstein [16] is a rather complex one. Here we use a simpler method by Shaked and Sutton [20].

Proposition 4.2 *For any discount factor $\delta \in (0, 1)$, the alternating offer bargaining game has a unique solution by backward induction, and its equilibrium path is that in period 1, player 1 proposes $\frac{1}{1+\delta}$ and player 2 chooses Yes. Therefore the game ends immediately with the equilibrium payoff vector $(\frac{1}{1+\delta}, \frac{\delta}{1+\delta})$.*

Proof Assume that this game has a solution by backward induction (a strategy combination). There may be multiple solutions, and let σ^H be the one that gives highest payoff to player 1. Let (u_H, v_H) be the payoff from σ^H, where $v_H \leqq 1 - u_H$. (The first coordinate is player 1's payoff. The sum of the players' payoffs is at most 1.)

Notice that this game looks the same from any odd period on, except the discounting of the payoffs. Hence the strategy combination σ^H is well-defined in any subtree starting at player 1's proposal decision node, and it must give the highest payoff for player 1 among the solutions by backward induction of the subtree.

Assume that if the game enters the third period, then the players follow σ^H thereafter. We replace the subtree with the resulting payoff of σ^H. Because the payoff generated in the subtree starting from the third period should be discounted two periods to be comparable in the entire game, the resulting payoff of σ^H is $(\delta^2 u_H, \delta^2 v_H)$. The reduced game is depicted in Fig. 4.11.

In the first period, player 1 has infinitely many actions from $x_1 = 0$ to $x_1 = 1$. This cannot be illustrated by arrows, and thus we use an arc with the two end points

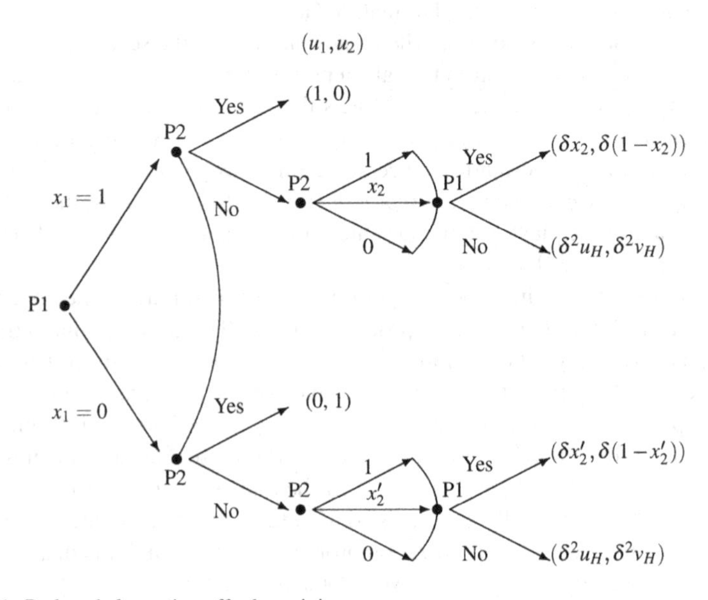

Fig. 4.11 Reduced alternating offer bargaining game

of $x_1 = 0$ and $x_1 = 1$ to show the continuum of actions. Each point on the arc corresponds to a proposal and an information set of player 2, from which two actions, Yes and No, are emanating. Because we cannot illustrate the infinitely many information sets of player 2 either, we show the end points after $x_1 = 0$ and $x_1 = 1$ as representatives. After player 2's No decision in the first period, player 2 also has infinitely many actions from $x_2 = 0$ to $x_2 = 1$. We display generic actions as x_2 and x_2' for two information sets.

To solve the reduced game by backward induction, consider player 1, who chooses between Yes and No at the end of the second period. As in the Ultimatum Game, there are two patterns of best responses to the proposals $x_2 \in [0, 1]$, and the one that allows the existence of player 2's optimal proposal is "Yes if and only if $x_2 \geqq \delta u_H$". (At this point, the history of first period is irrelevant because the payoffs do not depend on rejected proposals.) Given this action plan by player 1, player 2 compares the proposals in the region $x_2 \geqq \delta u_H$, which will be accepted, and others which will be rejected. Among the proposals that will be accepted, the best one for player 2 is $x_2 = \delta u_H$, which keeps player 1's share minimum in the range. Then player 2 gets the payoff $\delta(1 - \delta u_H)$. If player 2 proposes less than δu_H, it is rejected and he gets $\delta^2 v_H$ by the assumption. Since $\delta \in (0, 1)$, $\delta(1 - \delta u_H) > \delta^2(1 - u_H) \geqq \delta^2 v_H$ holds. Therefore, in any information set at the beginning of the second period, player 2's optimal action is $x_2 = \delta u_H$.

Let us go back to the first period. The last decision maker in the first period is player 2. In an information set after player 1's proposal x_1, if he says Yes to it, he gets $1 - x_1$, while if he says No, the game continues to the next period and he expects to receive $\delta(1 - \delta u_H)$, as both players will follow the solution by backward induction in the subtree. Therefore an optimal action plan for player 2 is Yes if and only if $1 - x_1 \geqq \delta(1 - \delta u_H)$. (Against the other optimal plan that rejects the proposal when the equality holds, player 1 does not have an optimal proposal.) Based on this, among the proposals such that $1 - x_1 \geqq \delta(1 - \delta u_H)$, the optimal proposal for player 1 is the one that maximizes x_1, or $x_1 = 1 - \delta(1 - \delta u_H)$. If player 1 proposes a higher share than this, she gets rejected and the players play the optimal strategy combination in the subtree starting in the second period. This gives $\delta^2 u_H$. Again, $\delta \in (0, 1)$ implies that $1 - \delta(1 - \delta u_H) = 1 - \delta + \delta^2 u_H > \delta^2 u_H$, and therefore $x_1 = 1 - \delta(1 - \delta u_H)$ is optimal.

We have found a unique solution by backward induction under the assumption that the players play σ^H if the game reaches the third period. Player 1's payoff is $1 - \delta(1 - \delta u_H)$. Recall that u_H was the highest payoff player 1 could get among solutions by backward induction, and $1 - \delta(1 - \delta u_H)$ is increasing in u_H. Thus it must be that $u_H = 1 - \delta(1 - \delta u_H)$, which yields $u_H = \frac{1}{1+\delta}$.

There may be multiple solutions by backward induction, and we can consider σ^L, which gives the lowest payoff to player 1 from the third period on. However, the reduced game is essentially the same as Fig. 4.11 with modifications of $(\delta^2 u_L, \delta^2 v_L)$ as the payoff vector replacing the subtree from the third period on, where $u_L = Eu_1(\sigma^L)$ and $v_L = Eu_2(\sigma^L)$. Therefore by an analogous argument, we get $u_L = 1 - \delta(1 - \delta u_L)$. In summary, the solution by backward induction is unique. □

In the unique equilibrium, player 1's equilibrium payoff $1/(1+\delta)$ is strictly greater than that of player 2, $\delta/(1+\delta)$. Therefore, it is advantageous to be the first proposer. Moreover, the equilibrium is efficient, because the bargaining ends immediately and the total payoff of the two players is not discounted by the delay.

This game and its unique-efficient equilibrium made a big impact in the bargaining analysis of non-cooperative games. The framework of alternating offers is realistic for wage negotiations between labor unions and firms, price negotiations between sellers and buyers, and so on. Theoretically, incorporating a bargaining process into some application models is also easy thanks to the uniqueness of the equilibrium. Thus, an avalanche of applications was created. (The book by Osborne and Rubinstein [13] contains many of the applied models and extensions of this game.)

Nonetheless, the equilibrium can be interpreted as a similar paradox to the chain store paradox, in the sense that the theoretical prediction is too stark and may seem unrealistic. Potentially, the players can bargain ad infinitum, and if player 2 commits to not agree immediately in order to delay the negotiation, player 1 reduces her/his demand. Knowing of this possible reduction of the demand, player 2 has an incentive to prolong the game. However, rational players settle with the immediate agreement. This paradoxical result is related to the *Coase conjecture* (Coase [5]) in economics.[12] In reality, we also seldom hear of immediate agreements between labor unions and employers. Bargaining between a seller and a buyer often takes some time, starting from a too-high price quote by the seller or a too-low price request by the buyer to find a mid point.

The "paradox" of the alternating-offer bargaining equilibrium can be resolved by introducing incomplete information of payoff functions, just like the chain store paradox (see Chap. 8 for the chain store paradox resolution). That is, there is a rational equilibrium with some delay in agreement when players are not sure about the payoff function of their opponents. For details, see the book by Osborne and Rubinstein [13].

4.8 Introduction of Nature

So far, we have assumed that the outcomes of a game are solely determined by the decisions of the players. However, there are a lot of strategic situations in reality that cannot be fully settled by players' choices. On one hand, if all external factors other than the players' decision making are completely known, then we can incorporate them into the game as a part of the rules or the payoff functions. An example is the effect of the day of the week on seller-buyer games. Sundays come regularly, and it is possible that buyers behave differently on Sundays than on other weekdays because they have more time to spend on shopping. This regular effect can be taken into account in the feasible action sets or in the payoff functions. On the other hand, if exogenous factors do not affect the outcomes in a specific way for sure, we need

[12] Gul et al. [9] gives a unified approach to encompass the potentially infinite horizon bargaining and the durable good sales problem considered in Coase [5].

to explicitly introduce *uncertainty* into the game. For example, random shuffling of the cards affects the outcome of a card game, and variable tastes of consumers or exchange rates influences the outcomes of firm competitions. These are factors that cannot be foreseen, but are critical for the outcomes of a game.

In extensive-form games, we can explicitly incorporate uncertainty by introducing an artificial player called *Nature*. During a game, when some realization of a random event determines the course of the game, we assume that Nature chooses among possible courses according to a (commonly known) probability distribution. Then we can derive the strategic equilibria of the other (rational) players in the same way as before. Nature is not a rational player but just determines the realization probabilities of the branches of a game, and thus its choice is not considered to be a part of an equilibrium. With uncertainty in the game tree, the terminal nodes realize stochastically, so we assume that each player maximizes her/his expected payoff by computing the probability of the terminal nodes given all players' strategies.

Take the very simple example of a one-person decision problem, in which a decision maker P chooses whether to take an umbrella with her or not in the morning, before knowing the weather of the day. The day has two possible courses, sunny or rainy. Suppose that the probability of becoming a sunny day is 0.3 and a rainy day is 0.7. Because the umbrella is bulky to carry all day, if the decision maker brings the umbrella but it did not rain, her payoff is -1. If it rains, her payoff is 1. But if she does not bring the umbrella and it rains, her payoff becomes -3, and if it becomes sunny, her payoff is 0. The tree diagram of this decision problem is depicted in Fig. 4.12. N is Nature.

Fig. 4.12 Introduction of risk

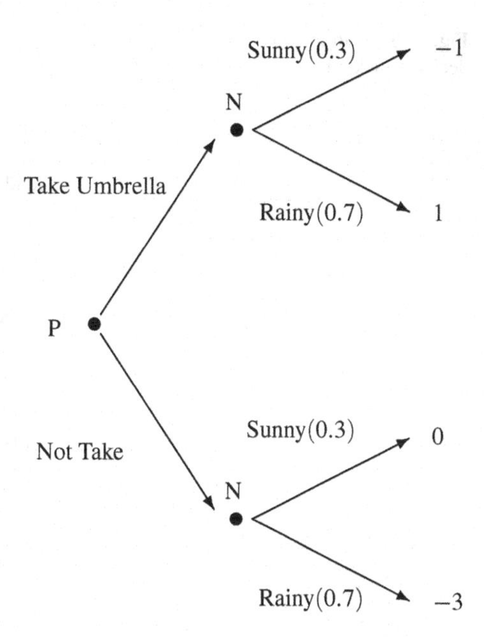

Because the actual weather realizes after the decision maker's choice, Nature moves after her. Nature would not change the probabilities of the different courses of weather depending on the decision maker P's choices, and thus the two decision nodes belonging to Nature have the same probability distribution over the two branches. The branches of Nature's moves include the possible event descriptions and their probabilities (usually in parentheses). All of Nature's decision nodes are singleton information sets by assumption.

Now, P's expected payoff from the strategy to take the umbrella is $(0.3) \cdot (-1) + (0.7) \cdot 1 = 0.4$, while the expected payoff from the strategy not to take it is $(0.3) \cdot 0 + (0.7) \cdot (-3) = -2.1$. Hence the optimal strategy is to take the umbrella with her.

For later reference, we give another simple example. Reconsider the Meeting Game in Table 3.1. Before playing this game, the two players will now flip a coin. The players can change their actions depending on the result of the coin flip. This makes a new extensive-form game. For example, a conditional action plan would be to choose action A if Heads comes up and choose action B if Tails comes up. This is a strategy of the new game. We do not analyze an equilibrium of this game now, but instead investigate Ann's decision-making problem when Bob is following this strategy. Ann does not have to follow the same strategy and can choose any action after each side is shown. (This is the fundamental characteristic of a non-cooperative game.) If we ignore mixed strategies, Ann's problem can be depicted as a tree, shown in Fig. 4.13. The payoffs are computed based on Bob's strategy "A if Heads, B if Tails".

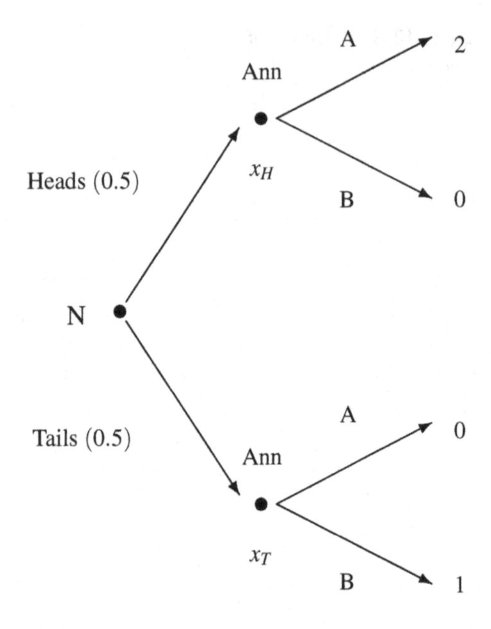

Fig. 4.13 Coin flip and decision making

Ann can choose different actions after each side, and hence her two decision nodes x_H and x_T belong to different information sets. At the information set $\{x_H\}$, she compares the two terminal nodes with payoffs of 2 and 0, and thus action A is optimal. Similarly, at $\{x_T\}$, action B is optimal. This means that Ann's optimal strategy is also "A if Heads, B if Tails", when Bob uses this strategy.

The same argument goes for Bob, and therefore the strategy "A if Heads, B if Tails" is mutually the best response, when both players observe the **same realization** of the coin flip. The outcome is (A, A) with probability $\frac{1}{2}$ and (B, B) with probability $\frac{1}{2}$. Their common expected payoff is $\frac{1}{2} \times 2 + \frac{1}{2} \times 1 = \frac{3}{2}$. Notice that this payoff combination $(\frac{3}{2}, \frac{3}{2})$ cannot be achieved by any independently mixed strategy combination by the two players. If Ann uses a mixed strategy $(p, 1-p)$ and Bob uses $(q, 1-q)$ (where the first coordinate is the probability of action A), their payoff combination takes the form

$$(2pq + (1-p)(1-q),\ pq + 2(1-p)(1-q)).$$

For any combination of p and q in $[0, 1]$, $(\frac{3}{2}, \frac{3}{2})$ is not feasible. In fact, the area of feasible expected payoff combinations as p and q move in $[0, 1]$ is approximately the meshed area in Fig. 4.14. (Each line is drawn by fixing p and moving q in $[0, 1]$.)

To see this fact from a different angle, the feasible probability distributions over the four outcomes (A, A), (A, B), (B, A) and (B, B) for independently mixed strategy combinations are special cases of all feasible probability distributions over the outcomes, which are possible if the two players can act according to a common probability device. A strategy using a common randomization device is called a *correlated strategy*. A correlated strategy combination can attach three different probabilities among the four cells in Table 4.2, while an independently mixed strategy combination attaches two parameters with the multiplication rule.

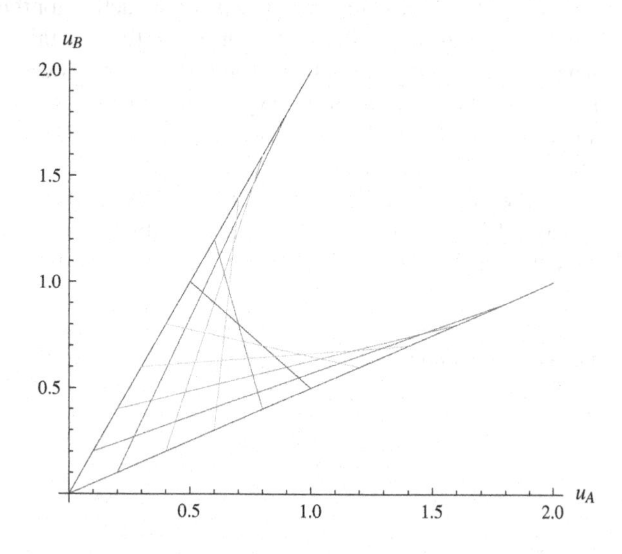

Fig. 4.14 Feasible payoff combinations under independent mixed strategy combinations

Table 4.2 Comparison of probability distribution structures

Ann \ Bob	A	B
A	pq	$p(1-q)$
B	$(1-p)q$	$(1-p)(1-q)$

Independently mixed strategy combination

Ann \ Bob	A	B
A	p	q
B	r	$1-p-q-r$

Correlated strategy combination

Therefore, if there is a randomization device that the players can jointly use, they can choose a probability distribution over **strategy combinations**, and it may improve their social efficiency. An equilibrium using a correlated strategy combination is a pair of a probability distribution and the strategy combinations for each possible realization, and is called a *correlated equilibrium* (see Aumann [3]).

4.9 Common Knowledge of Rationality and Backward Induction**

To conclude this chapter, we explain a problem in which solving an extensive-form game backwards may not be consistent with the common knowledge of rationality of all players.

Consider the extensive-form game in Fig. 4.15. This game is similar to the two-person harvest game in Sect. 4.2 and is a variant of Rosenthal's *centipede game* (Rosenthal [15]). Two players, 1 and 2, choose In and Out alternatingly. If one player chooses Out, the game ends. If a player chooses In, the game continues and the other player gets to move next. As the game continues, the sum of the two players' payoffs increases, but not individually.

There is a unique solution by backward induction for this game, namely to choose Out at every information set (decision node). To reach this solution, players anticipate what the other player (and herself, for player 1) will do in the future. For example, player 2 at the information set $\{x_2\}$ cannot choose an optimal strategy without reasoning about player 1's action at $\{x_{12}\}$. If player 2 thinks that player 1 is rational, then player 1 would choose Out at $\{x_{12}\}$. In that case, player 2 should choose Out at $\{x_2\}$. Now, if player 2 gets to move, it means that player 1 chose In at $\{x_{11}\}$. However, if player 1 is rational and knows the rationality of player 2, then player 1 would not have chosen In at $\{x_{11}\}$! In summary, at any information set following someone's

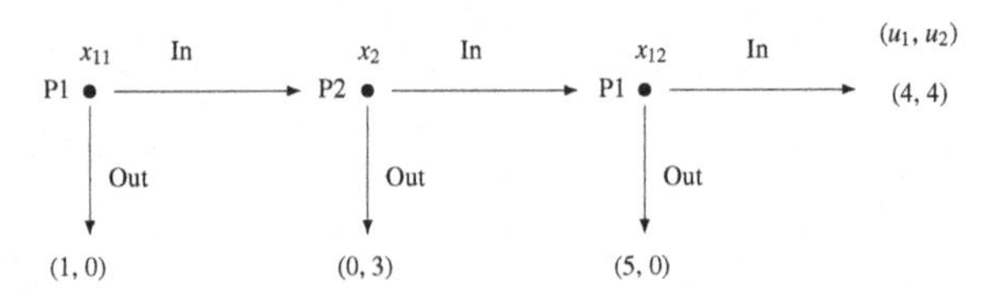

Fig. 4.15 A short centipede game

action "In", the common knowledge of rationality is violated. Therefore, it becomes doubtful if we should follow backward induction for games like this.

To solve this problem, there is an approach to make an equilibrium with a specification for how the players perceive the other players' rationality or objectives.[13] First, at information sets which are reachable if all players choose actions rationally, we can assume that all players believe in the others' rationality. Next, for information sets that are not reachable if all players are rational, we prepare a second scenario. For example, in the game of Fig. 4.15, when the game reaches the information set $\{x_2\}$, player 2 cannot believe that player 1 is rational any more. Consider a second scenario that player 1 is an egalitarian (who prefers (4, 4) over all other payoff combinations). With this second scenario, player 2's rational action is In. Moreover, if player 1 knows this second scenario of player 2 and his rationality, then the rational player 1 would choose In at $\{x_{11}\}$ (and Out at $\{x_{12}\}$). This solution has the property that both players are rational at all information sets, but the outcome is different from the one by backward induction.

Of course, there are other possible scenarios. For example, player 2 may think that it was player 1's mistake to reach $\{x_2\}$ and continues to believe that player 1 will make a rational choice at $\{x_{12}\}$. In this case, player 2 should choose Out. Given this, player 1 chooses Out at $\{x_{11}\}$ and the game ends there. This is another solution with the property that the players are rational at all information sets and turns out to be the same as the one by backward induction.

Overall, if we consider an equilibrium which takes into account how players update their beliefs about others' rationality, it is a combination of "optimal" strategies based on the updating process. For more details, see the book by Asheim [1] and references therein. In general, if we impose the assumption that for any information set (including the ones that are not reachable if players are rational), players believe in others' rationality, then we have solutions by backward induction. Otherwise, the resulting strategy combinations may not coincide with those found by backward induction.

Problems

4.1 Construct the induced normal form of the entry game of Fig. 4.9 and find all Nash equilibria.

4.2 Reconsider the two generals' game in Problem 3.3(d). Assume that General 2 uses a spy satellite to see how Country 1's army proceeds. General 1 does not use a spy satellite. In this case, General 2 can choose an action after learning the strategy of General 1.

---

[13]The following is a simplification of the argument in Brandenburger [4].

(a) Draw the game tree of this sequential decision game.
(b) General 1 discovered the spy satellite. Now General 1 knows that his strategy will be seen by General 2, so he knows the game of (a) as well. Interpreting this situation as a simultaneous-move game ex ante, construct the induced normal form and find all Nash equilibria in pure strategies.
(c) Find solutions by backward induction of the extensive-form game of (a), in pure strategies. Compare with the solutions in Problem 3.3(d).

4.3 Consider internet shopping. There is one seller and one buyer of a good. The seller gets a payoff of 300 if she does not sell the good and consumes it herself. If the buyer obtains the good and consumes it, he gets a payoff of 1000. Currently, the seller is posting the price of 500. Each player's total payoff is the sum of the consumption payoff and the monetary payoff (1 monetary unit is a payoff of 1). For example, if the buyer pays 500 and the seller sends the good, the buyer consumes the good so that his total payoff is $1000 - 500 = 500$, while the seller's total payoff is 500 from the money. If the buyer pays 500 but the seller does not send the good, the buyer's total payoff is -500 and the seller's is $300 + 500 = 800$.

(a) (Advance Payment Game) First, the buyer chooses whether to send the payment of 500 or not. If he does not send the money, the game ends. If he does, then the seller chooses whether to ship the good or not and the game ends. If she ships the good, the buyer consumes the good; otherwise the seller consumes the good. Draw the game tree of this game, specifying the payoff combinations at each terminal node, and find the solution by backward induction.
(b) (Advance Shipment Game) Assume that the buyer has ordered the good already. The first mover is the seller, who chooses whether to ship the good or not. If she does not ship it, the game ends. In this case the seller consumes the good and the buyer does not pay. If the seller ships the good, it is the buyer's turn to choose whether to send the payment of 500 or not and the game ends. If the buyer does not send the money, he consumes the good for free, and the seller does not consume the good nor receive the money, which results in 0 payoff. Draw the game tree of this game, specifying the payoff combinations at each terminal node, and find the solution by backward induction.
(c) Now assume that the seller has two units of the good. After one round of the Advance Payment Game, if and only if (send money, ship the good) was the outcome of the first round, they will play the Advance Payment Game again. The payoff is the sum of the one-shot payoffs over two rounds, if the game continues. Draw the game tree of this game and find the solution by backward induction.
(d) Reconsider the Advance Payment Game in (a). If the seller did not ship the good after the buyer sent the money, the buyer informs the Consumer Authority and the authority fines the seller. Half of the fine will be paid to the buyer, and the other half will be taken by the authority as fee.

(i) Let x be the fine. Modify the payoff combinations of the game tree you made in (a).
(ii) What is the minimum level of x which makes (send money, ship the good) the outcome of a solution by backward induction?

4.4 In the market for sofas, Firm A and Firm B are competing in price. Firm A's production cost is 100 per sofa, and Firm B's unit cost is 150, both regardless of the total number of sofas produced. This week, there is a demand for two sofas in the market. The firms' strategies are their prices, and they can choose from $\{300, 250, 200\}$. If the two firms choose the same price, each firm sells one sofa. If they choose different prices, the cheaper firm sells two sofas at that price and the more expensive firm sells nothing. A firm's payoff is the profit (the difference between the total sales and the total production cost). For example, if both firms chose the same price of 300, then Firm A's payoff is $300 \cdot 1 - 100 \cdot 1 = 200$ and Firm B's payoff is $300 \cdot 1 - 150 \cdot 1 = 150$. The firms produce after they get an order, so they incur no cost if they do not sell. Answer the following:

(a) Consider a simultaneous-move game in which the two firms choose their prices without knowing the other's price. Write down the matrix representation of the game and find all pure-strategy Nash equilibria.
(b) Consider a perfect-information extensive-form game such that Firm B chooses a price first, and after seeing it Firm A chooses its price. Find all pure-strategy solutions by backward induction (not the equilibrium path but a strategy combination).
(c) Consider a perfect-information extensive-form game such that Firm A chooses a price first, and after seeing it Firm B chooses its price. Find all pure-strategy solutions by backward induction.

4.5 Shop M is currently selling an item at a price of 100. Two buyers are considering purchasing this item, but they compare the prices between Shop M and Shop R in the neighborhood. If the two shops charge different prices, both buyers go to the cheaper shop. If the two shops charge the same price, one buyer goes to M and the other goes to R.

Shop R can choose between the prices 100 and 98. After seeing Shop R's price, Shop M chooses whether to change prices or not, and the game ends. Shop M also chooses between 100 and 98. Both shops get the item from the same wholeseller at the wholesale price of 50. (Assume that if a shop does not sell an item, it returns the item to the wholeseller so that there is no cost.) The payoff of a shop is its profit, which is the revenue (its price times the quantity sold) minus the cost (50 times the quantity sold).

(a) The game tree is as follows. Find a solution by backward induction.

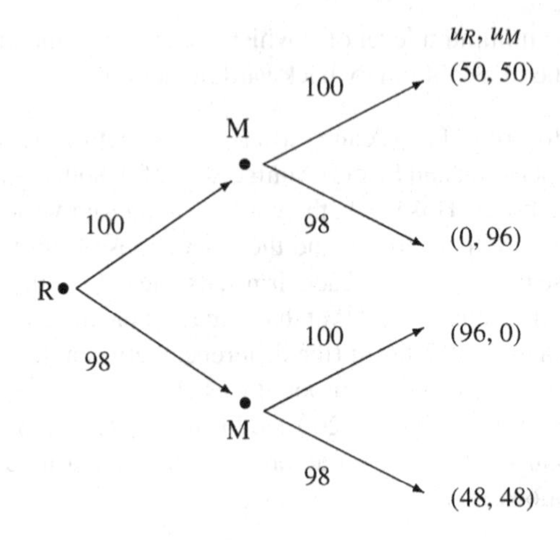

(b) Shop M started the "low price guarantee" policy. It will match its price to Shop R's price, if the latter's price is lower than Shop M's. Thus, after the price choice of Shop R, M has a unique option as in the new game tree below. Find a solution by backward induction and discuss the economic implications.

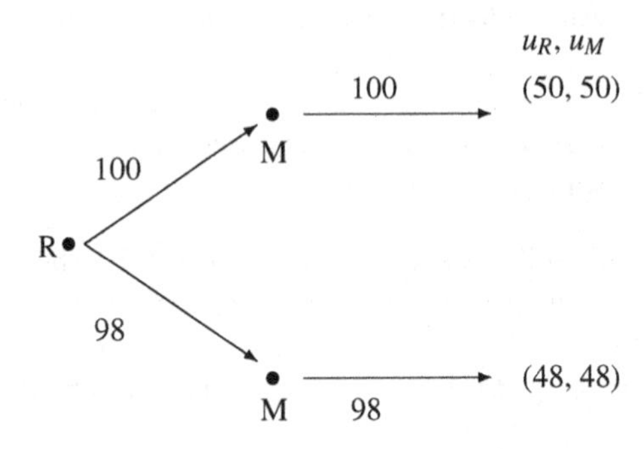

References

1. Asheim G (2006) The consistent preferences approach to deductive reasoning in games. Springer, Dordrecht
2. Aumann R (1964) Mixed and behavior strategies in infinite extensive games. In: Dresher, Shapley, Tucker (eds) Advances in game theory, annals of mathematic studies, vol 52. Princeton University Press, Princeton, pp 627–650
3. Aumann R (1987) Correlated Equilibrium as an expression of Bayesian rationality. Econometrica 55(1):1–18
4. Brandenburger A (2007) The power of paradox: some recent developments in interactive epistemology. Int J Game Theory 35(4):465–492
5. Coase R (1972) Durability and monopoly. J Law Econ 15(1):143–149

6. Cournot A (1838) Recherches sur les Principes Mathèmatiques de la Théorie des Richesses (Researches into the Mathematical Principles of the Theory of Wealth). Hachette, Paris (English translation, 1897.)
7. Fudenberg D, Tirole J (1991) Game theory. MIT Press, Cambridge
8. Gibbons R (1992) Game theory for applied economists. Princeton University Press, Princeton
9. Gul F, Sonnenschein H, Wilson R (1986) Foundations of dynamic monopoly and the coase conjecture. J Econ Theory 39(1):155–190
10. Kreps D (1990) A course in microeconomic theory. Princeton University Press, Princeton
11. Kreps D, Wilson R (1982) Sequential equilibrium. Econometrica 50(4):863–894
12. Kuhn H (1953) Extensive games and the problem of information. In: Kuhn, Tucker (eds) Contributions to the theory of games, vol 2. Princeton University Press, Princeton, pp 193–216
13. Osborne M, Rubinstein A (1990) Bargaining and markets. Academic Press, San Diego
14. Piccione M, Rubinstein A (1997) On the interpretation of decision problems with imperfect recall. Games Econ Behav 20(1):3–24
15. Rosenthal R (1981) Games of perfect information, predatory pricing and the chain-store paradox. J Econ Theory 25(1):92–100
16. Rubinstein A (1982) Perfect equilibrium in a bargaining model. Econometrica 50(1):97–109
17. Selten R (1965) Spieltheoretische Behandlung eines Oligopolmodells mit Nachfrageträgheit. Zeitschrift für die gesamte Staatswissenschaft 121:301–324, 667–689
18. Selten R (1975) Reexamination of the perfectness concept for equilibrium points in extensive games. Int J Game Theory 4(1):25–55
19. Selten R (1978) The chain store paradox. Theory Decis 9(2):127–159
20. Shaked A, Sutton J (1984) Involuntary unemployment as a perfect equilibrium in a bargaining model. Econometrica 52(6):1351–1364
21. von Stackelberg H (1934) Marktform und Gleichgewicht (Market Structure and Equilibrium). Springer, Vienna (English translation, 2011.)

Chapter 5
Subgame Perfect Equilibrium

5.1 Subgame Perfect Equilibrium

In this chapter, we extend backward induction to games with imperfect information in order to derive an equilibrium concept that works for any extensive-form game with complete information. This is one of the contributions of the seminal paper by Selten [21]. Consider, for example, the modified entry game in Fig. 5.1. There are two firms, E and M. Firm E chooses first whether to enter the market (action Enter) or not (action Out). If firm E does not enter, the game ends and firm E gets 0, while firm M gets 100. If firm E enters the market, two firms simultaneously choose designs for their products from two styles, A and B, and the game ends. The first coordinate of each payoff vector is the payoff for firm E and the second coordinate is the payoff for firm M.

How do we find a rational, stable strategy combination for this game? On the surface, firm E is the last decision maker, but the last information set contains two decision nodes. Depending on which decision node firm E is at, its optimal action

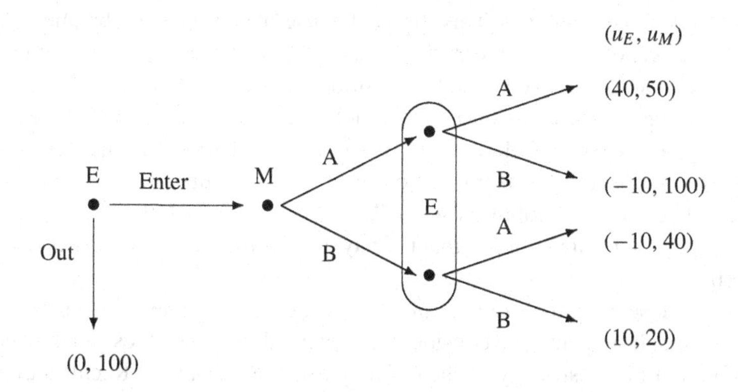

Fig. 5.1 A modified entry game

T. Fujiwara-Greve, *Non-Cooperative Game Theory*, Monographs in Mathematical Economics 1, DOI 10.1007/978-4-431-55645-9_5

Table 5.1 Second half of the modified entry game

E\M	A	B
A	40, 50	−10, 40
B	−10, 100	10, 20

differs. Therefore, we cannot apply backward induction directly. Notice, however, that the design choice part of the game after firm E enters is essentially a normal-form game as in Table 5.1 (where firm E is the row player so that the payoff vectors look the same as those in the game tree).

From Table 5.1, the unique rational action combination in the simultaneous-move part of the modified entry game is (A, A). Hence, rational firm E would expect (A, A) after entry and the payoff 40. Based on this belief, firm E compares entry with a payoff of 40 and staying out with a payoff of 0 to conclude that it is optimal to enter.

This way of thinking is a natural extension of backward induction and embeds a Nash equilibrium in the simultaneous-move part of the game. To generalize, Selten [21] divides a game into parts called *subgames* and requires a strategy combination to be a Nash equilibrium when it is restricted to the last subgames, and taking this into account, it is a Nash equilibrium in the second-to-last subgames, and so on. The resulting strategy combination is called a *subgame perfect equilibrium.*

Definition 5.1 A *subgame* of an extensive-form game is a part of the extensive-form game such that (i) it starts with a singleton information set, (ii) it includes all decision nodes after the starting information set and all actions of all of such decision nodes, and (iii) it does not cut any player's any information set.

A subgame is a part of the extensive-form game, which itself is a game and preserves the feasible actions and the information structure of the original game. Clearly, the original extensive-form game satisfies the definition of its subgame. We call the subgames other than the whole extensive-form game *proper subgames*.

Another way of looking at a subgame is that it is a collection of information sets with the above three properties. This approach is useful in writing a subgame formally. Let X be the set of all decision nodes of an n-person extensive-form game. They are partitioned among players, and the partition can be written as $(X_0, X_1, \ldots, X_n)$ where X_0 is the set of decision nodes which belong to Nature, and X_i is the set of decision nodes which belong to player i. For each player i, the collection of her/his information sets $\mathcal{H}_i$ is a partition of X_i. Then a subgame can be expressed as some collection of information sets $(\tilde{H}_0, \tilde{H}_1, \ldots, \tilde{H}_n)$ such that $\tilde{H}_i \subseteq \mathcal{H}_i$ for all $i = 0, 1, 2, \ldots, n$ (which implies that no player's information set is cut) and satisfies (i) and (ii).

As we have seen by the example of the modified entry game, we want to restrict strategies to each subgame and consider rational combinations of restricted strategies (the projection of the strategy of the entire game to a subgame). Recall that player i's (pure or behavioral) strategy s_i in an extensive-form game is a function which assigns an action distribution to each information set $H_i \in \mathcal{H}_i$.

Definition 5.2 For any player $i = 1, 2, \ldots, n$, the *restricted strategy* of a strategy s_i to a subgame $\tilde{H} = (\tilde{H}_0, \tilde{H}_1, \ldots, \tilde{H}_n)$ is denoted as $s_i \mid_{\tilde{H}}$, and it is a function defined on the collection of player i's information sets $\tilde{H}_i$ in the subgame such that the prescribed action (distribution) coincides with the original strategy's choice: for each $H_i \in \tilde{H}_i$, $s_i(H_i) = s_i \mid_{\tilde{H}} (H_i)$.

Let the payoff vector of a terminal node in any subgame correspond to the one in the original extensive-form game. Then it is meaningful to consider a Nash equilibrium of a subgame.

Definition 5.3 A strategy combination $s^* = (s_1^*, s_2^*, \ldots, s_n^*)$ of an extensive-form game is a *subgame perfect equilibrium* if, for each subgame $\tilde{H}$, its restricted strategy combination $(s_1^* \mid_{\tilde{H}}, s_2^* \mid_{\tilde{H}}, \ldots, s_n^* \mid_{\tilde{H}})$ to $\tilde{H}$ constitutes a Nash equilibrium of $\tilde{H}$.

For any perfect-information, extensive-form game with a finite length, Theorem 4.2 induces a Nash equilibrium for each subgame by backward induction. Thus the solution by backward induction is a subgame perfect equilibrium. The definition of a solution by backward induction for arbitrary length perfect-information games (Sect. 4.7) shows that it satisfies the definition of a subgame perfect equilibrium. Therefore, the solutions in Chap. 4 are all subgame perfect equilibria. There are many books and papers that combine solutions by backward induction and subgame perfect equilibria using only the concept of subgame perfect equilibria. In this book, however, we clarify the significance of the notion of subgame perfect equilibrium with separate chapters. Subgame perfection is not only backward induction but also a Nash equilibrium in parts, and the latter idea is effective when the game does not have perfect information.

5.2 Capacity Choice Game

To learn to use the concept of subgame perfect equilibrium, we analyze an extensive-form game which features a capacity choice of a firm, followed by a Cournot game. If a firm can choose its production capacity before a Cournot game, first it must solve a Nash equilibrium after each possible capacity choice and then determine the optimal capacity.

Firm 1 has decided to enter a market which has been monopolized by firm 2. Their products are the same good, and they face the same price, which is solely determined by the total supply quantity. Firm 1 can build either a large factory (action L) or a small factory (action S) to enter. A large factory costs 80,000 to build but can produce any amount of the product without additional cost. (Clearly, this is a simplification assumption. Imagine that costs are normalized.) A small factory costs 20,000 and can produce up to 100 units without additional cost, but no more. After the capacity choice of firm 1, which is observed by firm 2, the two firms choose production quantities of the same good simultaneously, and the game ends. Firm 2 has no capacity constraint and no cost to produce any amount of the good. If firm 1 produces q_1 units and firm 2 produces q_2 units, the market price of the good is $900 - q_1 - q_2$. The payoff of a firm is the profit. For firm 1, it is

$$\Pi_1(q_1, q_2) = (900 - q_1 - q_2)q_1 - C,$$

where C is either 80,000 or 20,000 depending on the capacity choice (and q_1 must be at most 100 if the small factory is chosen), and for firm 2,

$$\Pi_2(q_1, q_2) = (900 - q_1 - q_2)q_2.$$

This game has two proper subgames, starting after action L and action S of firm 1. Suppose that firm 1 selected action L. Then the simultaneous-move game part is the same as the Cournot game in Sect. 3.2 by setting $A = 900$ and $c_1 = c_2 = 0$. Hence there is a unique Nash equilibrium of the subgame after action L, $(q_1^*(L), q_2^*(L)) = (300, 300)$, and the payoff vector is (10,000, 90,000).

If firm 1 chooses action S, its actions in the proper subgame are restricted in $[0, 100]$, and we need to specify firm 1's best response to firm 2's action q_2 with this restriction. By differentiation,[1]

$$\frac{\partial \Pi_1}{\partial q_1} = -2q_1 + 900 - q_2.$$

Hence, if $q_2 \leqq 900$, the optimal q_1 that maximizes Π_1 is

$$q_1 = \frac{1}{2}(900 - q_2).$$

Notice that if $q_2 < 700$, then this quantity exceeds 100. This is not feasible in this subgame. Although a higher profit could be achieved if firm 1 could produce more than 100 units, the best response to $q_2 < 700$ is to produce 100 units. In summary,

$$BR_1(q_2) = \begin{cases} 0 & \text{if } 900 < q_2 \\ \frac{1}{2}(900 - q_2) & \text{if } 700 \leqq q_2 \leqq 900 \\ 100 & \text{if } q_2 < 700. \end{cases}$$

Therefore, as Fig. 5.2 shows, there is a unique Nash equilibrium of this subgame, $(q_1^*(S), q_2^*(S)) = (100, 400)$, and the payoff vector is (20000, 160000).

Finally, for the whole game, given the above Nash equilibria in the proper subgames, firm 1's optimal capacity choice is S. The unique subgame perfect equilibrium can be written as $(s_1^*, s_2^*) = ((\text{capacity choice}, q_1^*(L), q_1^*(S)), (q_2^*(L), q_2^*(S)) = ((S, 300, 100), (300, 400))$. An economic implication is that it is an equilibrium behavior that firm 1 would "tie its own hands" to commit to not producing a lot.

[1] This shows that after firm 1 has **sunk** the cost of the capacity, it does not affect decisions in later subgames.

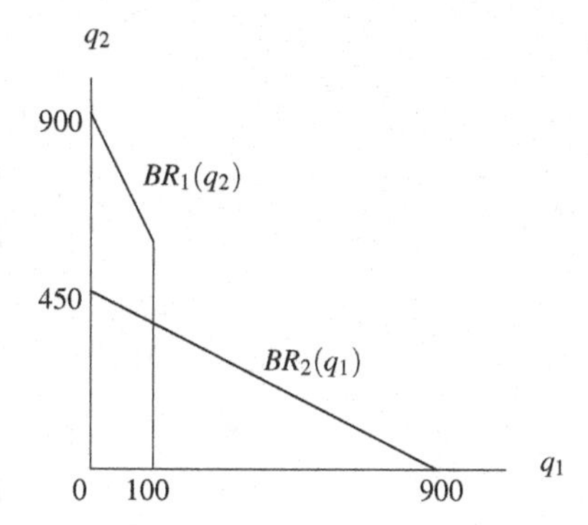

Fig. 5.2 Best response with capacity constraint

5.3 Prisoner's Dilemma of Neuroeconomics

Let us introduce another interesting game called the Prisoner's Dilemma of Neuroeconomics, introduced by Fujiwara-Greve [11]. One of the goals of recently rising literature of neuroeconomics is to discover a substance which "solves" problems like the Prisoner's Dilemma. (See for example, Kosfeld et al. [15]). We show, by a simple example, that such a discovery may not resolve the Prisoner's Dilemma but, rather, it produces a new Prisoner's Dilemma.

The game starts after a substance called X is discovered, which makes anyone be cooperative in the Prisoner's Dilemma. Two players, player 1 and 2, choose whether to take this substance before playing the Prisoner's Dilemma of Table 5.2, and after that, they play the Prisoner's Dilemma with or without the restriction on their feasible actions. In Table 5.2, the action C is Cooperation with the opponent and the action D is Defection against the opponent. A player who takes the substance X can only have one action C in the Prisoner's Dilemma, or her/his payoff function changes so that action D is strictly dominated by action C. There is no payoff (positive or negative) from taking the substance X.

Strictly speaking, depending on the order of intake decisions (simultaneous or sequential in some order), there are many possible formulations of this extensive-form game. However, in fact the outcome is essentially the same.

Table 5.2 General prisoner's dilemma ($g > c > d > \ell$)

P1\P2	C	D
C	c, c	ℓ, g
D	g, ℓ	d, d

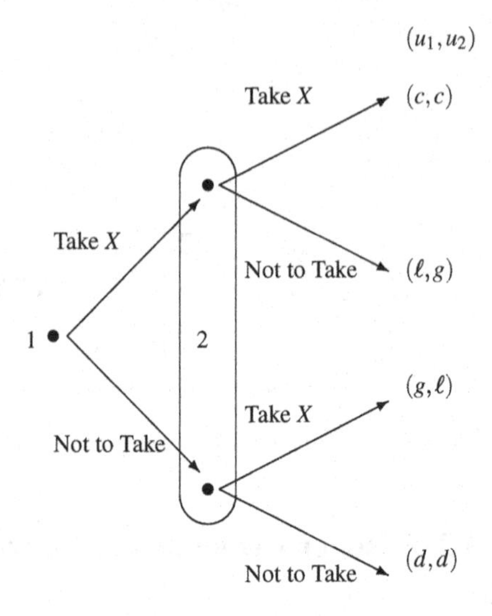

Fig. 5.3 Reduced game of simultaneous intake decisions

Let us consider the case that the two players simultaneously decide whether to take the substance X,[2] and then they play the Prisoner's Dilemma in Table 5.2. Let us find the Nash equilibria of the subgames after the intake decisions. If both players take X, they play (C, C) in the Prisoner's Dilemma and each player receives the payoff of c. If one player takes X and the other does not, then the player who took the substance plays C, but the other player chooses D, which is the dominant action. Hence the former receives ℓ, while the latter gets g. If neither of the players takes the substance X, then they face the Prisoner's Dilemma which has a unique Nash equilibrium of (D, D). Therefore in this case the players receive d. Taking these Nash equilibria of later subgames into account, we can reduce the whole game to the one illustrated in Fig. 5.3.

The reduced game is again a Prisoner's Dilemma. For each player, the strategy to take X is strictly dominated by the strategy not to take it, and hence this game has a unique subgame perfect equilibrium, in which no player takes the substance X. To be precise, in the unique subgame perfect equilibrium, each player does not take X, plays C if (s)he took X (regardless of the opponent's intake decision), and plays D otherwise (regardless of the opponent's intake decision).

Alternatively, we can consider a sequential decision game, and without loss of generality assume that player 1 first chooses whether to take X or not, and after observing her decision, player 2 chooses whether to take X or not. The reduced game of this sequential-decision game only separates the information set of player 2 in Fig. 5.3 into two. The Prisoner's Dilemma actions in the second stage are the same as for the simultaneous intake decision game. Therefore, at each information

[2]Whether the players observe the other's action afterwards or not does not matter, because a player who takes X always chooses C and the player who does not take X always chooses D in the subsequent subgames, regardless of the observation.

set of the intake decision stage, it is optimal for player 2 to not take X. Expecting this, player 1's optimal action is to not take X also. Overall, the sequential decision game also has a unique subgame perfect equilibrium, in which no player takes X.

Finally, we can also postulate that the two players may negotiate to choose the timing of their intake decisions (simultaneous or sequential in some order). In general, it is meaningful to investigate such an extended framework of a game to allow players to choose the timing of decision-making. This is because some games have the first-mover advantage or the second-mover advantage and, together with payoff asymmetry, it is not a priori clear who should move first to achieve an efficient outcome. (See, for example, Hamilton and Slutsky [12] and van Damme and Hurkens [7, 8]). However, in the current situation, regardless of the timing of the substance-taking decisions, there is no first-mover or second-mover advantage. Thus endogenizing the timing will not change the subgame perfect equilibrium.

In summary, rational players cannot voluntarily commit to cooperate in the Prisoner's Dilemma, because their commitment does not guarantee that the opponent changes behavior. A social problem like the Prisoner's Dilemma needs to be solved by the society, i.e., by an arrangement which **aligns players' incentives** towards a social goal. We turn to this idea in Sects. 5.5 and 5.6.

5.4 Finitely Repeated Games

An important class of extensive-form games is *repeated games*, in which the same game is repeated by the same set of players. The game that is repeated is called a *stage game* and is usually assumed to be a simultaneous-move game. A significance of this class of repeated games is that we can make a general theory which gives a property that holds for a large class of stage games. Moreover, we have a solution to the Prisoner's Dilemma (to sustain mutual cooperation by rational players in an equilibrium) in this class.

What causes the difference between a one-shot game and its repeated game? If players do not know the play of each round of the stage game (which we call a *period* of the repeated game) and choose actions independently in every period, of course the situation is essentially the same as a one-shot game. However, repetition makes it more plausible that players know some of the past plays of the game, on which they may base their future actions. (The strategies in the stage game are called *actions* to distinguish them from the strategies of the repeated game.) If players choose their actions based on the history of the repeated game, it becomes an extensive-form game, and it may generate a new equilibrium which involves actions that cannot be a part of any one-shot equilibrium.

In an extensive-form game, the information structure is crucial to the set of equilibria. After each period, we need to specify what is observed by the players. If all players observe the action profile of the stage game after each period, the information structure is called *perfect monitoring*. Otherwise, the information structure is called *imperfect monitoring*. For example, a repeated rock-paper-scissors game usually has

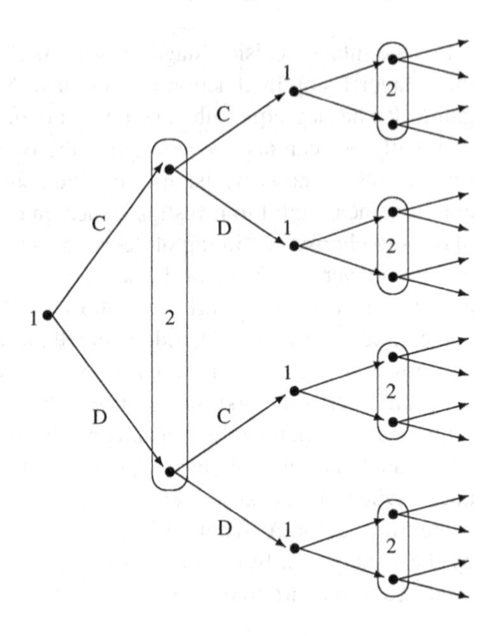

Fig. 5.4 The information structure of the twice-repeated prisoner's dilemma with perfect monitoring

perfect monitoring. A Bertrand game of simultaneous price setting may or may not have perfect monitoring, because firms might not observe their rival's secret price cut.

The analysis of subgame perfect equilibria is quite simple for a repeated game with perfect monitoring. This is because all subgames start at the beginning of some period and never start in the middle of a period, since the stage game is a simultaneous-move game. Moreover, a player's information sets are singletons at the beginning of a period, corresponding to each history of actions by all players until that period, thanks to perfect monitoring. For example, the twice-repeated Prisoner's Dilemma in Table 5.2 with perfect monitoring becomes a game tree shown in Fig. 5.4. (We omitted the payoff vectors and the action names in the second period, to outline only the tree.)

In this section, we consider *finitely repeated games*, denoted by G^T, such that the same set of players play a simultaneous-move game $G = (\{1, 2, \ldots, n\}, S_1, \ldots, S_n, u_1, \ldots, u_n)$ (with the interpretation that S_i is the set of actions) for a given finite number of T times. We first formally define the repeated game G^T by specifying the three components of the game.

The set of the players of G^T is the same as that of the stage game G, namely $\{1, 2, \ldots, n\}$. The set of pure strategies of a player in the repeated game depends on what is observable to each player. Even with perfect monitoring, we postulate that it is too much to assume that players can observe the probabilities of a mixed action played by opponents. Thus we assume *weak perfect monitoring*, such that each player observes only pure actions (or realizations of mixed actions) by all players in each period. Then, a pure strategy consists of action plans for each period of $t =$

$1, 2, \ldots, T$, such that all possible sequences of pure-action profiles (observations) up to period t are assigned an action of the stage game in period t.

To formalize, let us call a sequence of (pure) action profiles until period $t = 1, 2, \ldots, T$ a *history* until period t, and define the set of histories until period t as

$$H_t = (S_1 \times S_2 \times \cdots \times S_n)^{t-1}.$$

Note that the set of histories until the first period of the repeated game is degenerate. For notational convenience we use this notation. Note also that the set of histories is common to all players under perfect monitoring.

Definition 5.4 Under weak perfect monitoring, a pure strategy b_i of player i in G^T is a sequence $b_i = (b_{i1}, b_{i2}, \ldots, b_{iT})$ such that for each period $t = 1, 2, \ldots, T$,

$$b_{it} : H_t \to S_i.$$

Thus, a pure strategy is an action plan of each period based on the histories until that period. Moreover, since the "history until the first period" is degenerate, it specifies a single action in S_i in the first period. This definition of a pure strategy of a repeated game with weak perfect monitoring is the same for infinitely repeated games (Sect. 5.5), except that there will be no end period.

Once the pure strategies are defined, the definitions of mixed strategies and behavioral strategies should be obvious. The payoff of the repeated game G^T is defined as the sum of the stage game payoffs from G over the T periods.

Let us find the subgame perfect equilibria of a twice-repeated Prisoner's Dilemma game with weak perfect monitoring. Player i's pure strategy is $b_i = (b_{i1}, b_{i2})$; where $b_{i1}(\emptyset) \in \{C, D\}$ is an action in the first period and $b_{i2} : \{C, D\} \times \{C, D\} \to \{C, D\}$ is an action plan of the second period, contingent on the first period observation. Consider subgames of the last period, which are one-shot Prisoner's Dilemma games. There are four subgames, depending on the history until the second period. Although the first period payoffs are different for different histories, the second period payoffs in all four proper subgames are determined by the same G. This is illustrated by Fig. 5.5. (The first coordinate of each payoff vector belongs to player 1.)

As Fig. 5.5 shows, in each subgame starting at the beginning of the second period, no player can change the first period payoff. Hence, they are essentially playing the one-shot Prisoner's Dilemma game. This implies that in any subgame of the second period, there is a unique Nash equilibrium; which is to play (D, D).

Going back to the first period, we find a Nash equilibrium for the entire game. Because in any subgame in the second period the players play (D, D), the second period payoffs are the same after any action profile of the first period. Hence, in the entire game (given that each player plays the Nash equilibrium action in the future subgames), each player maximizes the first period payoffs. This means that the players are essentially playing the one-shot Prisoner's Dilemma game!

In summary, there is a unique subgame perfect equilibrium in the twice-repeated Prisoner's Dilemma game such that both players play D after any history, including

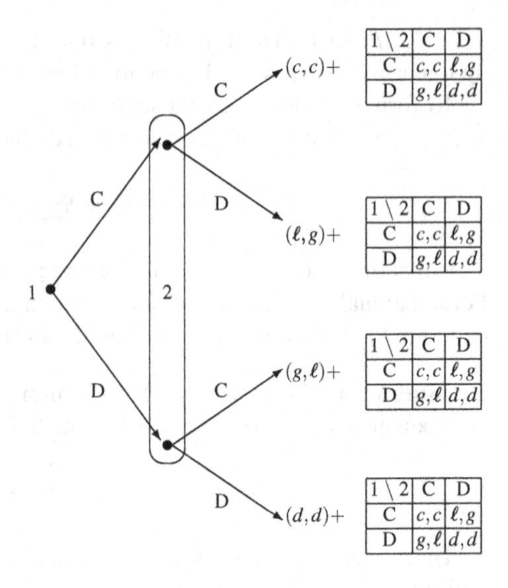

Fig. 5.5 Payoff structure of twice-repeated prisoner's dilemma game

the first period. Formally, for each player $i = 1, 2$, let $b^*_{i1}(\emptyset) = D$ and $b^*_{i2}(h) = D$ for all $h \in \{C, D\}^2$. Then the unique subgame perfect equilibrium is (b^*_1, b^*_2). Alternatively, we can express a strategy by a list of actions in the order of first period action, second period action for histories (C, C), (C, D), (D, C) and (D, D) respectively. Then the equilibrium strategy by each player is $b^*_i = (D, (D, D, D, D))$. Clearly, the dilemma is not resolved. Also, using an analogous argument, as long as the number of repetitions is finite, the players play (D, D) all the time. This logic is generalized below. (The proof is analogous to the above analysis, as well as that of the chain store paradox, and is thus omitted.)

Theorem 5.1 *For any stage game G with a unique Nash equilibrium and any $T < \infty$, there is a unique subgame perfect equilibrium of G^T such that all players play the stage-game Nash equilibrium action after any history.*

Therefore, if the stage game has only one Nash equilibrium, then the players do not behave differently from the one-shot game, no matter how many times they repeat the game, as long as the repetition ends at a finite period.

However, once the stage game possesses multiple Nash equilibria, the conclusion may be different. Consider a Business Meeting Game inspired by an example in Watson [25]. Two players, 1 and 2, need to make a business meeting. There are two possible meeting rooms that they can use; room A and room B. Player 2 has an additional option to use an internet meeting from home (action H). Player 1 does not have this option. Meeting room A has an internet connection, but room B is in a factory and does not have internet access. Therefore, the action combinations in which the two can meet are (A, A), (B, B) and (A, H) (Table 5.3).

Table 5.3 Business meeting game

1\2	A	B	H
A	5, 3	0, 0	1, 4
B	0, 0	3, 1	0, 0

This game has two Nash equilibria in pure actions, which are (A, H) and (B, B). Although the combination (A, A) attains the highest total payoff of the two players, it is not a one-shot Nash equilibrium.

Consider the twice-repeated Business Meeting Game. Player 1's pure strategy consists of a first-period action $b_{11}(\emptyset) \in \{A, B\}$ and an action plan for the second period $b_{12} : \{A, B\} \times \{A, B, H\} \to \{A, B\}$, while player 2's pure strategy consists of $b_{21}(\emptyset) \in \{A, B, H\}$ and $b_{22} : \{A, B\} \times \{A, B, H\} \to \{A, B, H\}$.

First, there are trivial subgame perfect equilibria in which the same one-shot Nash equilibrium is played after any history. For example, in all six subgames starting at the beginning of the second period, they play (A, H) and also in the first period. Formally, let player 1's strategy b_1^A be such that $b_{11}^A(\emptyset) = A$ and $b_{12}^A(h) = A$ for all $h \in \{A, B\} \times \{A, B, H\}$. Let player 2's strategy b_2^H be such that $b_{21}^H(\emptyset) = H$ and $b_{22}^H(h) = H$ for all $h \in \{A, B\} \times \{A, B, H\}$. Then (b_1^A, b_2^H) is a subgame perfect equilibrium of the twice-repeated Business Meeting Game. We can alternatively write the strategy combination by two lists of actions in the order of the first period, second period, after (A, A), (A, B), (A, H), (B, A), (B, B), and (B, H) are observed;

$$(b_1^A, b_2^H) = \big((A, (A, A, A, A, A, A)), (H, (H, H, H, H, H, H))\big).$$

Other trivial subgame perfect equilibria include the one playing (B, B) after any history and (A, H) in the first period. Basically, it is a subgame perfect equilibrium to separate the second period from the first period, to play the same one-shot Nash equilibrium regardless of the first period action profile, and some one-shot Nash equilibrium in the first period.

However, we have a non-trivial subgame perfect equilibrium in which (A, A) is played in the first period. Recall that (A, A) cannot be played if the Business Meeting Game is played only once. To motivate the players to play a non-Nash equilibrium action combination (A, A) in the first period, the second period Nash equilibria must be specified in such a way that the players (in particular player 2) want to play A in the first period, instead of the other action(s). (In all subgames starting at the beginning of the second period, a one-shot Nash equilibrium must be played. Hence, there is no subgame perfect equilibrium that plays (A, A) in the second period.) Using the fact that the two (pure) Nash equilibria give different payoffs to player 2, we can construct a carrot and stick mechanism: if player 2 "cooperates" with player 1 and (A, A) is played in the first period, they play the (A, H) equilibrium in the second period to "reward" player 2. However, if player 2 "betrays" and chooses action H in the first period, the players play (B, B) equilibrium in the second period to "punish" player 2.

Formally, let player 1's strategy b_1^* be such that $b_{11}^*(\emptyset) = A, b_{12}^*(A, A) = A$, and $b_{12}^*(A, H) = B$. For player 2, let b_2^* be a strategy such that $b_{21}^*(\emptyset) = A, b_{22}^*(A, A) = H$, and $b_{22}^*(A, H) = B$. After other histories, we only need to coordinate the two players' strategies to play the same one-shot Nash equilibrium. For example, let $b_{i2}^*(h) = B$ for both $i = 1, 2$ and any $h \neq (A, A)$ or (A, H).

We now prove that each player is playing a best response to the opponent's (restricted) strategy in each subgame. Consider player 2in the second period. If the history was (A, A), then player 1 would play A in the second period so that player 2's optimal action is H. Similarly, after other histories, player 1 would play B in the second period so that player 2 should play action B as well. Go back to the first period. Player 1 will play action A. If player 2 chooses H, he gets 4 in the first period but after the history of (A, H), the second period payoff is 1, from (B, B). Hence the total payoff is 5. By contrast, if player 2 plays A in the first period, he only gets 3 in the first period but receives 4 in the second period from (A, H). If player 2 plays B in the first period, not only does he get 0 in the first period but also only 1 in the second period, and thus it is not optimal. Therefore, given b_1^*, player 2's best response is to follow b_2^* in any subgame. It is easy to check that, given b_2^*, player 1's best response is (the restriction of) b_1^* in any subgame.

In summary, using two Nash equilibria of the stage game as a carrot and stick mechanism, we constructed a subgame perfect equilibrium in which a non-Nash action combination (A, A) is played on the equilibrium path. The possibility of enlarging the set of equilibrium actions beyond the Nash equilibria of the stage game is a significant property of repeated games. Benoit and Krishna [4] generalized the above logic to show that if for each player there are Nash equilibria of the stage game with different payoffs, then a large set of payoff vectors is sustainable by subgame perfect equilibria of sufficiently long but finitely repeated games of the stage game. Essentially, the stage game Nash equilibrium with a low payoff is used as the stick and one with a higher payoff is used as the carrot for each player. Under some conditions, the set of equilibrium payoffs of finitely repeated games corresponds to the sustainable set of infinitely repeated games of the same stage game as well. The latter set is characterized by the *Perfect Folk Theorem* in Sect. 5.7.

5.5 Infinitely Repeated Games

From this section until Sect. 5.7, we study *infinitely repeated games* such that the same stage game is repeated over the (discrete) infinite horizon, $t = 1, 2, \ldots,$ by the same set of players. The analysis is mathematically somewhat involved, but this class of games is very useful because, regardless of the stage game, there is a characterization[3] of the set of equilibrium payoff combinations sustainable by subgame perfect equilibria.

[3] A *characterization* of something is its necessary and sufficient condition.

When the game continues *ad infinitum*, a player receives an infinite sequence of payoffs. Then the sum of stage game payoffs does not make sense; because the sum of any sequence of positive numbers (resp. negative numbers) becomes the same $+\infty$ (resp. $-\infty$) so that such sequences are not distinguishable. We want a measure which indicates that, for example, the sequence of 3, 3, 3, ... is strictly better than the sequence of 1, 1, 1, ... In fact, there are multiple such measures, but we adopt the discounted sum as the total payoff of an infinitely repeated game.[4] This is to discount future payoffs by a *discount factor*, which is a real number $\delta \in (0, 1)$ common to all players, in such a way that the second period payoff is discounted by multiplying with δ, the third period payoff is multiplied with δ^2 and so on, and then to add them up over the entire horizon. Formally, if player i receives a sequence of stage game payoffs $u_i(1), u_i(2), \ldots$, then the payoff of the infinitely repeated game[5] is

$$\sum_{t=1}^{\infty} \delta^{t-1} u_i(t).$$

As long as the stage game payoffs are bounded (which is implicitly assumed throughout this book), the discounted sum is finite. For example, the discounted sum[6] of the sequence 3, 3, 3, ... is

$$3 + \delta \cdot 3 + \delta^2 \cdot 3 + \cdots = \frac{3}{1-\delta},$$

and this is strictly greater than the discounted sum $1/(1 - \delta)$ of the sequence $1, 1, 1, \ldots$, for any $\delta \in (0, 1)$.

The *average payoff* of the discounted sum of an infinite sequence of payoffs is defined by

$$(1-\delta) \sum_{t=1}^{\infty} \delta^{t-1} u_i(t).$$

Although the discounted sum of the payoff sequence can be very large when δ is close to 1, this average payoff is comparable to the stage game payoff for any $\delta \in (0, 1)$. For example, the average payoff of the payoff sequence 3, 3, 3, ... is 3. However, following the tradition of the dynamic programming, we assume that each player chooses strategies to **maximize the discounted sum** of the stage game payoffs of the infinitely repeated game.[7]

---

[4] Another way to evaluate infinite sequences of payoffs is *overtaking criterion*. See Rubinstein [19].

[5] Some authors think that the exponent $t-1$ is not "neat" and take the time horizon as $t = 0, 1, 2, \ldots$ so that the discounted sum is formulated as $\sum_{t=0}^{\infty} \delta^t u_i(t)$. In either way, the idea is the same in that the first period payoff is undiscounted and future payoffs are discounted exponentially.

[6] The sum of a geometric sequence starting with a and the common ratio r is derived as follows. Let X be the sum, so that $X = a + ar + ar^2 + ar^3 + \cdots$. Then by multiplying X with r, we obtain $r \cdot X = ar + ar^2 + ar^3 + \cdots$. Subtracting this from X, we have $X - r \cdot X = a$, i.e., $X = \frac{a}{1-r}$.

[7] The same analysis goes through if players maximize the average payoff.

There are at least three interpretations of the discount factor. One is a psychological factor that future payoffs evaluated at an earlier period seem to be less than their face value. Another important interpretation is that δ is the probability that the game continues, with the assumption that when the game ends (with probability $1 - \delta$), the future payoffs are 0. Then the total expected payoff becomes the above form. With this interpretation, we can justify the infinite horizon game against the possible criticism that no player (human or not) can live forever. The game ends after a finite period, but the ending date is uncertain.

Yet another interpretation is an economic one, in which the discount factor is $\delta = 1/(1+r)$ where r is the interest rate. The idea is as follows. If you have 1 dollar this year, you can deposit it in a safe asset (such as a bank deposit) to earn 1 plus the interest rate $r \in (0, 1)$ next year. By inverting this relationship, the future one dollar is worth $1/(1+r)$ dollar now. Hence we can set $\delta = 1/(1+r)$.

For any of the above interpretations, the greater the discount factor is, the less discouted the future payoffs are, which means that the players put a higher weight on future payoffs. Sometimes we say that the players become more patient.

Given a stage game G and a common discount factor $\delta \in (0, 1)$ for all players, we denote by $G^{\infty}(\delta)$ the infinitely repeated game with (weak) perfect monitoring in which the payoff function is the discounted sum of the stage game payoffs.

Let us analyze the subgame perfect equilibria of an infinitely repeated Prisoner's Dilemma game with the stage game as in Table 5.2. Recall that if it is repeated finitely many times, the unique subgame perfect equilibrium is to play (D, D) after any history. By contrast, when it is repeated infinitely many times or the end is uncertain, we can construct a totally different equilibrium using the fact that there is always the next period. For example, consider the following *grim-trigger strategy*, denoted as b^{GT}, played by both players, as a candidate for a symmetric subgame perfect equilibrium.

For any period $t = 1, 2, \ldots$ and any history $h \in [\{C, D\} \times \{C, D\}]^{t-1}$,

$$b_t^{GT}(h) = \begin{cases} C & \text{if } h = \emptyset \text{ or } (C, C)^{t-1}; \\ D & \text{otherwise.} \end{cases}$$

In words, the grim-trigger strategy prescribes the cooperative action C at the beginning of the game, and as long as no one deviated from (C, C). This is called the *on-path* behavior, meaning that repeated (C, C) is on the play path generated by (b^{GT}, b^{GT}). If someone has deviated from (C, C) in the past even once, the strategy selects the non-cooperative action D forever after, because the future history will be always in the second category. A deviation thus "triggers" the punishment, and the punishment never ends, which gives the name "grim" to the strategy. Note that, if both players obey (b^{GT}, b^{GT}), no deviation occurs, and thus the punishment is an off-path behavior. If (b^{GT}, b^{GT}) is a subgame perfect equilibrium, the punishment is credible.

When the discount factor δ is sufficiently high, i.e., players put a high weight on the future payoffs, the symmetric strategy combination (b^{GT}, b^{GT}) is in fact a subgame perfect equilibrium of the infinitely repeated Prisoner's Dilemma. To prove this result, we need to show that the grim-trigger strategy is optimal among an infinite number of feasible strategies of the game. Bellman's dynamic programming method [3] makes this optimization easy. For the completeness of this book, we provide formal results of dynamic programming in Appendix A.2, but let us sketch the idea below. (Another short introduction to dynamic programming is included in the Mathematical Appendix of Kreps [16]).

Focus on a player and fix other players' strategies. (The other players' actions are viewed as "states" for this player, which evolve according to the history of the game.) A strategy b satisfying the following property is called *unimprovable in one step*: for each period $t = 1, 2, \ldots$ and after any history until the tth period, the player cannot improve the total discounted payoff by changing the actions in one period/step, given that (s)he follows the strategy b from the next period on. It is then proved that an unimprovable strategy is an optimal strategy which maximizes the total expected payoffs over the infinite horizon, under discounting. This result is called the *one-step deviation principle*.

Thanks to the one-step deviation principle, we have the following routine to find a best response, given the opponents' strategy combination. (i) Make a candidate strategy of a best response. (ii) After any history and given that the player follows the candidate strategy from the next period on, compare the total discounted payoffs of one-period deviations and the candidate strategy's prescribed action. If the latter is not less than the former, then the candidate strategy is in fact a best response. That is, we do not need to vary the strategies over the infinite horizon to compare them. We only need to vary a strategy one period at a time. We also define a useful term, a continuation payoff, which focuses on the total payoff of the future.

Definition 5.5 For any infinitely repeated game $G^{\infty}(\delta)$, any (long-run) strategy combination $b = (b_1, b_2, \ldots, b_n)$ of $G^{\infty}(\delta)$, any period $t = 1, 2, \ldots$, and any subgame $\tilde{H}$ of $G^{\infty}(\delta)$ starting at the beginning of period t, let $(b_1 \mid_{\tilde{H}}, b_2 \mid_{\tilde{H}}, \ldots, b_n \mid_{\tilde{H}})$ be the restricted strategy combination of b to the subgame $\tilde{H}$. The sum of the discounted payoffs of a player i from $(b_1 \mid_{\tilde{H}}, b_2 \mid_{\tilde{H}}, \ldots, b_n \mid_{\tilde{H}})$ over the (infinite) horizon $t, t+1, t+2, \ldots$ is called the player's *continuation payoff* from b after the history reaching $\tilde{H}$.

At each period $t = 1, 2, \ldots$ and any history up to the tth period, the past payoffs are the same across one-period deviation strategies at t, and therefore in order to check if a strategy is unimprovable in one-step, we only need to compare the continuation payoffs. In fact, the notion of subgame perfect equilibrium has the continuation payoff comparison, when we check if a restricted strategy combination is a Nash equilibrium of a subgame.

Let us investigate whether (b^{GT}, b^{GT}) restricted in an arbitrary subgame is a Nash equilibrium of it. There are two classes of subgames; subgames after *on-path* histories that do not include action D and others (subgames after *off-path* histories). After an off-path history, (b^{GT}, b^{GT}) prescribes the one-shot Nash equilibrium of the stage game forever after. This is clearly a Nash equilibrium for any δ.

Consider a subgame after an on-path history of the form $(C, C)^{t-1}$, including the degenerate history (the first period). Given this, player i is about to choose an action in the tth period. If both players follow b^{GT} forever after, then player i's continuation payoff from tth period on is

$$u_i(b^{GT}, b^{GT}) = c + \delta c + \delta^2 c + \cdots = \frac{c}{1-\delta}.$$

If player i deviates to D, (s)he receives a high one-shot payoff g in this period, but (D, D) will be played forever after, following b^{GT} in the future. Therefore, the continuation payoff of this one-shot deviation is

$$g + \delta d + \delta^2 d + \cdots = g + \frac{\delta d}{1-\delta}.$$

Since the past payoffs are fixed, this one-step deviation does not improve the total discounted payoff of a player if and only if its continuation payoff is not greater than the one from b^{GT}:

$$\frac{c}{1-\delta} \geqq g + \frac{\delta d}{1-\delta} \iff \delta \geqq \frac{g-c}{g-d}.$$

(Because $g > c > d$, $(g-c)/(g-d)$ is strictly less than 1. That is, such δ exists.) Also, if this condition is satisfied, any mixed action of C and D followed by b^{GT} from the next period on would give the continuation payoff not greater than $c/(1-\delta)$. Therefore $\delta \geqq (g-c)/(g-d)$ is sufficient for the grim-trigger strategy to be unimprovable in one step, after any on-path history.

In summary, for any discount factor δ not less than $(g-c)/(g-d)$, the grim-trigger strategy combination is a subgame perfect equilibrium of the infinitely repeated Prisoner's Dilemma. On the equilibrium path, (C, C) is played every period, and the players' equilibrium average payoff combination is the efficient one (c, c). That is, the dilemma is resolved.

For general stage games, Friedman [9] first proved a *Folk Theorem* which characterizes the set of equilibrium payoff vectors by grim-trigger strategies, i.e., using a one-shot Nash equilibrium of the stage game as the punishment. The name Folk Theorem means that it is somehow known for a long time, like a folk tale, that in long-term relationships, a wide variety of payoff combinations are enforceable beyond one-shot (often inefficient) equilibrium payoff combinations. Let us formally state his result.

For convenience, we assume that correlated action profiles (see Sect. 4.8) are feasible in the stage game. That is, players can find a joint randomization device (such as a roulette) to coordinate their **action profiles** depending on the realizations of the device. Mathematically, a *correlated action profile* is a probability distribution over the set of action combinations of the stage game, $S = S_1 \times S_2 \times \cdots \times S_n$. Thus, the assumption is that for any probability distribution $\alpha \in \Delta(S_1 \times \cdots \times S_n)$, the players can find a randomization device which gives its outcomes with the probability distribution α and they can jointly observe a realization each period, before playing the stage game. With the existence of joint randomization devices, the set of feasible payoff vectors are defined as follows.

Definition 5.6 For a normal-form game $G = (\{1, 2, \ldots, n\}, S_1, \ldots, S_n, u_1, \ldots, u_n)$, a payoff vector $(v_1, v_2, \ldots, v_n) \in \Re^n$ is *feasible* if there exists a correlated action profile $\alpha \in \Delta(S_1 \times \cdots \times S_n)$ such that for any $i = 1, 2, \ldots, n$, $v_i = Eu_i(\alpha)$. Let V be the set of feasible payoff vectors.

As we have seen in Sect. 4.8, the set of payoff vectors which are attainable under correlated action profiles is often larger than the one under independently-mixed action profiles. Clearly, V is the largest set of possible payoff vectors of G.

Note, however, that even if the players cannot use a joint randomization device, it is still possible to approximate a feasible payoff vector $v \in V$ by designing a sequence of action profiles over time. For example, for the Prisoner's Dilemma in Table 5.2, in order to achieve $(\frac{c+d}{2}, \frac{c+d}{2})$ on average, one way is to correlate their action profiles (e.g., by flipping a coin) to play (C, C) and (D, D) with the equal probability, every period. Another way is to play (C, C) in odd-numbered periods and (D, D) in even-numbered periods. If δ is close to 1, then the latter sequence also achieves approximately $(\frac{c+d}{2}, \frac{c+d}{2})$ as the average payoff. Therefore, our assumption of the feasibility of correlated action profiles is purely for analytical convenience.

Recall that we have assumed only weak perfect monitoring. Hence, if a player uses an (individual) mixed action, then the underlying probability need not be observable to others. However, if a correlated action profile is used, all players are assumed to observe the realizations of the randomization device before playing the stage game, and hence they can see if any other player(s) had deviated from the specified correlated action profile.

Theorem 5.2 (Friedman's Folk Theorem) *For any normal-form stage game $G = (\{1, 2, \ldots, n\}, S_1, \ldots, S_n, u_1, \ldots, u_n)$, any Nash equilibrium σ^* of G, and any feasible payoff vector $v = (v_1, v_2, \ldots, v_n) \in V$ such that $v_i > Eu_i(\sigma^*)$ for all $i = 1, 2, \ldots, n$, there exists $\underline{\delta} \in (0, 1)$ such that for any $\delta \geqq \underline{\delta}$, there is a Nash equilibrium*[8] *of the infinitely repeated game $G^\infty(\delta)$ with (weak) perfect monitoring such that the average equilibrium payoff combination is v.*

In the proof, we construct grim-trigger strategy equilibria with a Nash equilibrium of the stage game as the punishment. The above theorem shows that for any Nash equilibrium of the stage game as the punishment, any (correlated) action profile can be repeated forever, if (i) the target action profile gives greater one-shot payoff than the punishment Nash equilibrium to all players, and (ii) the discount factor is sufficiently high. Notice that the minimum discount factor can depend on the target action profile. (The order of the statement, "for any v, there exists $\underline{\delta}$", is important.)

Proof Fix an arbitrary normal-form game G, its Nash equilibrium σ^*, and a feasible payoff combination $v \in V$. There exists a correlated action profile of G, $\alpha \in \Delta(S_1 \times \cdots \times S_n)$, such that $Eu_i(\alpha) = v_i$ for all $i = 1, 2, \ldots, n$. Consider a grim-trigger

[8]Because Friedman's work appeared before Selten's subgame perfect equilibrium, the original theorem uses Nash equilibrium as the equilibrium concept. Nowadays, we see that it is also a subgame perfect equilibrium.

strategy combination[9] $(b_1^{GT}, b_2^{GT}, \dots, b_n^{GT})$ such that, if no one has deviated from α in the past (including the first period), play α, and otherwise play σ^*.

After any history in which no player has deviated from α (including the first period), if all players follow the grim-trigger strategy combination, then the continuation payoff of player i is

$$Eu_i(\alpha) + \delta \cdot Eu_i(\alpha) + \delta^2 \cdot Eu_i(\alpha) + \cdots = \frac{v_i}{1-\delta}.$$

We compare this with the continuation payoff when player i deviates from α in one period and follows the grim-trigger strategy combination afterwards. Given that all others follow the grim-trigger strategy combination, player i's maximal one-shot payoff is $\max_{\sigma_i \in \Delta(S_i)} Eu_i(\sigma_i, \alpha_{-i})$, but after this period, σ^* will be played forever. Hence the maximal continuation payoff from a one-step deviation is

$$\max_{\sigma_i \in \Delta(S_i)} Eu_i(\sigma_i, \alpha_{-i}) + \frac{\delta \cdot Eu_i(\sigma^*)}{1-\delta}.$$

Therefore, the grim-trigger strategy is optimal if

$$\frac{v_i}{1-\delta} \geqq \max_{\sigma_i \in \Delta(S_i)} Eu_i(\sigma_i, \alpha_{-i}) + \frac{\delta Eu_i(\sigma^*)}{1-\delta}.$$

By computation, the above inequality is equivalent to

$$v_i - (1-\delta) \max_{\sigma_i \in \Delta(S_i)} Eu_i(\sigma_i, \alpha_{-i}) - \delta Eu_i(\sigma^*) \geqq 0$$

$$\Longleftrightarrow \delta \geqq \frac{\max_{\sigma_i \in \Delta(S_i)} Eu_i(\sigma_i, \alpha_{-i}) - v_i}{\max_{\sigma_i \in \Delta(S_i)} Eu_i(\sigma_i, \alpha_{-i}) - Eu_i(\sigma^*)}.$$

Let this bound be $\underline{\delta}$. Because $v_i > Eu_i(\sigma^*)$, $\underline{\delta} \in (0, 1)$. □

Two remarks are in order. First, as we noted in Footnote 8, the grim-trigger strategy combination is in fact a subgame perfect equilibrium. We just add that after any history that someone had deviated from α, the stage game Nash equilibrium is repeated forever, which is a Nash equilibrium of such subgames for any δ. Second, if we fix a discount factor first, then there can be unsustainable but feasible payoff vectors. This is because the minimum discount factor is dependent on the target payoff vector v.

Another well-known strategy for an infinitely repeated Prisoner's Dilemma game is the *Tit-for-Tat strategy*, which showed high performance in computer tournaments conducted by Axelrod [2]. It starts out with cooperation, but after that, it imitates the opponent's action of the previous period.

[9] Because we utilize a correlated action profile, it is not easy to "separately" define an individual player's strategy. The idea is that, in each period, each player chooses an action prescribed by α and the realization of the randomization device. If a player does not choose an action consistent with α and the realization of the randomization device, her/his action is a deviation.

Definition 5.7 The *Tit-for-Tat strategy* of the infinitely repeated Prisoner's Dilemma game with perfect monitoring is a strategy which plays C in the first period and the opponent's action from the previous period for all $t \geqq 2$.

The idea of the Tit-for-Tat strategy is simple. The player cooperates if the opponent cooperated in the previous period, but punishes one period, if the opponent did not cooperate. After that, reconciliation is possible if the opponent cooperates. If both players use the Tit-for-Tat strategy, under some payoff condition (see below) and sufficiently large discount factors, it is a Nash equilibrium of the infinitely repeated Prisoner's Dilemma and achieves repeated mutual cooperation. However, it may not be a subgame perfect equilibrium. Let us use the general parameter Prisoner's Dilemma in Table 5.2 as the stage game again, and assume that player 2 follows the Tit-for-Tat strategy.

Consider any (off-path) subgame starting with the previous period action profile (C, D), i.e., player 1 played C and player 2 played D. Following the Tit-for-Tat strategy yields an alternating action profile sequence $(D, C), (C, D), \ldots,$ so that the continuation payoff is

$$\begin{aligned} & g + \delta\ell + \delta^2 g + \delta^3\ell + \delta^4 g + \delta^5\ell + \cdots \\ & = (g + \delta\ell)(1 + \delta^2 + \delta^4 + \cdots) = \frac{g + \delta\ell}{1 - \delta^2}. \end{aligned}$$

If player 1 deviates to play C in this period and follows the Tit-for-Tat strategy afterwards, the players play (C, C) forever after. Hence the continuation payoff is $c/(1 - \delta)$. By computation,

$$\frac{c}{1-\delta} - \frac{g + \delta\ell}{1 - \delta^2} = \frac{1}{1 - \delta^2}\{(1 + \delta)c - g - \delta\ell\}.$$

The difference is non-negative in the range of $\delta \in (0, 1)$ if and only if

$$\delta \geqq \frac{g - c}{c - \ell}.$$

Thus, if $\frac{g-c}{c-\ell} < 1$ holds, then for any $\delta > \frac{g-c}{c-\ell}$, the Tit-for-Tat strategy is not a best response to itself in any subgame after (C, D).

Next, consider the initial period to find a sufficient condition for the Tit-for-Tat strategy to become a Nash equilibrium. There are in fact only two deviation strategies that should be compared with the Tit-for-Tat strategy. One is to play D in the first period (against C by the opponent) and to follow the Tit-for-Tat strategy from the second period on, which gives the sequence of payoffs $g, \ell, g, \ell, \ldots$. By the previous analysis, $\delta \geqq \frac{g-c}{c-\ell}$ guarantees that the Tit-for-Tat strategy is not worse than this deviation strategy. The other is to play D twice in the first and second period and then to follow the Tit-for-Tat strategy from the third period on. This deviation strategy generates a sequence of action combinations $(D, C), (D, D), (D, D), \ldots,$ as seen from Fig. 5.6, where double arrows indicate the prescribed actions by the

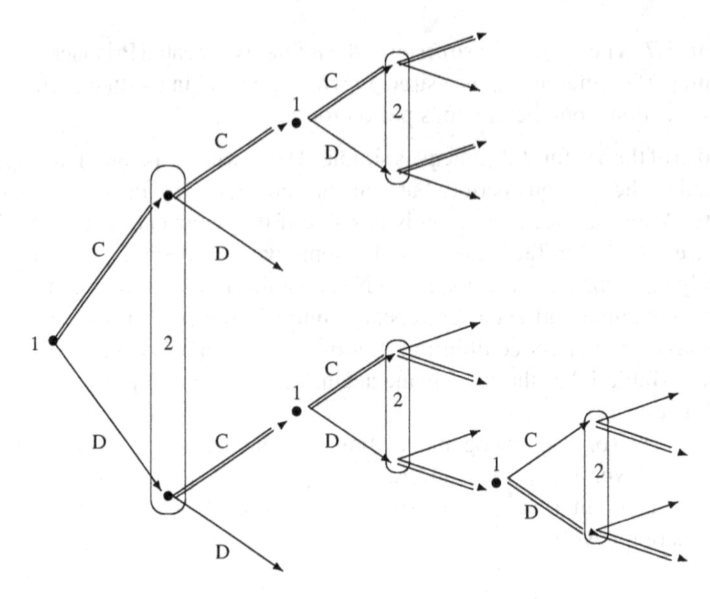

Fig. 5.6 Prescription by the Tit-for-Tat strategy

Tit-for-Tat strategy for both players. Hence the payoff sequence of this deviation strategy is $g, d, d, \ldots$.

Therefore, if

$$\frac{c}{1-\delta} \geqq g + \delta \frac{d}{1-\delta} \iff \delta \geqq \frac{g-c}{g-d}$$

also holds, then this deviation strategy is not better than the Tit-for-Tat strategy. If these two deviation strategies are not better than the Tit-for-Tat strategy, any mixed action deviations in the first two periods or longer periods deviations do not earn a higher total payoff than the Tit-for-Tat strategy, either. After any on-path history $(C, C), \ldots$, the players face the same deviation possibilities as the initial period. Hence, we conclude that if $\frac{g-c}{c-\ell} < 1$ and

$$\delta \geqq \max\{\frac{g-c}{c-\ell}, \frac{g-c}{g-d}\},$$

then the Tit-for-Tat strategy played by both players is at least a Nash equilibrium.

A strength of the Tit-for-Tat strategy is that, if some player chooses an unintended D by mistake, or if monitoring is imperfect so that C was taken as D by the opponent, the players can still reconcile in the future, if C is observed by the opponent. Therefore the loss of payoff can be small under small perturbations of actions/observations, unlike the grim-trigger strategy, which never stops the punishment once it is provoked.

5.6 Equilibrium Collusion

As we saw in Sects. 3.2 and 3.3, the Nash equilibria of duopoly competitions such as the Cournot game and the Bertrand game are not efficient. That is, even though only two firms are in the market, they cannot enforce a mutually beneficial outcome such as to share the monopoly profit by producing half of the monopoly quantity each or by charging the monopoly price.[10]

The story changes completely if two firms engage in an infinitely repeated Cournot or Bertrand game. The Folk Theorem implies that some of the efficient payoff combinations[11] are sustainable by subgame perfect equilibria, for sufficiently high discount factors (i.e., when the firms put a sufficiently high weight on future payoffs). In other words, collusion can become an equilibrium behavior.

To show this, let us take the example of the Cournot game with a symmetric marginal cost $c_1 = c_2 = c$ in Sect. 3.2 as the stage game. The unique one-shot Nash equilibrium is $(q_1^*, q_2^*) = (\frac{1}{3}(A-c), \frac{1}{3}(A-c))$, which yields the payoff combination $(\frac{1}{9}(A-c)^2, \frac{1}{9}(A-c)^2)$. We have seen that if the two firms restrict their production quantities as $(q_1^o, q_2^o) = (\frac{1}{4}(A-c), \frac{1}{4}(A-c))$, then each firm can earn $\frac{1}{8}(A-c)^2$. This is a half of the monopoly profit, and hence there is no symmetric payoff greater than this.

Consider a grim-trigger strategy for firm i, which produces $q_i^o = \frac{1}{4}(A-c)$ in the first period and as long as the action combination was (q_1^o, q_2^o) every period in the past. Otherwise, produce the Cournot-Nash equilibrium quantity $q_i^* = \frac{1}{3}(A-c)$. If both firms follow this grim-trigger strategy, a firm's continuation payoff after any on-path history is

$$\frac{1}{8}(A-c)^2 + \delta \cdot \frac{1}{8}(A-c)^2 + \delta^2 \cdot \frac{1}{8}(A-c)^2 + \cdots = \frac{(A-c)^2}{8(1-\delta)}.$$

If a firm deviates in one step, the optimal deviation is to produce the one-shot best response to q_j^o, $BR_i(\frac{1}{4}(A-c)) = \frac{1}{2}\{A - \frac{1}{4}(A-c) - c\} = \frac{3}{8}(A-c)$. This gives the one-shot payoff of $\{A-c-(\frac{3}{8}+\frac{1}{4})(A-c)\}\frac{3}{8}(A-c) = \frac{9}{64}(A-c)^2$. Following the grim-trigger strategy afterwards means that the two firms play the one-shot Cournot-Nash equilibrium, which gives the payoff $\frac{1}{9}(A-c)^2$ every period. Hence the continuation payoff of the optimal one-step deviation is

$$\frac{9}{64}(A-c)^2 + \frac{\delta}{1-\delta} \cdot \frac{(A-c)^2}{9}.$$

[10] Of course, the efficient outcome for the firms may not be good for the consumers in the market. The definition depends on the relevant stakeholders.

[11] For a complete characterization of the sustainable set, see Sect. 5.7.

The grim-trigger strategy is unimprovable in one step if and only if

$$\frac{(A-c)^2}{8(1-\delta)} \geqq \frac{9}{64}(A-c)^2 + \delta\frac{(A-c)^2}{9(1-\delta)} \iff \delta \geqq \frac{9}{17}.$$

We have established that, for any discount factor such that $\delta \geqq \frac{9}{17}$, the collusion to share the monopoly profit every period is sustainable as a subgame perfect equilibrium in the infinitely repeated Cournot game.

In reality also, many cartels seem to have a carrot and stick structure. Successful cartels are obviously very hard to detect, and thus clear empirical evidence of cartel behavior is limited. However, price wars have been observed in some markets, after periods of peaceful high prices (e.g., Bresnahan [6]). Thus, in order to prevent collusions, we can propose policies which destroy either the repeated game structure or the carrot and stick system. For example, policies to encourage entrants are effective in shaking the existing firms' long-term relationships, to make it difficult to agree on a mutually beneficial outcome among old and new firms, as well as to reduce the profit per firm, which shrinks carrots. Policies to hinder communication among firms may make deviations easier and collusion more difficult. However, in the spectrum auctions in the U.S., although many steps were taken to prevent participants from learning the identities of others and from communicating with them, some bidders seemed to have communicated via the daily bid announcements. Auction design is a very important application area of (repeated) game theory, and many governments are in fact seeking the advice of game theorists. See Klemperer [14] and Milgrom [18].

Recently, a "leniency system" has been introduced in many countries. This system re-structures the game among the firms by adding the option to defect to the authorities. If a cartel participant confesses to the authorities, it avoids prosecution and fines, but if another cartel member defects first, it can be punished. Hence the game becomes a Prisoner's Dilemma, and cartels are effectively prevented.

5.7 Perfect Folk Theorem*

There are many stage games for which the Folk Theorem by Friedman does not cover a wide range of feasible payoff vectors. For example, when the Nash equilibria of the stage game are on the frontier of feasible payoff vectors, then Friedman's theorem does not mean much. Fudenberg and Maskin [10] provided a "perfect" Folk Theorem by constructing a most stringent punishment to enlarge the set of equilibrium payoff vectors. The word "perfect" refers to both the subgame perfect equilibrium and the fact that we cannot enlarge the set of equilibrium payoff vectors more than theirs.

The point of their construction is not only to make a carrot-and-stick system but also to punish a player who does not punish. This layer of punishment is important to constitute a subgame perfect equilibrium, and in reality, efficient outcomes may not be sustainable because we cannot punish a non-punisher after a deviation. For example, if a firm in a price-fixing cartel deviates to a low price, other firms may not want to punish the deviator by starting a price war, because it also hurts them.

To determine the largest candidate set of equilibrium payoff vectors, we consider the minmax value in Chap. 2 again. For simplicity, we continue to assume that, in the stage game, players can use correlated action profiles and define the minmax value accordingly. (The minmax value depends on the set of feasible action profiles. See Problem 5.7).

Definition 5.8 Given a game $G = (\{1, 2, \ldots, n\}, S_1, \ldots, S_n, u_1, \ldots, u_n)$, player i's *minmax value* (under correlated action profiles) is

$$\underline{v}_i = \min_{\sigma_{-i} \in \Delta(S_{-i})} \max_{\sigma_i \in \Delta(S_i)} Eu_i(\sigma_i, \sigma_{-i}).$$

Let an action profile by players other than i to give player i the minmax value be

$$m^i_{-i} \in \arg\min_{\sigma_{-i} \in \Delta(S_{-i})} \max_{\sigma_i \in \Delta(S_i)} Eu_i(\sigma_i, \sigma_{-i}).$$

That is, if the other players play the (possibly mixed) action combination m^i_{-i}, then the best one-shot payoff player i can obtain is the minmax payoff $\underline{v}_i$. This action profile is the most stringent punishment towards player i. To see this, notice that the minmax value is never greater than the payoff from any Nash equilibrium of the stage game. (For the case of pure Nash equilibria, see Problem 5.12. One can generalize the argument in Problem 5.12 for mixed Nash equilibria). Moreover, the minmax value is the payoff that a player can guarantee unilaterally in any period, hence any strategy combination of the repeated game which gives each player less than her/his minmax value on average would not become a Nash equilibrium. Therefore, the minmax value gives the lower bound to possible average equilibrium payoffs.

To get an intuition of sustainable payoff vectors, consider the game in Problem 3.3(c) (reproduced in Table 5.4) as the stage game. The expected payoff of player 2 from the unique Nash equilibrium is 12/5, but player 1 can drag down player 2's payoff to 1 by choosing pure action D. This is player 2's minmax value.

Definition 5.9 A feasible payoff vector $v \in V$ satisfies *individual rationality* if for any $i = 1, 2, \ldots, n$, $v_i > \underline{v}_i$.

Table 5.4 Game for Problem 3.3(c) (reproduced)

P1\P2	L	R
U	0, 3	3, 2
M	5, 0	0, 4
D	1, 1	1, 1

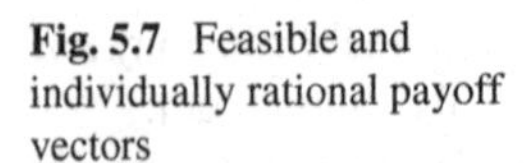

Fig. 5.7 Feasible and individually rational payoff vectors

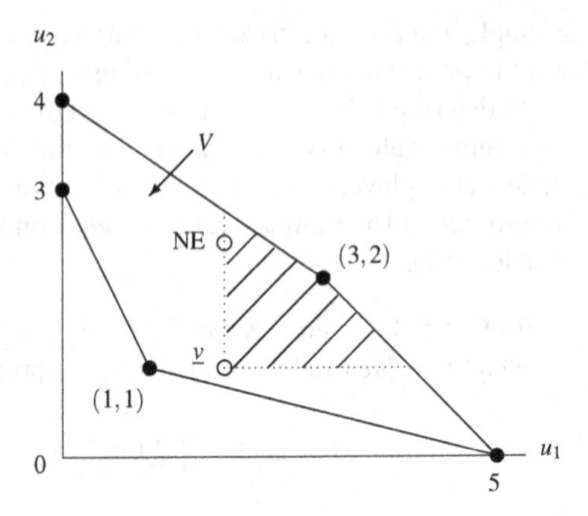

For the game in Table 5.4, the set of payoff vectors satisfying feasibility and individual rationality is the shaded area to the north-east of the minmax payoff vector $\underline{v}$ in the pentagon V surrounded by the five payoff vectors of pure action profiles (Fig. 5.7).

When two players repeat this stage game infinitely many times, to enforce an action profile that is not a one-shot Nash equilibrium in every period in the repeated game, one possible punishment is to play an action that imposes at most the minmax value to the deviator. If such a punishment is enforceable, then any payoff vector which gives more than the minmax value to all players is sustainable. The problem is that the action that "minmaxes" the opponent may also give the punisher a very low payoff, and thus the player would not play it *ad infinitum*, like a grim-trigger equilibrium. A solution to this problem is to punish a deviator for a finite number of periods, but long enough to wipe out any possible (one-step) deviation gain. (A careful reader may notice that this is like mechanism design in economics. The proof for a Folk Theorem constructs an incentive-compatible strategy combination).

When there are three or more players in the repeated game, there is another problem to sustain all feasible and individually rational payoff vectors. If all players have essentially the same payoff function (i.e., they have identical interest), then in subgames where players punish a deviator, all they can do is give a payoff higher than the minmax value to the deviator, in order for punishers not to deviate. See an example of this in Fudenberg and Maskin [10]. A solution to this problem is to exclude such cases where the payoff vectors are proportional to one another, by restricting attention to the stage games such that V has the same dimension as the number of players. (There is a weaker condition, or a larger class of stage games, which gives the same result. We discuss it after the proof of the perfect Folk Theorem.)

Theorem 5.3 (The perfect Folk Theorem by Fudenberg and Maskin) *Take any n-player normal-form game G such that the set of feasible payoff vectors V satisfies* $dim(V) = n$ *when* $n \geqq 3$. *(No condition is imposed when* $n = 2$*). Assume (weak) perfect monitoring. For any feasible payoff vector* $v \in V$ *that satisfies individual* *rationality, there exists $\underline{\delta} \in (0, 1)$ such that, for any $\delta \geqq \underline{\delta}$, there exists a subgame perfect equilibrium of $G^{\infty}(\delta)$ with the average payoff vector v.*

Proof We divide the proof into the case of $n = 2$ and $n \geqq 3$. Fix a feasible payoff vector $v \in V$ that satisfies individual rationality. By the definition of V, there exists a correlated action profile $\alpha \in \Delta(S_1 \times \cdots \times S_n)$ such that $(Eu_1(\alpha), \ldots, Eu_n(\alpha)) = v$. We construct a strategy combination to play α on the equilibrium path every period. When $n = 2$: we make an additional notation. When the two players "minmax" each other by playing (m_1^2, m_2^1), let the payoff of player i be $w_i = Eu_i(m_1^2, m_2^1)$. Choose a positive integer $T < \infty$ such that

$$(T+1)v_i > \max_{\sigma \in \Delta(S)} u_i(\sigma) + Tw_i, \quad \forall\, i = 1, 2.$$

Because $v_i > \underline{v}_i \geqq w_i$, such T exists.

Consider the following *trigger strategy* of $G^{\infty}(\delta)$. (Because the punishment phase is finite, it is simply called a trigger strategy.)

Phase 1 (equilibrium path): in the first period of the repeated game and after any history in which no single player has deviated[12] from α, play α. If a player has deviated from α, go to Phase 2.

Phase 2 (punishment phase): play (m_1^2, m_2^1) for T consecutive periods. If no single player has deviated, go back to Phase 1. If a single player deviated during this phase, restart Phase 2.

Let us show that this strategy combination is a Nash equilibrium in any subgame. The subgames are classified into two classes, those in which the players are in Phase 1 and those in which the players are in Phase 2.

Consider player i in Phase 1. If the opponent follows the trigger strategy, following the trigger strategy gives

$$\frac{v_i}{1-\delta}$$

as the continuation payoff. By the one-step deviation principle, the trigger strategy is optimal if no one-period deviation followed by the trigger strategy would give a continuation payoff greater than the above. If player i changes actions in one period, the maximal one-shot payoff is $\max_{\sigma \in \Delta(S)} u_i(\sigma)$. If (s)he follows the trigger strategy afterwards, player i receives w_i for T periods, and then v_i forever after. Thus the continuation payoff of any one-period deviation is at most

$$\max_{\sigma \in \Delta(S)} u_i(\sigma) + (\delta + \cdots + \delta^T)w_i + \frac{\delta^{T+1}}{1-\delta}v_i.$$

---

[12]Any equilibrium notion based on a Nash equilibrium is defined with respect to unitary deviations. Hence for histories in which more than one player has deviated, no punishment is needed.

By subtracting this from $v_i/(1-\delta)$, we have

$$(1+\delta+\cdots+\delta^T)v_i - \max_{\sigma\in\Delta(S)} u_i(\sigma) - (\delta+\cdots+\delta^T)w_i,$$

and as δ approaches to 1, this value difference converges to

$$(T+1)v_i - \max_{\sigma\in\Delta(S)} u_i(\sigma) - Tw_i.$$

Therefore, by the choice of T, there exists a lower bound to δ above which this value difference is non-negative, i.e., player i follows the trigger strategy.

Next, consider player i in Phase 2. Suppose that the punishment must continue for t more periods. Following the trigger strategy gives the continuation value of

$$(1+\delta+\cdots+\delta^{t-1})w_i + \frac{\delta^t}{1-\delta}v_i.$$

In particular, this value is smallest when $t = T$, i.e., the first period in Phase 2. Thus it suffices to prove that player i would not deviate when $t = T$. Recall that the opponent plays the minmaxing action for player i. Thus any one-period deviation gives at most $\underline{v}_i$ for player i, and the punishment restarts for T more periods. Hence the maximal continuation value of a one-period deviation is

$$\underline{v}_i + (\delta+\cdots+\delta^T)w_i + \frac{\delta^{T+1}}{1-\delta}v_i.$$

The value difference (at $t = T$) is

$$w_i + \delta^T v_i - \underline{v}_i - \delta^T w_i = \delta^T(v_i - w_i) - (\underline{v}_i - w_i),$$

which converges to $(v_i - w_i) - (\underline{v}_i - w_i) = v_i - \underline{v}_i > 0$ when δ goes to 1. Hence there exists another lower bound to δ above which no player deviates in Phase 2. By taking the maximum of the lower bounds of Phase 1 and Phase 2 over the two players, the theorem is proved.

When $n \geqq 3$: we need more preparation. Let $m^i = (m^i_i, m^i_{-i})$ be the action profile in which player i is "minmaxed", so that $Eu_i(m^i_i, m^i_{-i}) = \underline{v}_i$. (This will be the punishment action profile, and it is different from the one for the two-player case). Let $w^i_j = Eu_j(m^i)$ be player j's one-shot payoff when the players punish player i.

By the assumption that $\dim(V) = n$, there exist $v' \in int(V)$ and $\epsilon > 0$ such that, for any i,

$$\underline{v}_i < v'_i < v_i;$$

and

$$v'(i) = (v'_1 + \epsilon, \ldots, v'_{i-1} + \epsilon,\ v'_i,\ v'_{i+1} + \epsilon, \ldots, v'_n + \epsilon) \in V.$$

This payoff combination $v'(i)$, or an underlying (correlated) action profile $\alpha(i) \in \Delta(S_1 \times \cdots \times S_n)$ such that $(Eu_1(\alpha(i)), \ldots, Eu_n(\alpha(i))) = v'(i)$, is used to give "rewards" to players other than i after successfully punishing player i. This is to motivate the players to punish deviators.

Pick $T < \infty$ such that

$$\max_{\sigma \in \Delta(S)} u_i(\sigma) + T\underline{v}_i < \min_{\sigma \in \Delta(S)} u_i(\sigma) + Tv'_i, \quad \forall i = 1, 2, \ldots, n.$$

Later, it will become clear that this is the sufficient punishment length to wipe out any one-shot deviation gain for any player.

With the above preparations, we can introduce the following three-phase trigger strategy combination.

Phase 1 (equilibrium path): in the first period of the repeated game and after any history in which no player unilaterally deviated from α, play α. If a single player j deviated from α, go to Phase 2(j).

Phase 2(j): play m^j for T periods. During this phase, if a single player i deviated from m^j, start Phase 2(i). If no player deviated for T consecutive periods, go to Phase 3(j).

Phase 3(j): play $\alpha(j)$ forever. If a single player i deviated, go to Phase 2(i).

Fix an arbitrary player i and we show that in any subgame, player i's total expected payoff from one-step deviations would not be greater than that from the trigger strategy combination. We write the analysis from the easiest case, not the order of the phases.

Phase 1: if no player has deviated from α (the first period is also this case), and if player i follows the trigger strategy combination, the continuation payoff is

$$\frac{v_i}{1-\delta}.$$

If player i chooses a different action in one period and follows the trigger strategy combination afterwards, the one-shot payoff is at most $\max_{\sigma \in \Delta(S)} u_i(\sigma)$. After a deviation, the play shifts to Phase 2(i) and then to Phase 3(i). Hence the continuation payoff (including this period) of any one-shot deviation is at most

$$\max_{\sigma \in \Delta(S)} u_i(\sigma) + \delta \frac{1-\delta^T}{1-\delta}\underline{v}_i + \frac{\delta^{T+1}}{1-\delta} v'_i. \tag{5.1}$$

To compare these, multiply both by $(1-\delta)$ and subtract the latter from the former. We have

$$v_i - \{(1-\delta) \max_{\sigma \in \Delta(S)} u_i(\sigma) + \delta(1-\delta^T)\underline{v}_i + \delta^{T+1} v'_i\}.$$

As δ approaches to 1, this converges to $v_i - v_i'$. That is, for very patient players, only the Phase 3 payoff v_i' matters. But this is strictly less than v_i. Therefore no one-step deviation is better than the trigger strategy for sufficiently high δ's in this phase. (That is, there is a lower bound to δ such that for any δ not less than the lower bound, following the trigger strategy is optimal for any player in this phase.)
Phase 3(j) such that $j \neq i$: in this phase, player i is to receive "rewards" after successfully punishing player j. If player i deviates from $\alpha(j)$ for one period and follows the trigger strategy combination afterwards, the continuation payoff including this period is again bounded by the Eq. (5.1). By the same logic for Phase 1, for sufficiently high δ's we have

$$\frac{v_i' + \epsilon}{1-\delta} > \max_{\sigma \in \Delta(S)} u_i(\sigma) + \delta \frac{1-\delta^T}{1-\delta} \underline{v}_i + \frac{\delta^{T+1}}{1-\delta} v_i'.$$

Therefore no one-step deviation is better than the trigger strategy in this phase.
Phase 3(i): this is the phase after i has been punished and the other players get rewards. If player i follows the trigger strategy combination, the continuation value is

$$\frac{v_i'}{1-\delta}.$$

By contrast, if player i deviates for one period, the maximal continuation value including this period is given by (5.1) again. (5.1) can be expanded as

$$\max_{\sigma \in \Delta(S)} u_i(\sigma) + (\delta + \cdots + \delta^T)\underline{v}_i + \frac{\delta^{T+1}}{1-\delta} v_i'.$$

Subtracting this from $v_i'/(1-\delta)$ yields

$$\{v_i' + (\delta + \cdots + \delta^T)v_i'\} - \{\max_{\sigma \in \Delta(S)} u_i(\sigma) + (\delta + \cdots + \delta^T)\underline{v}_i\}.$$

When δ goes to 1,

$$\begin{aligned} &\lim_{\delta \to 1}[\{v_i' + (\delta + \cdots + \delta^T)v_i'\} - \{\max_{\sigma \in \Delta(S)} u_i(\sigma) + (\delta + \cdots + \delta^T)\underline{v}_i\}] \\ &= (T+1)v_i' - \max_{\sigma \in \Delta(S)} u_i(\sigma) - T\underline{v}_i > v_i' - \min_{\sigma \in \Delta(S)} u_i(\sigma), \end{aligned}$$

where the last inequality follows from the definition of T. Hence for sufficiently high δ's, the payoff difference is non-negative.
Phase 2(j) such that $j \neq i$: suppose that the players must punish player j for t more periods. Following the trigger strategy combination gives player i

$$(1 + \delta + \cdots + \delta^t)w_i^j + \frac{\delta^{t+1}}{1-\delta}(v_i' + \epsilon)$$

as the continuation value. If player i deviates, the maximal continuation value including this period is

$$\max_{\sigma \in \Delta(S)} u_i(\sigma) + (\delta + \cdots + \delta^T)\underline{v}_i + \frac{\delta^{T+1}}{1-\delta} v_i'.$$

For sufficiently high δ's, the reward $v_i' + \epsilon$ in Phase 3(j) is strictly greater than the payoff v_i' in Phase 3(i). Thus, the trigger strategy is better than making one-step deviations.

Phase 2(i): suppose that player i needs to be punished t more periods. Following the trigger strategy combination gives

$$(1 + \delta + \cdots + \delta^t)\underline{v}_i + \frac{\delta^{t+1}}{1-\delta} v_i'.$$

In this phase, the other players impose the minmax value on player i so that a one-period deviation gives at most $\underline{v}_i$, and Phase 2(i) restarts. Therefore, no deviation is better than the trigger strategy. □

Fudenberg and Maskin assumed the full dimensionality condition, $\dim(V) = n$, for repeated games with three or more players. This is not a necessary condition for a Folk Theorem. Abreu et al. [1] proposed a weaker sufficient condition for a Folk Theorem of repeated games with three or more players.

Definition 5.10 Game G satisfies the *Non-Equivalent Utilities condition* if for any pair of players i and j, there are no real numbers c and d such that

$$u_i(s) = c + d u_j(s), \quad \forall s \in S_1 \times \cdots \times S_n.$$

The full dimensionality condition and the NEU condition restrict the stage games, while Wen [26] adjusted the set of sustainable payoff vectors. He introduced the *effective minmax payoff* such that for any n-player repeated game, the Folk Theorem holds in the sense that any feasible payoff vector that gives more than the effective minmax payoff to all players is sustainable for sufficiently large discount factors.

5.8 Repeated Games with Non-simultaneous Moves*

In the standard repeated game framework, the stage game is assumed to be a simultaneous-move game. A natural question is whether the Folk Theorem holds if the stage game is not a simultaneous-move game. There are many possible formulations of repeated games with non-simultaneous moves. For example, an extensive-form game can be repeated. Another example is a dynamic game in which one player moves in odd-numbered periods and the other plays in even-numbered periods. There are other more complex timing structures that one may want to analyze.

Table 5.5 A pure coordination game

P1\P2	A	B
A	2, 2	0, 0
B	0, 0	1, 1

In this section we introduce an alternating decision game by Lagunoff and Matsui [17]. This game is interesting because the set of sustainable payoff vectors dramatically changes by only staggering the decision timing of the two players. Assume that two players engage in the game in Table 5.5 over an infinite horizon. The game is a *coordination game* such that it is the best for the two players to play the same action and, moreover, the pure-action Nash equilibria are ranked by the strict inequality of their payoff vectors. In particular, the game in Table 5.5 is called a *pure coordination game*, because if the actions do not match, both players receive 0.[13]

In the game in Table 5.5, both players prefer the (A, a) equilibrium to the (B, b) equilibrium. However, since the action combination (B, b) is a Nash equilibrium of the stage game, it can be played every period of a subgame perfect equilibrium of the ordinary repeated game in Table 5.5. For example, player 1 (resp. player 2) can use a strategy to play action B (resp. b) regardless of the history. For any discount factor $\delta \in (0, 1)$, this strategy combination is a subgame perfect equilibrium. In other words, we cannot exclude the inefficient average payoff (1,1) to be sustained if the simultaneous-move game in Table 5.5 is repeated forever.

Let us stagger the players' decision timing. Before the game starts, player 1 must choose the "initial action" from $S_1 = \{A, B\}$, and this will be played in the first period. In the first period, only player 2 chooses an action from $S_2 = \{a, b\}$, which will be played in both the first and second period. In the second period, only player 1 chooses an action from S_1, which will be played in the second and third period. The game continues this way *ad infinitum*.

Therefore, each player's action choice is valid for two periods, or they commit to play the same action for two periods. For example, suppose that player 1's initial action choice is A, player 2's first-period action choice is b, then player 1 chooses action B in the second period, and finally, player 2's action choice is again b in the third period. In this case, the first period action combination is (A, b), the second period action combination is (B, b), and the third period action combination is again (B, b) (see Fig. 5.8).

In this way, player 2 chooses an action in odd-numbered periods and player 1 chooses an action in even-numbered periods. Each player maximizes the total discounted payoff over the infinite horizon with a common discount factor $\delta \in (0, 1)$.

Let us consider the "always B (b)" strategy combination again, in which player 1 (resp. 2) chooses B (resp. b) after any history including the initial choice. It turns out that this strategy combination is not a subgame perfect equilibrium. That is, there

[13]In addition, in the game of Table 5.5, the payoffs of the two players are the same for all action combinations. Such a game is called a *common interest game*.

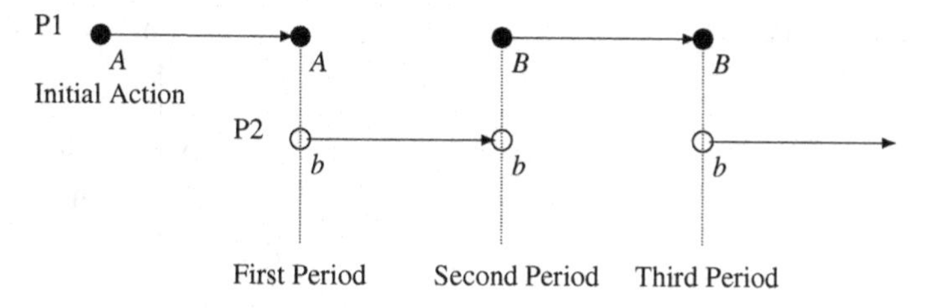

Fig. 5.8 Asynchronous choice game

is a subgame in which one of the players gets a higher total discounted payoff by deviating from the above strategy combination.

Consider a subgame in which player 1 has chosen A in the previous period and it is player 2's turn to choose an action. If both players follow the always-B (b) strategy combination, the action combination sequence will be $(A, b), (B, b), (B, b), \ldots$. The continuation payoff of player 2 is the sum of 0 in this period and $\delta/(1-\delta)$, which is $\delta/(1-\delta)$.

By contrast, if player 2 deviates to action a in this period and from the next period on both players follow the always-B (b) strategy combination, then the action combination sequence will be $(A, a), (B, a), (B, b), (B, b), \ldots$. Player 2 receives 2 in this period and 0 in the next period, but after that, (s)he receives 1 forever. The total discounted sum of this sequence is $2 + \delta \cdot 0 + \delta^2/(1-\delta)$. To compare with the continuation payoff from the always-b strategy,

$$2 + \frac{\delta^2}{1-\delta} - \frac{\delta}{1-\delta} = 2 - \delta > 0.$$

Therefore, the deviation gives a higher total discounted payoff. (Similarly, in subgames where player 2 has chosen a, it is better for player 1 to deviate from the always-B strategy.)

We have shown that a subgame perfect equilibrium in the ordinary repeated game framework is not a subgame perfect equilibrium in this model. The reason is that, if the opponent has committed to an action and the game is a coordination game, then it is better to match one's action with the opponent's. By generalizing this logic, Lagunoff and Matsui [17] showed an *anti-Folk Theorem* in which for any pure coordination game, only the most efficient payoff vector is sustainable by a subgame perfect equilibrium in the alternating choice repeated game.

Yoon [27] analyzed a more general class of asynchronously repeated games and pointed out that the pure coordination game of Lagunoff and Matsui [17] does not satisfy the NEU condition (although it is a two-person game). He showed that, if the stage game satisfies the NEU condition, then a Folk Theorem holds for a class of asynchronously repeated games including the alternating choice game of Lagunoff and Matsui [17].

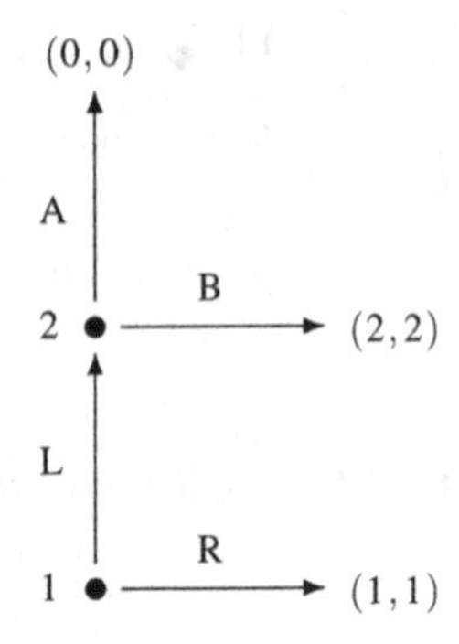

Fig. 5.9 A stage game

Table 5.6 The induced normal form of Fig. 5.9

P1\P2	A	B
R	1, 1	1, 1
L	0, 0	2, 2

On the other hand, Takahashi and Wen [24] showed that Yoon's Theorem depends on the timing structure of the dynamic game, and made a counter example to a Folk Theorem such that the average sustainable payoff vector may be lower than the minmax value. Their three-player game example has the structure where in a period when a player cannot change her/his actions, the other players can push down this player's payoff below the *maxmin value*[14] of the stage game. In general, the maxmin value is less than the minmax value, and thus an equilibrium average payoff can be less than the minmax value of the stage game.

There are Folk Theorems for repeated games of extensive-form stage games. (See Rubinstein and Wolinsky [20] and Sorin [23]). However, the Folk Theorem set of the induced normal-form game and the directly derived set of average equilibrium payoff vectors for the repeated extensive-form game do not coincide in general. For example, consider the infinitely repeated game of the extensive-form game in Fig. 5.9, presented in Rubinstein and Wolinsky [20].

In each period, player 1 first chooses between L and R, and if L is chosen, then player 2 gets to choose between A and B. What is the minmax value, for the infinitely repeated game of this extensive-form stage game?

One way is to convert the extensive-form game in Fig. 5.9 into the induced normal form (see Table 5.6) and compute the minmax value. Then, within the range of mixed actions, both players' minmax values are 1.

Hence, by the Folk Theorem of the normal-form games, any payoff vector on the line segment connecting $(1, 1)$ and $(2, 2)$ must be sustainable by some subgame perfect equilibrium, when δ is sufficiently high. However, the infinitely repeated game of the extensive form game in Fig. 5.9 has a unique subgame perfect equilibrium for any $\delta < 1$ such that (L, B) is repeated every period. To see this, let m be the infimum of a player's average discounted payoff among all subgame perfect equilibria. If

[14] The definition of the maxmin value over mixed actions is analogous to the one given in Sect. 2.6.

player 1 chooses action L in the first period, by choosing action B and following the strategy combination that yields m on average afterwards, player 2 can guarantee the total discounted payoff of $2+\frac{\delta}{1-\delta}m$ at least. Because of the common payoff function, player 1's total discounted payoff is also at least this value as well. Hence any average payoff y of a subgame perfect equilibrium must satisfy $y \geqq (1-\delta)\{2+\frac{\delta}{1-\delta}m\}$, which implies that the infimum also satisfies $m \geqq (1-\delta)\{2+\frac{\delta}{1-\delta}m\}$. By the feasibility condition, we must have $m = 2$.

Note, however, that this stage game is a *common interest game* and does not satisfy the full dimensionality condition, although it is a two-person game. The above argument hinges on player 1 having the same payoff as player 2. If the full dimensionality condition is imposed, Rubinstein and Wolinsky [20] showed a Folk Theorem for extensive-form stage games, by constructing trigger strategy combinations.

5.9 Repeated Games with Overlapping Generations*

So far we have only dealt with games in which players participate until the (possibly uncertain) end of the game. There are games that continue *ad infinitum*, but each player stays in the game only for a finite number of periods. In this section we consider overlapping generation games in this class.[15]

Assume that each player lives only two periods, and call players in their first period of their lives as "young" and those in their second period as "old". In each period of the game, one young and one old player play the Prisoner's Dilemma game in Table 5.7. After an old player dies, the current young player becomes old and remains in the game, and a new young player enters the game. This way, two overlapping generations of players play the game forever.

Each player maximizes the sum of the stage game payoffs over the two periods. Assume that all players observe all past players' actions, and let us show that the following grim-trigger strategy played by all players is a subgame perfect equilibrium.

When young, play C if the history consists only of (C, D), where C is played by the young and D is played by the old in each period. Otherwise, play D. When old, play D after any history.

Clearly, when a player is old, it is her/his last period, and hence playing D after any history is optimal. Suppose that a young player deviates from the above strategy and plays D after a history consisting only of (C, D). In this period, the young player gets 0 instead of -1, but in the next period the new young player punishes

Table 5.7 Prisoner's dilemma for the OLG model

Young\old	C	D
C	2, 2	−1, 3
D	3, −1	0, 0

[15] This section follows Smith [22]. Kandori [13] also made a similar analysis, independently from Smith [22].

this player by playing D, which gives this player 0 again. If this player follows the grim-trigger strategy, (s)he gets -1 in this period but 3 in the next period. Hence the total payoff is $2 > 0$. This implies that no young player deviates on the play path. In other subgames, (D, D) is the prescribed action combination and it is the Nash equilibrium of the stage game. Therefore, in this case, a young player does not deviate either.

In summary, although each player plays only two periods, C can be played in a subgame perfect equilibrium. In another (obvious) equilibrium where (D, D) is played every period, the average payoff over the two periods for each player is 0, but in the above equilibrium, the total payoff over the two periods is 2 and the average is 1.

However, as Bhaskar [5] points out, the above grim-trigger equilibrium is based on a very strong informational assumption. If players can only observe the previous period action combination, the above strategy is not feasible. A possible alternative strategy is the "reciprocal strategy" that young players play C if and only if the previous young player chose C. (For any informational structure, old players should play the unique dominant action D in any equilibrium). However, this reciprocal strategy is not a best response to itself played by the next young player, in some subgames. The problem is that even if the current old player did not play C when young, it is better for the young player not to play D, because the young player in the next period cannot know whether the D action this period is to punish the old player or the deviation by the young player.

More formally, assume that the next period young player uses the reciprocal strategy, and consider a subgame after the current old player has deviated in the previous period. For the current young player, playing D to punish the old (following the reciprocal strategy) generates (D, D) this period and (D, D) in the next period. Thus the total payoff is 0. If the young player does not punish the old and plays C, then this period payoff is -1 but the next period payoff is 3. This means that the reciprocal strategy is not a best response to the reciprocal strategy by the next period's young player.

Bhaskar [5] constructed randomized punishments to obtain a cooperative equilibrium[16] under the one-period memory, but this equilibrium unravels if the payoff functions are perturbed a little.[17] Therefore we conclude that the informational assumption is crucial for the existence of cooperative equilibria in the overlapping generation games.

[16]Because this game has imperfect monitoring, the equilibrium notion is not a subgame perfect equilibrium but a sequential equilibrium (see Chap. 8).

[17]For the definition of perturbation, see Sect. 6.5. Although the OLG game is not a normal-form game, this result is in contrast to Harsanyi's Purification Theorem (Theorem 6.2).

Problems

5.1 Two brothers are facing three candies to be shared. They cannot cut a candy into smaller pieces. Each player wants to maximize the number of the candies he gets.
(a) Consider a simultaneous-move game such that each player declares a number from $\{0, 1, 2, 3\}$. If the sum of the two numbers is not more than 3, then each player receives candies according to the number he declared. If the sum exceeds 3, then both players get 0 candy. Construct the matrix representation of this game and find all pure-strategy Nash equilibria.
(b) Consider an extensive-form game, in which the older brother declares the number he wants from the set $\{0, 1, 2, 3\}$ and, knowing this, the younger brother declares the number he wants from the set $\{0, 1, 2, 3\}$. The payoff structure is the same as in (a) so that if the sum of the two numbers does not exceed 3, each player receives candies according to the number he declared. Otherwise, both players get 0. Draw the game tree of this game and find all pure-strategy subgame perfect equilibria.

5.2 There are two cafes in a mall. Call them (or their managers) A and B. It turns out that the people of this town do not go to cafes so much, and if two cafes exist, neither of them earns enough profit. Both cafes start to consider closing the business. If the rival closes first, then the remaining cafe becomes a monopolist and can earn enough profit. If the rival does not shut down, then it is better to close one's own cafe. Assume that the managers cannot know each other's decision, i.e., it is a simultaneous-move game. The set of strategies for the players is the same, and $S_A = S_B = \{S, E\}$, where S means stay and E means exit. If a player chooses E, then regardless of the opponent's strategy, the payoff is 0. If a player chooses S and the opponent also chooses S, the payoff is -10 each. If a player chooses S and the opponent chooses E, then the staying player gets a payoff of 5.
(a) Find all Nash equilibria of this game, including mixed-strategy ones.

The manager of cafe A considers renovating the shop before making the stay-exit decision. The renovation costs -20 but increases customers. Assume that if cafe A renovates and chooses S, then its net payoff increases by 5 (after subtracting the cost). That is, (S, S) (where the first coordinate is A's action) gives $-10 + 5 = -5$ and (S, E) gives $5 + 5 = 10$ to player A. However, if A exits after renovation, the payoff is -20 regardless of the opponent's stay-exit decision.

Consider this new situation as an extensive-form game. Player A chooses whether to renovate or not first. A's renovation decision is observable to player B. If A does not renovate, the players play the simultaneous-move game of (a) and the game ends. If A renovates, they play the simultaneous-move game with the new payoff function described above (where player B's payoff function is the same as the one in (a)) and the game ends.
(b) Draw the game tree of the extensive-form game.
(c) Find all pure-strategy subgame perfect equilibria of the extensive-form game. Discuss the economic implications.

5.3 Find all pure-strategy subgame perfect equilibria of the following 3-player extensive-form games. The payoffs are written in the order of player 1, 2, and 3 from the top.

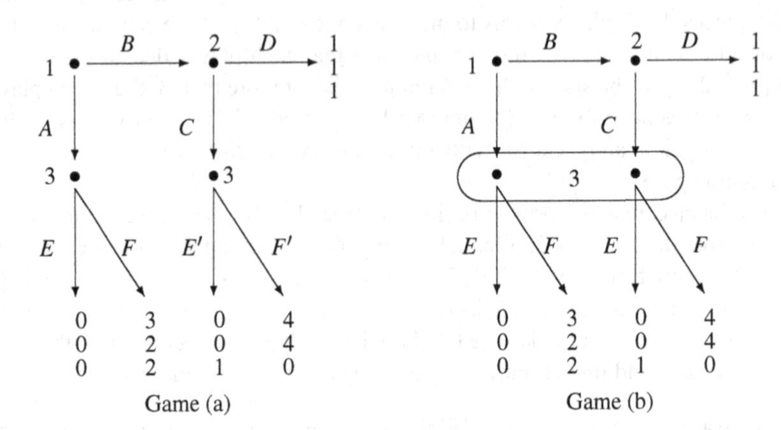

5.4 Find all subgame perfect equilibria of the following 3-player extensive-form game.

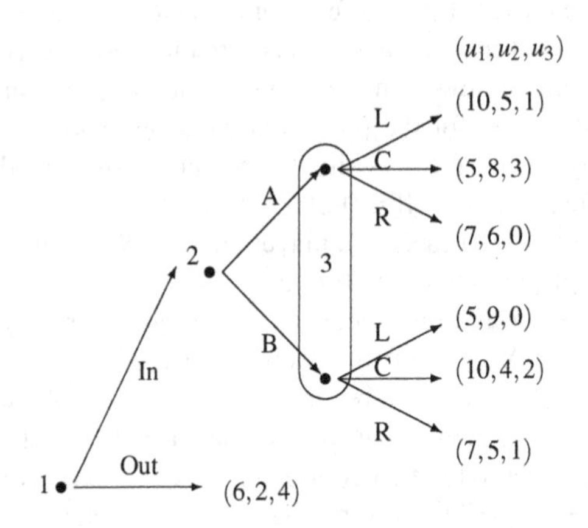

5.5 Consider a twice-repeated game of the stage game in Table 5.8. The payoff is the sum of the stage game payoffs over the two periods. Construct a subgame perfect equilibrium such that (B, B) is played in the first period. Explain why the strategy combination is a subgame perfect equilibrium. (Write the strategy combination precisely, for any history).

5.6 Consider the simultaneous-move game G in Table 5.9.
(a) Find all Nash equilibria (including mixed-strategy ones) of G.
(b) Assume that players can use correlated strategies. Draw the set of feasible payoff combinations of G. (Let the horizontal axis be the payoff of player 1). Illustrate the Nash equilibrium payoff combinations derived in (a) in the same figure.

Table 5.8 Stage game for Problem 5.5

P1\P2	A	B	C
A	2, 2	6, 0	1, 1
B	0, 0	5, 5	0, 6
C	1, 0	2, 1	4, 4

Table 5.9 Stage game for Problem 5.6

P1\P2	A	B
A	3, 2	0, 0
B	0, 0	1, 4

Table 5.10 Game for Problem 5.7

P1\P2	L	R
U	0, 3	1, 0
M	1, 0	0, 2
D	2, 5	4, 4

Consider the infinitely repeated game $G^{\infty}(\delta)$ of G with the discount factor $\delta \in (0, 1)$. For simplicity, assume not only perfect monitoring but also that players can observe mixed actions by all players.
(c) Let a strategy combination **s** of $G^{\infty}(\delta)$ be as follows.
In odd periods (including $t = 1$): after any history, P1 chooses action A and P2 chooses action a.
In even periods: after any history, P1 plays action B and P2 plays b.

Compute each player's total discounted payoff, the average payoff (the total discounted payoff times $(1 - \delta)$), and the limit average payoff as δ converges to 1 from the strategy combination **s**. (Hint: Consider two periods as one unit of time). Illustrate the limit average payoff in the figure for (b).
(d) Consider a subgame after an arbitrary history. Prove that if the opponent follows the strategy combination **s**, then no player earns a strictly higher payoff than that from **s** by any one-step deviation.

5.7 Consider the 2-player simultaneous-move game G in Table 5.10.
(a) Find all Nash equilibria (including mixed-strategy ones) of G.
(b) We now compute the minmax value of G for each player.
(i) Compute

$$v_1 := \min_{y \in \{L,R\}} \max_{x \in \{U,M,D\}} u_1(x, y)$$

$$v_2 := \min_{x \in \{U,M,D\}} \max_{y \in \{L,R\}} u_2(x, y)$$

in the range of pure strategies.

Table 5.11 Stage game for Problem 5.8

P1\P2	A	B
A	3, 3	0, 4
B	4, 0	1, 1

(ii) Next, allowing mixed strategies, compute

$$\underline{v}_1 := \min_{y \in \Delta\{L,R\}} \max_{x \in \Delta\{U,M,D\}} Eu_1(x, y)$$

$$\underline{v}_2 := \min_{x \in \Delta\{U,M,D\}} \max_{y \in \Delta\{L,R\}} Eu_2(x, y)$$

and compare with the result in (i). Write down player 1's (mixed) strategy which imposes $\underline{v}_2$ on player 2.
(c) Draw the set of feasible payoff combinations of G. (Let the horizontal axis be player 1's payoff.)

5.8 Consider the infinitely-repeated game with the stage game as in Table 5.11 and the discount factor δ. Each player (player 1 and 2) maximizes the total discounted payoff.

Consider the following grim-trigger strategy s^G (common to both players): for any history including the first period, if no player has deviated from (A, A), play A. If someone has deviated in the past, play B.

Because the game is symmetric, without loss of generality focus on player 1. Assume that player 2 follows the grim-trigger strategy, and consider a history where no one has deviated from (A, A). Let us show that if the one-step deviation to the pure action B (and following the grim-trigger strategy afterwards) does not give a higher total discounted payoff than that of s^G, then any one-step deviation to a mixed action $pA + (1 - p)B$ will not give a higher total discounted payoff than that of s^G.

First, the statement that the one-step deviation to the pure action B (and following the grim-trigger strategy afterwards) does not give a higher total discounted payoff than the one from s^G is formulated as the inequality

$$\frac{3}{1-\delta} \geqq (x) + \delta \frac{1}{1-\delta}. \tag{5.2}$$

Next, we formulate the second statement that any one-step deviation to a mixed action $pA + (1 - p)B$ will not give a higher total discounted payoff than that of s^G. Note that with probability p, this period action combination is (A, A). In this case, the players continue to play (A, A) afterwards. With probability $(1 - p)$, the action combination is (B, A), and then they play (B, B) forever after. Therefore the second statement is formulated as

$$\frac{3}{1-\delta} \geqq p\{(y) + \delta \frac{3}{1-\delta}\} + (1-p)\{(z) + \delta \frac{1}{1-\delta}\}. \tag{5.3}$$

(a) Fill in x, y, and z with appropriate numbers.
(b) Prove that for any $0 \leqq p < 1$, if Eq. (5.2) holds, then (5.3) holds.
(c) Find the smallest δ such that the grim-trigger strategy played by both players is a subgame perfect equilibrium.

5.9 Consider an extensive-form game between player P and player A. First, P offers a contract (w, b). The real number w is the basic wage which P pays to A for sure, if the contract is accepted by A. The real number b is the bonus payment, which P pays if and only if the sales amount is high. P can choose any non-negative real number pair (w, b) as an offered contract.

Next, after seeing the offered contract (w, b), A chooses whether to accept it (action Yes) or not (action No). If A does not accept the contract, the game ends and both players receive 0 payoff. If A accepts the contract, A starts working. A can choose whether to make an effort (action E) or to shirk (action S). If A shirks, the sales amount is low for sure, and A cannot get the bonus. If A makes an effort, with probability $1/2$, the sales amount is high and A gets the bonus. With probability $1/2$, the sales amount is low and A cannot get the bonus. Making an effort gives a disutility of -1 to A. (Shirking has no disutility.)

When the sales amount is high, P gets a revenue of 10. When the sales amount is low, P gets a revenue of 2. P's payoff is (when a contract (w, b) is accepted) the expected value of the revenue minus the payment to A. A's payoff (when a contract (w, b) is accepted) is only the basic wage if he shirks and the expected payment from P minus 1 if he makes an effort.
(a) Draw the game tree of this game.

Let us find the equilibrium path of a subgame perfect equilibrium such that A accepts a contract and makes an effort, backwards.
(b) Consider A's decision node after he has accepted a contract. Find A's optimal action as a function of (w, b).
(c) Given (b), in the set of (w, b) under which A's optimal action is E, what is the additional condition on (w, b) so that A accepts the contract?
(d) In the set of (w, b) under which A's optimal action is S, what is the additional condition on (w, b) so that A accepts the contract?
(e) From the above analysis, find the optimal contract (w, b) that maximizes P's payoff and so that A makes an effort. (You can assume that if A is indifferent between E and S, A chooses E.)

5.10 We consider a managerial problem. First, a worker chooses between making an effort or not. The revenue of this business hinges on the worker's effort: it is 10 if the worker makes an effort and 1 if not. The worker incurs a disutility of -3 if she makes an effort. After the revenue is determined, the manager demands $x \in [0, 1]$ of the revenue for himself. The worker can say either Yes or No to this demand.

The payoff combinations are determined as follows.

When the worker makes an effort and the manager demands x: if the worker says Yes, the payoff combination is $(10(1-x)-3, 10x)$. (The first coordinate is the payoff of the first-mover of the game, i.e., the worker). If the worker says No, they go to court, and the usual settlement gives 90 % to the manager. Hence their payoff combination is $(1 - 3, 9) = (-2, 9)$.

When the worker does not make an effort and the manager demands y: if the worker says Yes, the payoff combination is $(1 - y, y)$. If the worker says No, the settlement is $(0.1, 0.9)$.

Assume that the game has perfect information. Find a pure-strategy subgame perfect equilibrium. This is an example of the *Hold-up problem.*

5.11 Two animals are facing their prey. (Or two firms are operating in a declining industry). If one animal gets the prey, it receives a payoff of v, which is greater than 1. The game potentially continues over the discrete infinite horizon $t = 1, 2, \ldots$ and v is constant over time. As long as the two animals remain in the game, they both choose whether to Fight or Stop simultaneously in each period. In a period, if both choose to Fight, both pay the cost of -1 but cannot determine who gets the prey, so the game continues to the next period. If at least one animal chooses to Stop, the game ends. The one who chooses Stop gets 0, and the one who chooses Fight (if any) gets v without the cost.

The payoff of a player from this potentially infinite-length game is the total discounted sum of the payoffs from the playing periods. For example, if player 1 chooses to Stop in period T (but until then chooses to Fight) and player 2 chooses to Stop in period $T' > T$ (and until then chooses to Fight), then the game ends at T and player 1's total discounted payoff is

$$(-1)(1 + \delta + \delta^2 + \cdots + \delta^{T-2}) = \frac{-(1 - \delta^{T-1})}{1 - \delta}$$

and player 2's total discounted payoff is

$$(-1)(1 + \delta + \delta^2 + \cdots + \delta^{T-2}) + \delta^{T-1}v.$$

Find the unique subgame perfect equilibrium of this game and prove that it is so. (Hint: In each period, each player randomizes). This is a discrete time version of a game called the *War of Attrition*.

5.12 For an arbitrary 2-person normal-form game $G = (\{1, 2\}, S_1, S_2, u_1, u_2)$, prove the following statement.
For each player $i = 1, 2$, the minmax value within the range of pure strategies is not more than the payoff from any pure-strategy Nash equilibrium.

(In formula, this is equivalent to: for any $i = 1, 2$, and any pure-strategy Nash equilibrium $(s_1^*, s_2^*) \in S_1 \times S_2$ of G,

$$u_i(s_i^*, s_j^*) \geqq \min_{s_j \in S_j} \max_{s_i \in S_i} u_i(s_i, s_j).)$$

References

1. Abreu D, Dutta P, Smith L (1994) The folk theorem for repeated games: a NEU condition. Econometrica 62(4):939–948
2. Axelrod R (1984) The evolution of cooperation. Basic Books, New York
3. Bellman R (2003) Dynamic programming. Reprint. Dover Publications, Mineola, NY
4. Benoit J, Krishna V (1985) Finitely repeated games. Econometrica 53(4):905–922
5. Bhasker V (1998) Informational constraints and the overlapping generations model: folk and anti-folk theorems. Rev Econ Stud 65(1):135–149
6. Bresnahan T (1987) Competition and collusion in the American automobile industry: the 1955 Price War. J Ind Econ 35(4):457–482
7. van Damme E, Hurkens S (1999) Endogenous stackelberg leadership. Games Econ Behav 28(1):105–129
8. van Damme E, Hurkens S (2004) Endogenous price leadership. Games Econ Behav 47(2): 404–420
9. Friedman J (1971) A non-cooperative equilibrium for supergames. Rev Econ Stud 38(1):1–12
10. Fudenberg D, Maskin E (1986) The folk Theorem in repeated games with discounting or with Incomplete Information. Econometrica 54(3):533–554
11. Fujiwara-Greve T (2010) The prisoner's dilemma of neuroeconomics. Keio University, Mimeograph
12. Hamilton J, Slutsky S (1990) Endogenous timing in duopoly games: stackelberg or cournot equilibria. Games Econ Behav 2(1):29–46
13. Kandori M (1992) Repeated games played by overlapping generations of players. Rev Econ Stud 59(1):81–92
14. Klemperer P (2004) Auctions: theory and practice. Princeton University Press, Princeton, NJ
15. Kosfeld M, Heinrichs M, Zak P, Fischbacher U, Fehr E (2005) Oxytocin increases trust in humans. Nature 435:673–676
16. Kreps D (1990) A course in microeconomics theory. Prentice Hall, Upper Saddle River, NJ
17. Lagunoff R, Matsui A (1997) Asynchronous choice in repeated coordination games. Econometrica 65(6):1467–1477
18. Milgrom P (2004) Putting auction theory to work. Cambridge University Press, Cambridge
19. Rubinstein A (1979) Equilibrium in supergames with the overtaking criterion. J Econ Theory 21(1):1–9
20. Rubinstein A, Wolinsky A (1995) Remarks on infinitely repeated extensive-form games. Games Econ Behav 9(1):110–115
21. Selten R (1975) Reexamination of the perfectness concept for equilibrium points in extensive games. Int J Game Theory 4(1):25–55
22. Smith L (1992) Folk Theorems in overlapping generations games. Games Econ Behav 4(3):426–449
23. Sorin S (1995) A note on repeated rxtensive games. Games Econ Behav 9(1):116–123
24. Takahashi S, Wen Q (2003) On asynchronously repeated games. Econ Lett 79(2):239–245
25. Watson J (2007) Strategy: an introduction to game theory, 2nd edn. Norton, New York
26. Wen Q (1994) The "Folk Theorem" for repeated games with complete information. Econometrica 62(4):949–954
27. Yoon K (2001) A folk theorem for asynchronously repeated games. Econometrica 69(1): 191–200

References

Chapter 6
Bayesian Nash Equilibrium

6.1 Formulation of Games with Incomplete Information

We now introduce incomplete information, where some part of the game structure is not common knowledge among the players. This can mean that some players may not completely know the payoff function of certain player(s), or the strategy set of certain player(s), or even who the players are. Then the players need to form their own beliefs about the components of the game, which are called first order beliefs. In addition, to choose a strategy, the players need to form beliefs about how the others choose their strategies based on their first order beliefs. These are beliefs of beliefs, thus they are second order beliefs. Furthermore, players realize that the others also make decisions based on their second order beliefs, which requires third order beliefs...In this way, a game with incomplete information requires an infinite hierarchy of (mutual) beliefs for decision making.

Another big problem is the origin of these beliefs. Would the players form beliefs totally subjectively? Or, is there some common component to the beliefs among the players? If so, where does the common component come from? It is very difficult to even formulate incomplete information games with generality and precision, let alone to define a convincing equilibrium concept.

In this chapter, we explain the *Bayesian game* framework by Harsanyi [5–7]. Harsanyi's idea is that the beliefs of players are "mutually consistent", in the sense that they are conditional probability distributions derived from a common basic probability distribution (*common prior*) over the elements of the game, and that players have at least common knowledge of the set of possible games (or the parts that are not common knowledge, such as someone's payoff function) and the common prior. Each player maximizes the expected payoff using the updated beliefs by Bayes' rule when additional information becomes available.

With this setup, we can identify[1] the model of an infinite hierarchy of beliefs in a game with **imperfect information** such that Nature chooses the elements of

[1] To be precise, mathematical equivalence between the infinite hierarchy of beliefs model and Harsanyi's 'type' model needs to be proved. See Mertens and Zamir [12].

T. Fujiwara-Greve, *Non-Cooperative Game Theory*, Monographs in Mathematical Economics 1, DOI 10.1007/978-4-431-55645-9_6

a game probabilistically according to the common prior, and the players do not know the choice of Nature perfectly. That is, **we convert a game with incomplete information into a game with imperfect but complete information**. Because of the common knowledge assumption of the converted game (which is called the *Bayesian game* generated from the originally incomplete information situation), this method cannot address all incomplete information situations where such common knowledge is doubtful, but it is a very convenient formulation, under which we can use the analytical tools developed for complete information games. There is another formulation for normal-form games with incomplete information, by Carlsson and van Damme [2], which is explained in Chap. 9.

In this chapter we consider the case where the payoff functions are not common knowledge. In fact, the case where the set of strategies are not common knowledge can be transformed into the case where the domains of the payoff functions are not common knowledge. If the rules of the game are not common knowledge, we can also interpret this as the case where the possible strategies (the domain of the payoff functions) are not common knowledge. Furthermore, if someone's informational structure is not common knowledge, this can be expressed as the relevant player's set of strategies is not common knowledge as well. Thus, most incomplete information situations can be converted into the incomplete information of someone's payoff function.

6.2 Bayesian Games

We consider a famous economic problem called *adverse selection* as an example of incomplete information.[2] There is a seller S and a buyer B looking at a used car. The main problem of used car trading is the unknown quality. In this case, the seller who owns the used car knows the quality of the car, but the buyer does not. Hence, the buyer does not know his payoff until after he purchases the car.

For simplicity, assume that there are only two possibilities, that the car is in good condition or a "lemon" (i.e., it is problematic). There is also a known market price $P > 0$ for this model and the year of the car. The choices for the seller are whether or not to sell the car at the market price, and the choices of the buyer are whether or not to buy the car at the market price.

If the car is in good condition, the buyer gets a benefit worth 60, but if the car is a lemon, he gets only 30 (these are in monetary units). The payoff for the buyer is thus $60 - P$ if he buys a good car and $30 - P$ if he gets a lemon. The seller can keep the car and get a payoff of 55 (again in monetary units) if the car is in good condition but 0 if the car is a lemon. The seller gets a payoff of P if he sells the car and the buyer buys it. The two possible payoff functions (but not the normal-form game, since the buyer does not know which function he has) are shown in Table 6.1.

[2]This section is inspired by Watson [14]. The seminal paper on the lemon problem is Akerlof [1].

Table 6.1 Two payoff functions

Good			Lemon		
S \ B	Trade	Not	S \ B	Trade	Not
Trade	$P, 60 - P$	55, 0	Trade	$P, 30 - P$	0, 0
Not	55, 0	55, 0	Not	0, 0	0, 0

We assume that the two possible payoff functions are common knowledge. We still need to specify how the buyer assesses the likelihood of the two possible qualities of the car, or his payoff functions. Imagine that, by some publicly available market statistics, the probability is 0.3 that a used car is a lemon, and this is common knowledge. That is, this probability is the *common prior*. The assumption that the seller knows the quality of the car can be formulated as getting additional information. In effect, the knowledge difference between the two players is this additional information.

Mathematically, the seller S updates his belief about the quality of the car from the common prior using the additional information, via the *Bayes' rule*.[3] Then the seller uses the *posterior* probability distribution to compute his expected payoff. (This is the reason that the game is called a Bayesian game.) The buyer B does not have this additional information, hence he uses only the prior probability distribution for the expected payoff computation. We also assume that the fact that the seller has the additional information and the buyer does not is common knowledge. (This much of common knowledge is needed to make the Bayesian game with complete information.)

In the current example, the additional information for the seller S reveals the quality of the car, so that the seller S puts a posterior probability of 1 on one of the payoff functions. The buyer B assigns the probability 0.7 to the payoff function of the good car and 0.3 for lemon.

In general, suppose that a normal-form game $G = (1, 2, \ldots, n, S_1, S_2, \ldots, S_n, u_1, u_2, \ldots, u_n)$ has incomplete information regarding the payoff function of some player(s). (Hence $u_1, u_2, \ldots, u_n$ are not common knowledge, and there are other possible payoff functions.) To convert this into a Bayesian game, we assume that **each** player is identified by an *attribute vector*, which totally describes the player's characteristics (summarized by a payoff function, which includes her/his information structure) and her/his infinite hierarchy of beliefs. In the above example, an attribute vector includes the payoff function depending on the quality of the car and the common knowledge of what the players know about the car. A value of an attribute vector is called the *type* of a player. Since a type includes the player's informational

[3]Bayes' rule generates the conditional probability of an event A given that the event B has occurred from the prior probability distribution Pr in such a way that

$$Pr(A \mid B) = \frac{Pr(A \cap B)}{Pr(B)}.$$

structure, it is enough to say that each player knows only her/his type. Let T_i be the set of possible types of player $i \in \{1, 2, \ldots, n\}$. For the moment, assume that each T_i is finite. The set of type combinations is written as $T := \times_{j=1}^{n} T_j$. With a slight abuse of notation, in the Bayesian game of G with the type space T, player i has the (extended) payoff function $u_i : S \times T \to \Re$, which not only depends on the strategy combination but also the type combination of all players. (Recall that $S = \times_{i=1}^{n} S_i$.)

The true type combination $t = (t_1, \ldots, t_n) \in T$ is supposed to be chosen by Nature at the beginning of the Bayesian game, with the common prior probability distribution $p : T \to [0, 1]$. For simplicity, assume that any type combination $t \in T$ has a positive prior probability. For each possible realization $t^* = (t_1^*, \ldots, t_n^*)$, each player only learns her/his own type t_i^* and computes her/his posterior probability distribution $p(\cdot \mid t_i^*) : \times_{j \neq i} T_j \to [0, 1]$ of the others' type combinations by the Bayes' rule.

In the used car trading example, the seller S's type can be simply written as $T_S = \{G, L\}$, where G is the payoff function when the car is in good condition and L is the payoff function when the car is a lemon. This is because the payoff functions include the quality and the knowledge of the quality of the car. By contrast, the buyer has a unique type because he does not know the quality of the car, and this is common knowledge. Let us write the set of B's type as $T_B = \{GL\}$. The common prior assigns probability 0.7 that Nature chooses the type combination $t^* = (G, GL)$ and 0.3 on $t^* = (L, GL)$. (The first coordinate is the type of the seller.)

After Nature's move, player S learns either $t^* = (G, GL)$ is realized or $t^* = (L, GL)$ is realized. Hence he assigns (the posterior) probability 1 to one of the payoff functions in Table 6.1. Player B only learns his unique type anyway, which means that his "posterior" probability is the same as the common prior and that he assigns 0.7 to the good condition payoff function and 0.3 to the lemon payoff function. The tree diagram of the Bayesian game is depicted in Fig. 6.1. (The payoff vectors are omitted for simplicity.) Player S has two singleton information sets, but player B has a unique information set H_{BGL}.

A strategy of player i in a Bayesian game is a function from her/his type set T_i to S_i, i.e., a behavioral strategy. Note, however, that the original incomplete information situation implies that there must be a player who does not know Nature's choice. Therefore, although it is an extensive-form game, there is no proper subgame. This should be clear from Fig. 6.1. Hence, a meaningful equilibrium concept is a Nash equilibrium. A Nash equilibrium of a Bayesian game is called a *Bayesian Nash equilibrium*.

Definition 6.1 Given a normal-form game $G = (\{1, 2, \ldots, n\}, S_1, S_2 \ldots, S_n, u_1, u_2, \ldots, u_n)$ and its Bayesian game with finite type sets[4] $T_1, T_2, \ldots, T_n$ and a common prior p, a combination of behavioral strategies $(b_1, \ldots, b_n)$ such that for each $i \in \{1, 2, \ldots, n\}$, $b_i : T_i \to S_i$ is a *Bayesian Nash equilibrium* if, for each player $i \in \{1, 2, \ldots, n\}$, b_i maximizes the ex-ante expected payoff among functions from T_i to S_i, given other players' strategies b_{-i}. That is, for any $g_i : T_i \to S_i$,

[4] It is possible to extend the Bayesian framework and the Bayesian Nash equilibrium to infinite type spaces with appropriate probability measures.

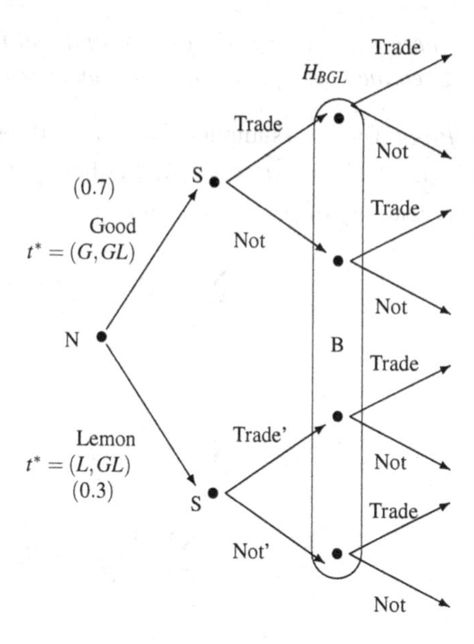

Fig. 6.1 The structure of the Bayesian game based on the lemon's problem

$$\sum_{t_i \in T_i} \sum_{t_{-i} \in T_{-i}} p(t_i, t_{-i}) u_i(b_i(t_i), b_{-i}(t_{-i}), t_i, t_{-i})$$
$$\geqq \sum_{t_i \in T_i} \sum_{t_{-i} \in T_{-i}} p(t_i, t_{-i}) u_i(g_i(t_i), b_{-i}(t_{-i}), t_i, t_{-i}), \quad (6.1)$$

where $t_{-i} \in T_{-i} := \times_{j \neq i} T_i$ is a combination of types of players other than i.

We should be careful that the expected payoff must be computed before Nature chooses a type combination. This is because a Nash equilibrium compares payoffs **before the game starts**. There are players who will get additional information, but they are supposed to be informed after Nature chooses the "true" type combination. Therefore, all players compare the ex-ante expected payoff in an equilibrium.

Nonetheless, the above ex-ante optimization is in fact equivalent to the type-dependent (ex-post) optimization, when all type combinations have a positive prior probability.

Lemma 6.1 *Assume that, for any* $t \in \times_{j=1}^{n} T_j$, $p(t) > 0$. *Then for any* $i \in \{1, 2, \ldots, n\}$ *and any* $g_i : T_i \to S_i$, $(b_1, \ldots, b_n)$ *satisfies (6.1) if and only if for any* $i \in \{1, 2, \ldots, n\}$, *any type* $t_i \in T_i$, *and any* $s_i' \in S_i$,

$$\sum_{t_{-i} \in T_{-i}} p(t_{-i} \mid t_i) u_i(b_i(t_i), b_{-i}(t_{-i}), t_i, t_{-i})$$
$$\geqq \sum_{t_{-i} \in T_{-i}} p(t_{-i} \mid t_i) u_i(s_i', b_{-i}(t_{-i}), t_i, t_{-i}),$$

where $p(t_{-i} \mid t_i)$ is the ex-post probability of the other players' type combination given the own type t_i and the common prior p.

Proof By the assumption that $p(t) > 0$ for any $t \in \times_{j=1}^{n} T_j$ and the Bayes' rule, $p(t_{-i} \mid t_i) = p(t_i, t_{-i})/p(t_i)$ (where $p(t_i) := \sum_{t'_{-i} \in T_{-i}} p(t_i, t'_{-i})$ is the marginal probability of $p(t)$ on T_i), and

$$
\begin{aligned}
& \sum_{t_{-i} \in T_{-i}} p(t_{-i} \mid t_i) u_i(b_i(t_i), b_{-i}(t_{-i}), t_i, t_{-i}) \\
& \geqq \sum_{t_{-i} \in T_{-i}} p(t_{-i} \mid t_i) u_i(s'_i, b_{-i}(t_{-i}), t_i, t_{-i}) \ \forall t_i \in T_i \\
\Longleftrightarrow & \sum_{t_{-i} \in T_{-i}} \frac{p(t_i, t_{-i})}{p(t_i)} u_i(b_i(t_i), b_{-i}(t_{-i}), t_i, t_{-i}) \\
& \geqq \sum_{t_{-i} \in T_{-i}} \frac{p(t_i, t_{-i})}{p(t_i)} u_i(s'_i, b_{-i}(t_{-i}), t_i, t_{-i}) \ \forall t_i \in T_i \\
\Longleftrightarrow & \sum_{t_{-i} \in T_{-i}} p(t_i, t_{-i}) u_i(b_i(t_i), b_{-i}(t_{-i}), t_i, t_{-i}) \\
& \geqq \sum_{t_{-i} \in T_{-i}} p(t_i, t_{-i}) u_i(s'_i, b_{-i}(t_{-i}), t_i, t_{-i}) \ \forall t_i \in T_i. \qquad (6.2)
\end{aligned}
$$

Therefore, by adding these inequalities over all $t_i \in T_i$, we obtain (6.1).

Conversely, assume that for any i and any $g_i : T_i \to S_i$, (6.1) holds. For each $t_i \in T_i$ and $s'_i \in S_i$, there exists g_i such that $g_i(t_i) = s'_i \neq b_i(t_i)$ and $g_i(t'_i) = b_i(t'_i)$ for other $t'_i \in T_i$ such that $t'_i \neq t_i$. The fact that (6.1) holds for this g_i implies (6.2). □

In other words, because the behavioral strategy implies that each type of player can choose actions independently, we can view each type as a different "player". In the next two subsections we solve the used car trading Bayesian game in two different but equivalent ways, from the perspective of the ex-ante comparison of expected payoffs and by the type-players' optimization.

6.2.1 Ex-Ante Optimization

Let us make a matrix representation of the used car trading Bayesian game, in terms of the ex-ante expected payoff of behavioral strategies. A pure (behavioral) strategy of player S (the seller) has two components, what to do if the car is in good condition and what to do if it is a lemon. By taking the initials of the actions in this order, let us write the set of pure strategies as $S_S = \{TT', TN', NT', NN'\}$. Similarly, the pure strategies of player B (the buyer) are T and N. If player B chooses N, then regardless

Table 6.2 Ex-ante payoff matrix for the lemon problem

S \ B	T		N
TT'	$P,$	$51 - P$	38.5, 0
TN'	$(0.7)P,$	$(0.7)(60 - P)$	38.5, 0
NT'	$38.5 + (0.3)P,$	$(0.3)(30 - P)$	38.5, 0
NN'	38.5,	0	38.5, 0

of S's strategy no transaction results, so that player S's ex-ante expected payoff is $(0.7) \times 55 + (0.3) \times 0 = 38.5$. Recall that we need to compare payoffs before the Bayesian game starts so that the expectation is taken before Nature chooses the quality of the used car. Player B's ex-ante expected payoff from strategy N and an arbitrary strategy by player S is 0. In addition, if player S chooses NN', regardless of B's strategy, S's expected payoff is 38.5 and B's is 0.

Next, suppose that player B chooses T. If player S chooses TT', then the used car is sold regardless of the quality. Hence, S's expected payoff is $Eu_S(TT', T) = (0.7)P + (0.3)P = P$. Player B's expected payoff is $Eu_B(TT', T) = (0.7) \times (60 - P) + (0.3) \times (30 - P) = 51 - P$. If player S chooses TN', then only the good car is traded, and S's expected payoff is $Eu_S(TN', T) = (0.7)P + (0.3) \times 0 = 0.7P$, while B's is $Eu_B(TN', T) = (0.7)(60 - P) + (0.3) \times 0 = (0.7)(60 - P)$. Finally, if S chooses NT', then only a lemon is traded, and S's expected payoff is $Eu_S(NT', T) = (0.7) \times 55 + (0.3)P$, while B's expected payoff is $Eu_B(NT', T) = (0.7) \times 0 + (0.3)(30 - P)$. These are summarized in Table 6.2. (The first coordinate is player S's payoff.)

Let us investigate if there is a (pure) Bayesian Nash equilibrium such that player B chooses strategy T.[5] Because $P > 0$, we have that $P > (0.7)P$ and $38.5 + 0.3P > 38.5$. Therefore, player S's pure strategy best response to T is either TT' or NT'. Among these, TT' is a best response if and only if

$$P \geqq 38.5 + 0.3P \iff P \geqq 55.$$

However, when P is so high that $P \geqq 55$, player B's best response to TT' is N. Therefore, there is no Bayesian Nash equilibrium under $P \geqq 55$ such that trade occurs. Notice that if the game has complete information, a good car can be traded for P not less than 55 as long as it is not greater than 60. But such a trade is impossible when information is not complete.

When $P < 55$, a best response of player S to T is NT'. If, in addition, $P \leqq 30$ holds, then a best response of player B to NT' is also T. Therefore, we have a Bayesian Nash equilibrium (NT', T). However, if $30 < P < 55$, then this strategy combination is not an equilibrium.

In summary, under incomplete information, only low prices $P \leqq 30$ admit a Bayesian Nash equilibrium in which trade occurs, and moreover, in that equilibrium,

[5] The combination (NN', N) is a Bayesian Nash equilibrium for any P.

only a lemon is traded. This is an example of an economic phenomenon called *adverse selection*, because it is an opposite consequence to the usual natural selection in the market, where good products remain after low quality products are expelled by consumer choice. When information is not complete, only bad products may remain in the market.

6.2.2 Optimization by 'Type' Players

We solve the same game from the perspective that there are two types of sellers corresponding to the two possible qualities. Denote by S_1 the seller player who owns a good car and by S_2 the seller who owns a lemon. We can re-interpret the extensive-form game in Fig. 6.1 such that the two information sets of the seller belong to different players. Thus, it becomes a three-player game.

When the price is $P \geqq 55$ and player B chooses strategy T, player S_1 compares the payoff P from his pure strategy T and 55 from strategy N. Hence it is optimal for player S_1 to trade. Analogously, player S_2 compares P and 0 and again it is optimal to trade. (Note that we are comparing non-expected payoffs, given a type of the player.) Let us write the combination of optimal strategies TT'. Player B, who does not know whether the seller is type G or type L, still compares the ex-ante expected payoff. Hence the expected payoff of strategy T is $Eu_B(TT', T) = (0.7)(60 - P) + (0.3)(30 - P)$, and this is negative when $P \geqq 55$. Therefore, player B's best response is N. That is, (TT', T) is not a Bayesian Nash equilibrium.

When $30 < P < 55$ and player B chooses T, player S_1 compares the payoffs P (from strategy T) and 55 (from strategy N), so that the best response is now N. Player S_2's best response is T'. In this case, player B's expected payoff is $Eu_B(NT', T) = (0.7)0 + (0.3)(30 - P) < 0$ so that (NT', T) is not a Bayesian Nash equilibrium.

Finally, when $P \leqq 30$ and player B chooses T, S_1's best response is N and S_2's best response is T'. Then $Eu_B(NT', T) = (0.7)0 + (0.3)(30 - P) \geqq 0$ implies that (NT', T) is a Bayesian Nash equilibrium. Thus, we have reached the same outcome as in the previous subsection.

In this example, the number of types is small and the optimization computation is not so different between the ex-ante payoff comparison and the type-separated comparison. However, when there are many types or many strategies, it is often the case that type-separated optimization is easier. The next section gives such an example.

6.3 Cournot Game with Incomplete Information

Reconsider the Cournot game in Sect. 3.2, in which two firms choose production quantities simultaneously. Let us now assume that firm 1's marginal cost c_1 is not common knowledge. There are two possibilities, c_h or c_ℓ (where $c_h > c_\ell$), and the

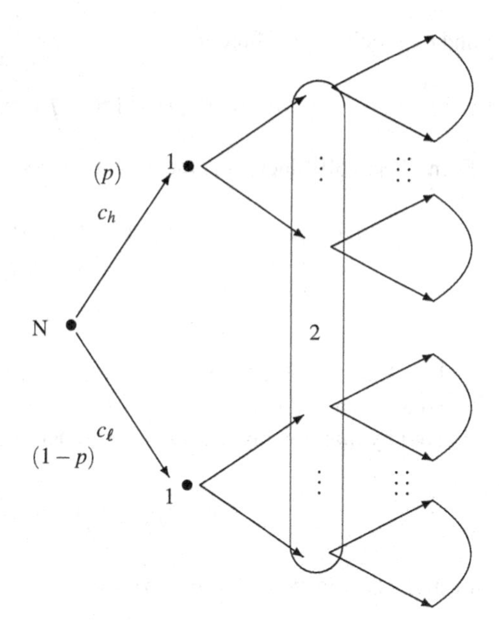

Fig. 6.2 Cournot game with two types

common prior probability is that the marginal cost is c_h with probability p, while it is c_ℓ with probability $1-p$. Only firm 1 knows which is true, but the above information structure is common knowledge.

To formulate the Bayesian game of this model, Nature chooses one of the types (the marginal cost parameters of firm 1) according to the prior probability distribution. Then firm 1 learns Nature's choice, but firm 2 does not. After that, the two firms choose quantities (q_1, q_2) simultaneously and the game ends. The outline of the Bayesian game is shown in Fig. 6.2. (Imagine that the information set of firm 2 contains an infinite number of actions of both types for firm 1.)

A (behavioral) strategy of firm 1 who has the cost information is to choose a quantity for each information set, and a strategy of firm 2 is to choose one quantity q_2 at its unique information set. In this example, it is easier to solve firm 1's optimization problem as two type-separated players choosing one variable each to maximize (non-expected) payoff, instead of varying two-dimensional quantity vectors to maximize the ex-ante expected payoffs. Therefore, we focus on type-players.

As in Sect. 3.2, firm i's payoff is its profit, and the price is determined by the formula $A - q_1 - q_2$. Let q_{1h} be the quantity choice by c_h-type of firm 1 (called 1_h), and $q_{1\ell}$ be the one by c_ℓ-type (called 1_ℓ). When firm 2's quantity is q_2, c_h-type has the payoff function

$$u_{1h}(q_{1h}, q_2) = (A - q_{1h} - q_2)q_{1h} - c_h \cdot q_{1h}, \tag{6.3}$$

and c_ℓ-type's payoff function is

$$u_{1\ell}(q_{1\ell}, q_2) = (A - q_{1\ell} - q_2)q_{1\ell} - c_\ell \cdot q_{1\ell}. \tag{6.4}$$

Firm 2's payoff function is the ex-ante expected profit:

$$\begin{aligned} Eu_2(q_{1h}, q_{1\ell}, q_2) =& p\{(A - q_{1h} - q_2)q_2 - c_2 \cdot q_2\} \\ &+ (1-p)\{(A - q_{1\ell} - q_2)q_2 - c_2 \cdot q_2\} \\ =&(A - q_2 - c_2)q_2 - q_2\{p \cdot q_{1h} + (1-p)q_{1\ell}\}. \end{aligned}$$

In a Bayesian Nash equilibrium, the three players, 1_h, 1_ℓ and 2, choose a best response to the others' strategies. The best response of player 1_j $(j = h, \ell)$ to q_2 is derived by maximizing the quadratic functions of (6.3) and (6.4) so that

$$q_{1j} = \frac{1}{2}(A - c_j - q_2).$$

Firm 2's best response to $(q_{1h}, q_{1\ell})$ is

$$q_2 = \frac{1}{2}[A - c_2 - \{p \cdot q_{1h} + (1-p)q_{1\ell}\}].$$

Solving these three simultaneous equations, we obtain

$$\begin{aligned} q_{1h}^* =& \frac{1}{6}\{2A + 2c_2 - (3+p)c_h - (1-p)c_\ell\} \\ q_{1\ell}^* =& \frac{1}{6}\{2A + 2c_2 - pc_h - (4-p)c_\ell\} \\ q_2^* =& \frac{1}{3}\{A - 2c_2 + pc_h + (1-p)c_\ell\} \end{aligned}$$

as the unique Bayesian Nash equilibrium.

This game has some economic implications. Suppose that the true marginal cost of firm 1 is c_h, and it is common knowledge. Then, in the Cournot Nash equilibrium, firm 2 would produce $\frac{1}{3}(A - 2c_2 + c_h)$. Because $pc_h + (1-p)c_\ell < c_h$, the above Bayesian Nash equilibrium quantity q_2^* is less than the complete information equilibrium quantity. Firm 2 is "conservative" in production, because it entertains the possibility of a lower cost parameter for the rival (which means a stronger rival). From firm 1's perspective, it benefits from the incomplete information. The Bayesian Nash equilibrium payoff of firm 1_h is

$$u_{1h}(q_{1h}^*, q_2^*) = \frac{1}{36}\{2A + 2c_2 - (3+p)c_h - (1-p)c_\ell\}^2.$$

This is a decreasing function of p. Hence, the common knowledge case of $p = 1$ gives a lower payoff than that in the incomplete information case. If the true cost parameter is c_ℓ, however, firm 2 would be more aggressive, because firm 1 may be weaker than the truth, and this hurts firm 1.

6.4 Auctions

An important application of Bayesian games is the analysis of auctions. In many auctions, such as art auctions, it is plausible that the benefit a bidder receives from the object is independent from other bidders' benefits and is private information. In such *private value auctions*, it is natural to formulate the bidding competition as a Bayesian game. There are also *common value auctions*, in which any bidder receives the same value after getting the object, such as auctions of extraction rights of an oilfield. Even in common value auctions, the game may have incomplete information, for example when the bidders do not know the true (common) value of the object before winning the auction. As in the used car problem, the true amount of the oil from the oilfield may be uncertain to bidders. Then no one knows the payoff functions completely. In this way, it is often the case that auctions should be formulated as games with incomplete information. Many auction procedures, such as sealed-bid auctions, are essentially simultaneous-move games, and we can use Bayesian games to predict an equilibrium.

So far, we have analyzed examples that yield fundamentally different equilibria or equilibrium payoffs depending on whether the game has complete information or not. However, auctions may not behave in the same way. Rather, an interesting property of auctions is that the result (such as the revenue for the seller) does not change even if the procedure or the information structure changes.

Consider, for example, the second-price, sealed-bid auction we studied in Problem 2.4. In Problem 2.4., we implicitly assumed complete information and proved that it is a weakly dominant strategy for a bidder to bid her/his true valuation of the object. The proof can be modified to fit an incomplete information game, where bidders do not know the other bidders' valuations (use the type-players). Hence, it is a weakly dominant strategy to bid one's true valuation in the auction game with incomplete information, and the strategy combination where all players bid their true valuation is a Bayesian Nash equilibrium.

For other auction procedures, good references are Klemperer [11] and Milgrom [13]. Among the many interesting findings in auction theory, we show the most important Revenue Equivalence Theorem below, which serves as an example of a Bayesian game with infinitely many types.

Auctions can take a variety of forms. The most commonly used bidding procedures include not only the sealed-bid method but also the well-known *ascending-bid auction* (or, the *English auction*,[6] where bidders gradually raise the price until only

---

[6] Also called the open or oral auction.

one bidder is left), and the *descending-bid auction* (also called the *Dutch auction*, where the seller starts with a very high price and lowers it until someone offers to buy). An ascending-bid auction is an extensive-form game because players (bidders) change their prices as the auction goes on. A descending-bid auction can be formulated as a simultaneous-move game because, although the actual time passes, the essence of strategic decision making is the highest price that a bidder is ready to accept. In addition, the payment rules can vary. A common rule is that the winner pays her/his bid, which is the highest bid (a *first-price auction*), but in some cases the winner pays the second highest bid (a *second-price auction*) or an even lower ranked bid. Furthermore, there is *all-pay auction*, in which all bidders must pay some participation fee. It may seem odd that bidders must pay some cost even though they know that most of them will lose the auction. However, win-or-lose competitions, such as R & D competitions and lobbying activities, can be viewed as all-pay auctions.

Despite the wide variety of auction forms, many of them yield the same revenue to the seller in equilibrium. This is the amazing result of the Revenue Equivalence Theorem. Following Klemperer [10] or Chap. 1 of Klemperer [11], we give the theorem for auctioning a single object where valuations among bidders are independent and can take any value in an infinite set. (Multiple-object and/or inter-dependent value cases are generally much more complex.)

Theorem 6.1 (Revenue Equivalence Theorem) *Assume that there are n bidders for a single object to be auctioned. Each bidder's valuation of the object is private information and is independently drawn from a common continuous distribution over $[\underline{v}, \bar{v}]$. The payoff of a bidder is the benefit (the valuation if the bidder obtains the object and 0 otherwise) minus the payment, and bidders choose strategies in order to maximize the expected payoff. Then, any auction mechanism in which (i) the object is always given to the bidder with the highest (realized) valuation, and (ii) any bidder with the lowest-feasible valuation gets 0 expected payoff, yields the same expected revenue to the seller in a Bayesian Nash equilibrium. (Or, any type bidder receives the same expected payoff in a Bayesian Nash equilibrium of such auctions.)*

For each bidder, a type is the realized valuation and a strategy is a function from types to bids. We briefly explain the idea of the proof of the theorem, given in Klemperer [10].

Fix an arbitrary auction mechanism and a bidder i. Consider a symmetric equilibrium so that all bidders use the same (increasing) function $s : [\underline{v}, \bar{v}] \to \Re$. For each $v \in [\underline{v}, \bar{v}]$, let $P_i(v)$ be the probability of this bidder i getting the object when (s)he follows the strategy $s(v)$. Let $E_i(v)$ be the expected payment. Then the expected payoff is

$$U_i(v) = P_i(v)v - E_i(v).$$

(Note that we do not specify how E_i depends on $P_i(v)$ so that we allow payment regardless of getting the object.)

If bidder i with type v deviates from $s(v)$ and uses the strategy of the type $v + dv$, the probability of winning the auction becomes $P_i(v+dv)$ and the expected payment is $E_i(v + dv)$. Hence, the expected payoff becomes

$$P_i(v + dv)v - E_i(v + dv). \tag{6.5}$$

Using the expected payoff of a bidder with the true type $v + dv$,

$$U_i(v + dv) = P_i(v + dv)(v + dv) - E_i(v + dv),$$

we can rewrite (6.5) as

$$P_i(v + dv)v - E_i(v + dv) = U_i(v + dv) - (dv)P_i(v + dv).$$

Hence, the condition that a bidder with type v does not deviate to the strategy of $v + dv$ is

$$U_i(v) \geqq U_i(v + dv) + (-dv)P_i(v + dv), \;\; \forall dv. \tag{6.6}$$

Conversely, for the type $v + dv$ not to deviate to the strategy of type v,

$$U_i(v + dv) \geqq U_i(v) + (dv)P_i(v)$$

must hold. Combining with (6.6), we obtain

$$P_i(v + dv) \geqq \frac{U_i(v + dv) - U_i(v)}{dv} \geqq P_i(v).$$

Letting $dv \to 0$, we have

$$\frac{dU_i}{dv} = P_i(v).$$

By integration, for any $v \in [\underline{v}, \bar{v}]$,

$$U_i(v) = U_i(\underline{v}) + \int_{\underline{v}}^{v} P_i(x)dx.$$

That is, for any auction mechanism, as long as $U_i(\underline{v})$ and $P_i(\cdot)$ are the same, each type's expected payoff is the same in a Bayesian Nash equilibrium. From this, the seller's expected revenue must be also the same. Condition (ii) implies that for any auction mechanism and any i, $U_i(\underline{v}) = 0$. Let F be the type distribution function. Then the probability that v is the highest realized valuation is $P_i(v) = \{F(v)\}^{n-1}$ (independent of i), and condition (i) guarantees that with this probability the bidder type v obtains the object, regardless of the auction mechanism.

6.5 Harsanyi's Purification Theorem*

Another important contribution by Harsanyi to game theory is to give a foundation for mixed-strategy Nash equilibria of normal-form games with complete information. Mixed-strategy equilibria are often hard to grasp intuitively, and people criticize that real players (such as firms) would not randomize their actions. It is also not clear why players must follow a particular probability distribution when multiple pure strategies give the same expected payoff in a mixed-strategy equilibrium.

Harsanyi [8] proved that, based on a normal-form game, we can construct a sequence of Bayesian games and its Bayesian Nash equilibria to converge to the original game's mixed-strategy Nash equilibrium. Therefore, a mixed-strategy equilibrium is approximately a pure-strategy equilibrium.

To formulate a sequence of Bayesian games "close" to the original game, we introduce perturbed games.

Definition 6.2 For any $\epsilon > 0$ and an n-player normal-form game $G = (\{1, 2, \ldots, n\}, S_1, S_2, \ldots, S_n, u_1, u_2, \ldots, u_n)$, a *perturbed game* $G(\epsilon)$ is a normal-form game $G(\epsilon) = (\{1, 2, \ldots, n\}, S_1, S_2, \ldots, S_n, v_1, v_2, \ldots, v_n)$ such that for each player $i \in \{1, 2, \ldots, n\}$, player i's payoff function is

$$v_i(s) = u_i(s) + \epsilon \cdot \theta_i(s),$$

for some random variable $\theta_i(s)$ that assigns a value in $[-1, 1]$ for each strategy combination $s \in S$, independently from the other players' $\theta_j(\cdot)$.

The perturbation of the payoffs can be interpreted as uncertainty from the other players' point of view, that is, a belief held commonly by the other players about i's payoff function. The perturbation range $[-1, 1]$ is just a normalization. As ϵ converges to 0, the players' beliefs converge to the payoff function of G so that the game converges to the complete information game of G.

The Bayesian game is such that Nature chooses all players' perturbed part $\{\theta_i(\cdot)\}_{i \in \{1,2,\ldots,n\}}$, and player i only learns her/his $\theta_i(\cdot)$. After that, the players simultaneously choose an action from S_i and the game ends. Let $b^*(\epsilon)$ be a Bayesian Nash equilibrium of $G(\epsilon)$. The analysis is on the limit of $b^*(\epsilon)$ when ϵ converges to 0, i.e., when the uncertainty disappears.

Consider the example of perturbed Matching Pennies, shown below in Table 6.3. For each player i, x_i follows the uniform distribution over $[-\epsilon, \epsilon]$. (That is, $\theta_1(s)$ is the uniform distribution over $[-1, 1]$ when $s = (H, H)$ and a degenerate distribution $\{0\}$ for other s, and $\theta_2(s)$ is the uniform distribution over $[-1, 1]$ if $s = (T, H)$ and $\{0\}$ otherwise.)

The unperturbed Matching Pennies game (when $x_1 = x_2 = 0$) has a unique Nash equilibrium such that both players use each pure strategy with probability $1/2$, i.e., a strategy combination $((\frac{1}{2}, \frac{1}{2})), (\frac{1}{2}, \frac{1}{2}))$.

In the perturbed game as interpreted as a Bayesian game, player i only knows the realization value of x_i and that the opponent's x_j follows the uniform distribution

Table 6.3 Perturbed matching pennies

1 \ 2	H	T
H	$1+x_1, -1$	$-1, 1$
T	$-1, 1+x_2$	$1, -1$

over $[-\epsilon, \epsilon]$. Thus, player i's type set is the set of values of x_i, and a pure strategy is a function that assigns H or T to each x_i. Consider a symmetric strategy combination such that, for each $i = 1, 2$,

$$b_i(x_i) = \begin{cases} H \text{ if } & x_i \geqq 0 \\ T \text{ if } & x_i < 0. \end{cases}$$

If the opponent follows the strategy b_j, the probability that the opponent plays H is the probability that $x_j \geqq 0$, which is $1/2$. Given b_j, the expected payoff of player i of type x_i when (s)he plays H is

$$Eu_i(H, b_j; x_i) = \frac{1}{2}(1 + x_i) + \frac{1}{2}(-1) = \frac{1}{2}x_i,$$

while the expected payoff of T is

$$Eu_i(T, b_j; x_i) = \frac{1}{2}(-1) + \frac{1}{2} \cdot 1 = 0.$$

Therefore, H is optimal if $x_i \geqq 0$. In other words, b_i is a best response to the same strategy by the opponent and (b_1, b_2) is a pure-strategy Bayesian Nash equilibrium. The combination (b_1, b_2) induces an ex-ante probability distribution $((\frac{1}{2}, \frac{1}{2})), (\frac{1}{2}, \frac{1}{2}))$ over the strategies $\{H, T\}$ of G for the two players. In this sense, the sequence of pure-strategy Bayesian Nash equilibria (trivially) converges to the mixed-strategy equilibrium of the underlying game G as $\epsilon \to 0$.

In general, for each mixed-strategy equilibrium of G, we can construct an appropriate sequence of perturbed games and their Bayesian Nash equilibria which induce a sequence of probability distributions over the strategy combinations S of G and can make the probability distribution converge to the target mixed-strategy equilibrium (as the perturbation disappears). This is the idea of the Purification Theorem below. Its proof is quite technical and thus is omitted.

Theorem 6.2 (Harsanyi's Purification Theorem) *For any generic*[7] *n-player normal-form game G and any mixed-strategy Nash equilibrium of G, there exist a sequence of perturbed games*[8] $\{G(\epsilon)\}$ *and a sequence of pure-strategy Bayesian Nash equilibria*

---

[7]Generic games is a set of games (interpreted as $n \times |S|$-dimensional payoff vectors) which excludes measure 0 sets in the $n \times |S|$-dimensional Eucledian space. They are not **all** games, and therefore counter examples exist. For more details, see Harsanyi [8], Chap. 6 of Fudenberg and Tirole [3], and Govindan et al. [4].

[8]To be precise, we need to restrict the perturbation structure as well as the target Nash equilibrium.

$b^*(\epsilon)$ of $G(\epsilon)$ such that the action distribution of $b^*(\epsilon)$ converges to the distribution of the mixed-strategy Nash equilibrium of G, as ϵ converges to 0.

Recall that the above result shows only the existence of "nearby" perturbed games and their Bayesian Nash equilibria which converge to the target mixed-strategy equilibrium. A more general question is whether distributions "close to" a Nash equilibrium of a complete information game are Bayesian equilibria of **any** "nearby" perturbed games. Kajii and Morris [9] gives a negative answer to this question. A robust equilibrium in this sense is for example the p-dominant equilibrium in Chap. 9. That is, only a subset of Nash equilibria are robust for any sufficiently small value of incomplete information.

Problems

6.1 Player 1 and 2 are thinking about investing in a firm's stock. This firm's performance, which influences its dividends and stock prices, is determined by Nature. First, Nature chooses good or bad performance. The prior problability is 0.3 that the performance is good. After that, players decide whether to buy the stock or not. The payoffs are dependent on Nature's choice as in Table 6.4.
(a) Assume that the game has complete information. Both players learn Nature's choice and then choose Buy or Not simultaneously. For each simultaneous-move game after Good and Bad performance, derive all Nash equilibria including mixed-strategy ones.
(b) Assume that the game has incomplete information. Only player 1 learns Nature's choice (and this is common knowledge) and the two players choose Buy or Not simultaneously. Write down the (ex-ante) matrix representation of the two-player Bayesian game.
(c) Find all pure-strategy Nash equilibria of the two-person Bayesian game in (b). (Note: you can do this part using the 'type' players as well.)

6.2 There are two bidders for an art auction. Player 1's valuation of the artwork is 1.5, and this is common knowledge. Player 2's valuation is not completely known. It is either 1 or 2 with the same problability.

The auction mechanism is *first-price, sealed-bid*. That is, the two players offer prices simultaneously, and the one who gave the highest price gets the artwork at that price.

Table 6.4 Two payoff functions

Good

1 \ 2	Buy	Not
Buy	10, 10	5, 0
Not	0, 5	0, 0

Bad

1 \ 2	Buy	Not
Buy	1, 1	−5, 0
Not	0, −5	0, 0

For simplicity, player 1 chooses an offer between the two prices 1.1 and 1.5. Player 2 chooses between the two prices 0.9 and 1.6. Each player's payoff is her/his valuation of the artwork minus the payment if (s)he wins and 0 otherwise. For example, if player 1 bids 1.1 and player 2 bids 0.9, player 1 wins and gets a payoff of $1.5 - 1.1 = 0.4$ and player 2's payoff is 0 regardless of the true valuation. We analyze this incomplete information game as a Bayesian game.
(a) Player 2's behavioral strategy is to choose a bid for each of his valuations. Write down all of his pure strategies.
(b) Write the (ex-ante) matrix representation of the two-player Bayesian game.
(c) Find all Bayesian Nash equilibria in pure (behavioral) strategies. (Note: you can do this part using the 'type' players as well.)

6.3 Consider an incomplete information game between a fruit seller S and a customer C. The seller has a very expensive melon, but the customer does not know its quality.
(a) Consider a one-person decision problem for the customer. Nature chooses the melon's quality from Ripe and Unripe. Without knowing Nature's choice, the customer chooses between Buy and Not. If the customer buys a melon of Ripe quality, her payoff is 1. If the quality is Unripe, the payoff is -2, and if she does not buy, her payoff is 0 regardless of the quality of the melon. The prior problability is 0.5 that the melon is Ripe. What is the optimal strategy for the customer?

Consider a Bayesian game where only the fruit seller S learns Nature's choice (i.e., the quality of the melon), and the seller chooses between Sell and Not Sell, while the customer chooses between Buy and Not, simultaneously. The payoff function of the customer is the same as in (a). The payoff of the seller S is 2 if a Ripe melon is sold, it is 1 if a Ripe melon is not sold, it is 2 if an Unripe melon is sold, and finally it is 0 if an Unripe melon is not sold. (The melon is sold if and only if the seller S chooses Sell and the customer C chooses Buy, simultaneously.) The prior problability is 0.5 that the melon is Ripe.
(b) Write down the (ex-ante) matrix representation of the two-player Bayesian game.
(c) Find all Bayesian Nash equilibria in pure (behavioral) strategies. (Note: you can do this part using the 'type' players as well.)

6.4 Two firms, X and Y, are producing the same good. When the production quantity of firm X (resp. Y) is x (resp. y) units, the market price of this good is determined by the formula $1800 - x - y$. Both firms can choose a production quantity from non-negative real numbers. Firm X maximizes its sales (the market price times its quantity). This is common knowledge. However, firm X does not know whether firm Y's payoff function is its sales or its profit. Firm Y's production cost is $300y$ when it produces y units. Therefore, when the production quantity of firm X (resp. Y) is x (resp. y) units, the profit of firm Y is $(1800 - x - y - 300)y$.

Let the prior problability be $2/3$ that firm Y's payoff function is its sales and $1/3$ that firm Y's payoff is its profit. This is common knowledge. Interpret this game with incomplete information as a Bayesian game where Nature chooses Y's payoff functions and the firms choose their quantities simultaneously. Find all Bayesian Nash equilibria in pure (behavioral) strategies.

Table 6.5 Two payoff functions

Type 1		
L \ S	First half	Second half
First half	3, 3	0, −1
Second half	0, −1	3, 3

Type 2		
L \ S	First half	Second half
First half	0, 3	0, −1
Second half	2, −1	2, 3

6.5 Tomorrow is the final exam of the game theory course. The student S has not studied at all. All he can do now is to memorize half of the course material with the remaining time, so he must choose the first half of the course or the second half. These are his feasible actions. The lecturer L of the game theory course can be either type 1, who is happy if the student does well on the exam, or type 2, who only cares about the course contents and puts emphasis on the second half in the exam. The student does not know the lecturer's type, but her reputation says she is of type 1 with problability 0.4. This is common knowledge. The lecturer's feasible actions are to give the final based on the first half of the course or based on the second half. The type-dependent payoff functions of the lecturer L and the unique payoff function of the student S are shown in Table 6.5. The players choose actions simultaneously.
(a) Interpret this incomplete information game as a Bayesian game and write down all of the lecturer's pure (behavioral) strategies.
(b) Find all Bayesian Nash equilibria in pure (behavioral) strategies.

6.6 Consider a Meeting Game with uncertainty as follows. Player 1 and 2 are trying to meet each other either at a movie theater or on a beach. Their payoffs depend on whether it is sunny or rainy and where they end up (Table 6.6).

The game starts with Nature choosing between Sunny and Rainy with the equal problability. Only player 1 learns the weather. After that the two players simultaneously choose Movie or Beach, and the game ends. The game tree looks like below.
(a) Complete the game tree by adding payoff vectors to all terminal nodes.

Player 2 has a unique information set, and thus he has two pure (behavioral) strategies, Movie and Beach. Player 1 has four pure strategies to be expressed as (action when it is sunny, action when it is rainy).

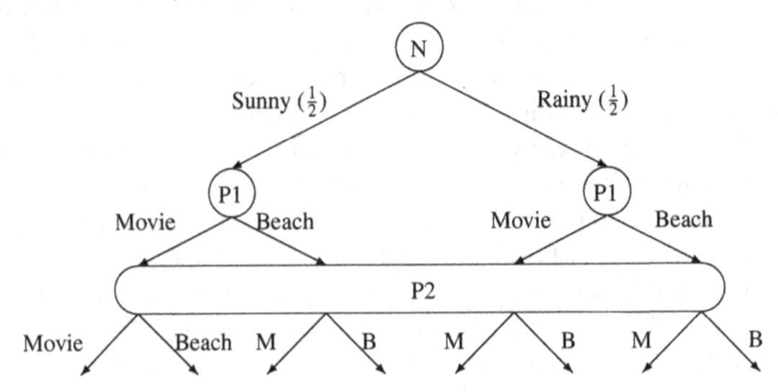

Table 6.6 Two payoff functions

Sunny		
P1 \ P2	Movie	Beach
Movie	3, 3	3, 0
Beach	0, 0	5, 5

Rainy		
P1 \ P2	Movie	Beach
Movie	3, 3	3, 0
Beach	0, 0	1, 1

(b) A pure strategy for player 1 is (Beach, Movie). Compute the expected payoff of Movie strategy and Beach strategy for player 2, and find his best response. (Hint: You do not need to worry about mixed strategies.)
(c) Prove that the pure strategy (Beach, Movie) for player 1 is a best response to the strategy you found in (b), by comparing the expected payoffs of four pure strategies for player 1.
(d) Similarly, prove that (Movie, Movie) for player 1 and a best response to it for player 2 constitute a (Bayesian) Nash equilibrium.

References

1. Akerlof G (1970) The market for lemons: quality uncertainty and the market mechanism. Q J Econ 84(3):488–500
2. Carlsson H, van Damme E (1993) Global games and equilibrium selection. Econometrica 61(5):989–1018
3. Fudenberg D, Tirole J (1991) Game theory. MIT Press, Cambridge, MA
4. Govindan S, Reny P, Robson A (2003) A Short proof of Harsanyi's purification theorem. Games Econ Behav 45(2):369–374
5. Harsanyi J (1967) Games with incomplete information played by Bayesian players, part I: the basic model. Manage Sci 14(3):159–182
6. Harsanyi J (1968) Games with incomplete information played by Bayesian players, part II: Bayesian equilibrium points. Manage Sci 14(5):320–334
7. Harsanyi J (1968) Games with incomplete information played by Bayesian players, part III: the basic probability distribution of the game. Manage Sci 14(7):486–502
8. Harsanyi J (1973) Games with randomly disturbed payoffs: a new rationale for mixed-strategy equilibrium points. Int J Game Theory 2(1):1–23
9. Kajii A, Morris S (1997) The Robustness of equilibria to incomplete information. Econometrica 65(6):1283–1309
10. Klemperer P (1999) Auction theory: a guide to the literature. J Econ Surv 13(3):227–286
11. Klemperer P (2004) Auctions: theory and practice. Princeton University Press, Princeton, NJ
12. Mertens J-F, Zamir S (1985) Formulation of bayesian analysis for games with incomplete information. Int J Game Theory 14(1):1–29
13. Milgrom P (2004) Putting auction theory to work. Cambridge University Press, Cambridge, UK
14. Watson J (2007) Strategy: an introduction to game theory, 2nd edn. Norton, New York, NY

Chapter 7
Perfect Bayesian Equilibrium

7.1 Extensive-Form Games with Incomplete Information

In this chapter, we consider extensive-form games in which the payoff functions are not common knowledge. As equilibrium concepts, not only the perfect Bayesian equilibrium that we introduce in this chapter, but also the sequential equilibrium in Chap. 8 is useful. Although the latter was developed before the former, the perfect Bayesian equilibrium is mathematically simpler, and thus we explain it first.

The basic idea of the equilibrium analysis is the same as in the case of a normal-form game with incomplete information. We extend the underlying extensive-form game into a Bayesian form. All components of a game tree, except (some of) the payoff vectors at the end of the tree, are assumed to be common knowledge. It is also assumed that a set of possible payoff function combinations (of all players) and a common prior distribution on this set are common knowledge. Hence, we can formulate a new extensive-form game such that Nature chooses one of the payoff function combinations, and some players are informed of Nature's choice, while others are not, before playing the underlying extensive-form game.

An important departure from a normal-form Bayesian game is that proper subgames may exist when the underlying game is an extensive-form. Therefore, a refinement of the Nash equilibrium concept must be used to incorporate rational decisions **during the game**.

We have already learned some extensive-form oriented refinements of Nash equilibria: backward induction and subgame perfect equilibrium. A common idea of these is to require optimality of strategies starting at each information set in the extensive-form game as much as possible (i.e., when the information sets are singletons). The concept of the perfect Bayesian equilibrium is developed to apply this *sequential rationality* to the Bayesian formulation of extensive-form games with incomplete information.

To get an intuition of the sequential rationality under a Bayesian framework, consider the example of a generalized entry game in Chap. 4, introduced by Kreps and Wilson [7], depicted in Fig. 7.1. The payoffs are generalized by parameters

T. Fujiwara-Greve, *Non-Cooperative Game Theory*, Monographs in Mathematical Economics 1, DOI 10.1007/978-4-431-55645-9_7

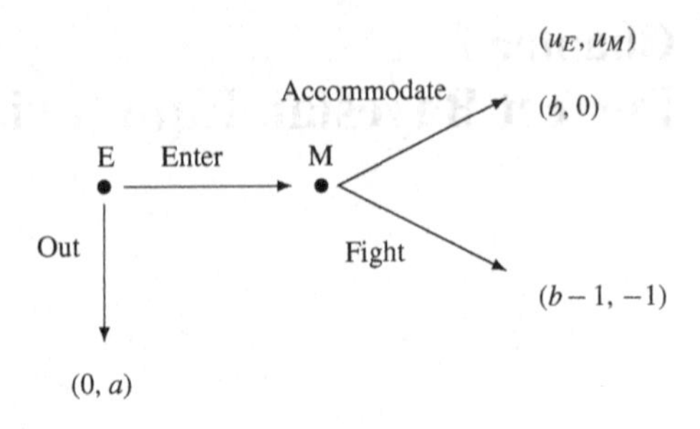

Fig. 7.1 An entry game ($a > 1, 1 > b > 0$)

a and b. (Because of this parameter name and also following Kreps' usual notation, we use π for a behavioral strategy.) We also rename the actions according to Kreps and Wilson [7] such that the monopolist's low price action is to Fight and the high price action is to Accommodate. The first coordinate of a payoff vector is the first mover E's payoff.

As we studied in Chap. 4, the extensive-form game in Fig. 7.1 has a unique solution by backward induction, or a unique subgame perfect equilibrium, which is (Enter, Accommodate). Let us now consider a case with incomplete information such that the entrant firm E does not completely know the payoff function of the monopolist M. For example, E assesses that there are two possible payoff functions that M might have, the one in Fig. 7.1, and another under which Fight is an optimal action when E enters the market. Let us call the monopolist with the payoff function in Fig. 7.1 a Rational type, and with the latter payoff function a Tough type. Let the common prior probability of the Rational type be $1 - \delta$, and the Tough type be δ. To convert the incomplete information model into a complete but imperfect information game, assume that Nature chooses one of the types first, and only the monopolist M is informed of its type. After that, the two players play the entry game. The new extensive-form game is depicted in Fig. 7.2. (As in Fig. 7.1, the first coordinate of each payoff vector is the payoff of firm E. For concreteness, we reversed the payoffs for Rational type to make the Tough type payoffs, but that is only one example.)

Ex-post, the monopolist M can assign probability 1 to one of its information sets. Hence, the type-contingent optimal action is

$$\pi_M = \begin{cases} \text{Fight} & \text{if Tough (at information set } M_T) \\ \text{Accommodate} & \text{if Rational (at } M_R). \end{cases}$$

The uninformed entrant E uses the common prior δ to assess the probability of being one of the decision nodes. The entrant can also reason by backward induction that M will use the type-contingent strategy π_M. Therefore, the expected payoff from each pure strategy is

$$Eu_E(\text{Out}, \pi_M; \delta) = 0$$
$$Eu_E(\text{Enter}, \pi_M; \delta) = \delta(b-1) + (1-\delta)b = b - \delta.$$

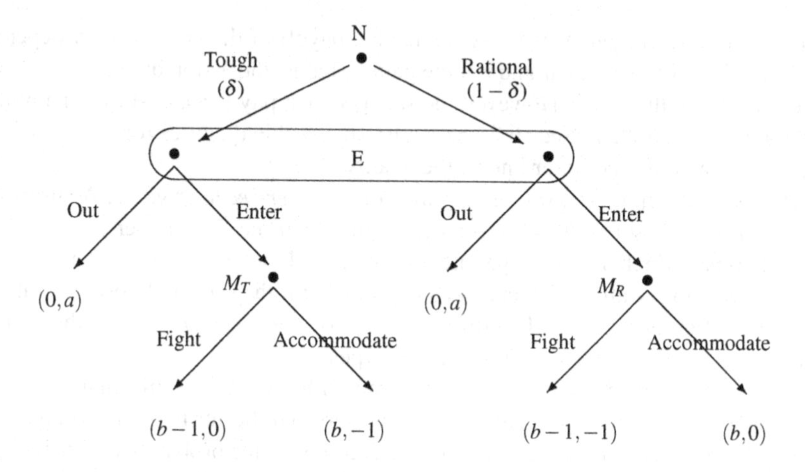

Fig. 7.2 Entry game with two types

Thus, Out is an optimal strategy if the probability δ that M is of Tough type is not less than b. An economic interpretation of this equilibrium is that, if the monopolist can make an entrant believe that M is tough with a sufficiently high probability, then M can deter entry, which is not possible if the game has complete information.

There are two important factors in constructing the above equilibrium. First, the players must form a *belief* at each information set, which is a probability distribution over decision nodes, based on the common prior and the additional information including the history of past actions, using Baye's rule as much as possible. The updating rule is already implicitly used in the derivation of a Bayesian Nash equilibrium using the ex-ante payoff optimization, and is called (weak) *consistency*. (Therefore, an information set of M has the ex-post probability 1 or 0, which is the belief of M.) With an extensive-form game after Nature's choice, there can be many information sets which require consistent belief formation.

Second, each player should choose continuation strategies starting at each information set, given the consistent beliefs and continuation strategies of all other players (starting at their later information sets). This is the formal definition of *sequential rationality*. A strategy combination satisfying these two properties constitutes a rational equilibrium, including optimization during the game. In fact, a subgame perfect equilibrium also requires the same two properties, but on the limited class of singleton information sets. (Hence consistency is ignored.) The concept of *perfect Bayesian equilibrium* (PBE) is a generalization of subgame perfect equilibrium, which requires the above two conditions for all information sets.

Although not clear in the game in Fig. 7.2, a consistent belief must take into account the strategies of players, if the history leading to the relevant information set includes actions by the players. To see this, consider a variant of the used car problem in Sect. 6.2. Suppose that the seller S who owns a used car is certain to put it on sale, and he chooses either to Invest in re-painting the car or Not. After observing whether the car is re-painted or not, player B decides whether to Trade or Not. This is an extensive-form game. We assume that the payoff of the buyer B only depends on the quality of the used car and not the paint. That is, the action by the seller S is a pure "signal" to the buyer. However, the seller S must pay some cost to re-paint the car. As in the example in Sect. 6.2, there are two possible qualities for the car, Good or Lemon, and only the seller knows the quality.

This is an example of a *signaling game*, or a *sender-receiver game*. A signaling game is a two-person extensive-form game in which the first mover (sender) has multiple types which affect the payoffs of the second mover (receiver). The actions of the sender do not affect the receiver's payoffs, but if the sender chooses a strategy such that different types lead to different actions, the receiver can use the action observation to deduce (some of) the hidden types.

Let us transform this extensive-form game with incomplete information into a complete but imperfect information game, as Harsanyi did for normal-form games. Nature chooses one of the qualities by the common prior probability distribution. Assume that a Good car is chosen with probability 0.7 and a Lemon is chosen with probability 0.3. The seller S observes Nature's choice and chooses whether to Invest or Not, depending on the quality (which is his type). The buyer B only observes the actions of the seller but not of Nature. Hence, the buyer B has two information sets corresponding to the seller's actions, and each of them has two possible decision nodes corresponding to the possible qualities of the car. At each information set, the buyer B chooses whether to Trade or Not. This extensive-form game can be illustrated as in Fig. 7.3, following a graphical invention by Cho and Kreps [2]. We omit the payoff vectors first, and explain how to read the figure.

The origin of the game tree is in the center of Fig. 7.3: Nature's decision node marked with N. After Nature's choice, the seller S has two singleton information

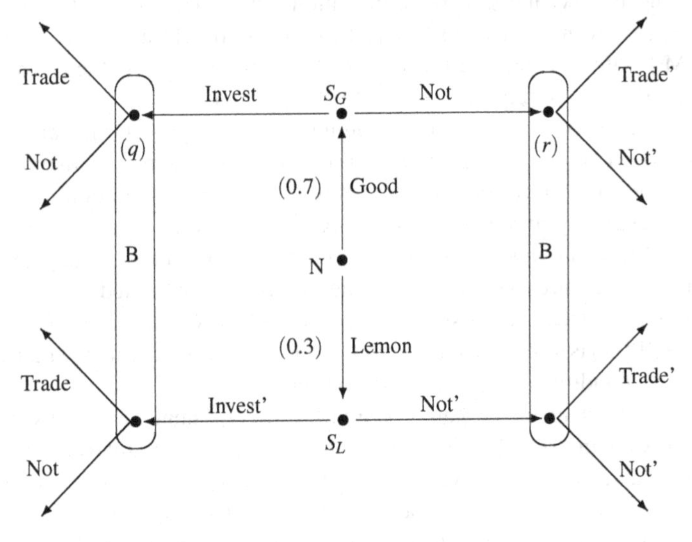

Fig. 7.3 Outline of the signaling game for a used car

sets, for each quality/type. By contrast, the buyer B does not see Nature's choice but whether the seller painted the car or not. Hence, the buyer's information sets are separated by the seller's actions but connected in Nature's choices.

To find an optimal behavioral strategy for the buyer, we need to specify the payoff vectors, and the buyer needs to form a belief at each information set about the probabilities of the decision nodes. In Fig. 7.3, we denote by q and r respectively the belief probability that the seller is of Good type, after seeing the seller's action Invest and Not, respectively. (A common custom is to put parentheses around probability variables to distinguish them from the names of actions, players etc.) In equilibrium, the seller's strategy is correctly anticipated by the buyer, which influences what the buyer deduces after seeing one of the actions.

For example, suppose that the buyer expects the seller to use a behavioral strategy such that the good car owner paints the car but the lemon owner does not (this kind of strategies are called *separating strategies*). Then the buyer's consistent belief after seeing the Invest action is that the seller is Good type for certain, which is formally computed by updating the prior probability by Bayes' rule, as follows.

$$
\begin{aligned}
q = Pr(G \mid I) &= \frac{Pr(G \cap I)}{Pr(I)} \\
&= \frac{Pr(G \cap I)}{Pr(G \cap I) + Pr(L \cap I)} = \frac{(0.7) \times 1}{(0.7) \times 1 + (0.3) \times 0} = 1,
\end{aligned}
$$

where G stands for Good type, L stands for Lemon type, and I is the seller's action to Invest in re-painting. Analogously, the consistent belief after seeing that the seller did Not invest is

$$
r = \frac{Pr(G \cap \text{Not})}{Pr(G \cap \text{Not}) + Pr(L \cap \text{Not})} = \frac{(0.7) \times 0}{(0.7) \times 0 + (0.3) \times 1} = 0.
$$

In this way, the consistent beliefs must depend not only on the common prior, but also on the seller's equilibrium strategy.

To derive the buyer's optimal behavioral strategy explicitly, let $P = 59$, and add payoff vectors to complete the game tree. The extensive-form game is in Fig. 7.4.

When the seller S does not invest (in the right-side information set), the payoff vectors are the same as in Table 6.1. The cost of painting the car is assumed to be dependent on the quality of the car. It is 2 if the car is Good, and it is 60 if it is a Lemon. Therefore, the seller's payoffs from the action Invest are all after subtracting this cost. The buyer's payoffs only depend on the quality of the car (or the type of the seller), and hence are the same after both information sets.

Let us investigate if there is an equilibrium in which the seller chooses the separating strategy (Invest, Not') to signal the quality of the car. As we have seen, in this case the buyer's consistent beliefs are $q = 1$ and $r = 0$. Based on these beliefs, the buyer can determine the optimal behavioral strategy (Trade, Not'), where the first action is after seeing the seller's action Invest, and the second action is after seeing Not.

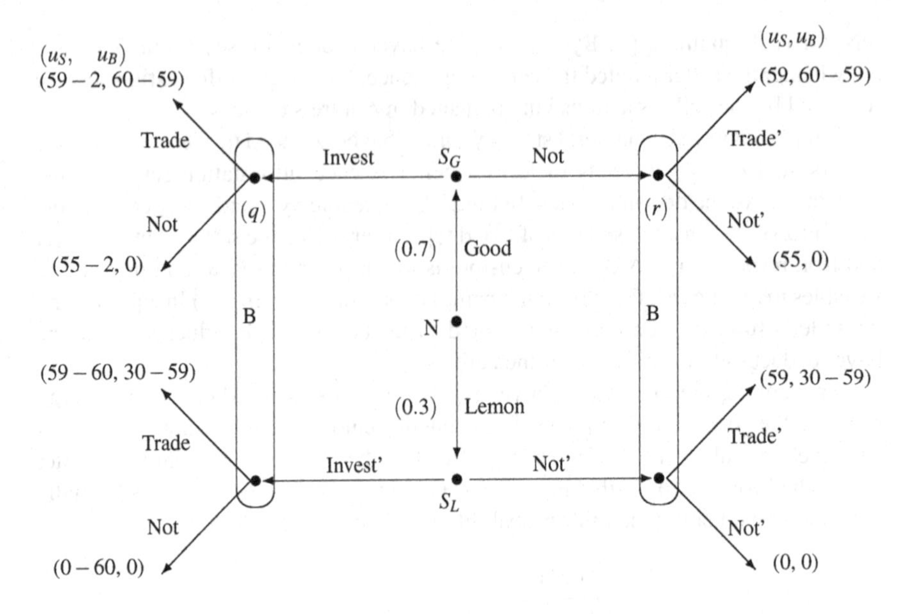

Fig. 7.4 Signaling game of a used car ($P = 59$)

To make an equilibrium, the seller's separating strategy must be a best response to the buyer's strategy. The Good type seller S_G gets $59 - 2 = 57$ if he chooses Invest and 55 if he chooses Not. Hence, it is optimal for this type seller to Invest. If the Lemon type seller S_L chooses to Invest, the car is sold but he gets $59 - 60 = -1$. If S_L chooses action Not', then although he keeps the car, his payoff is 0. Therefore, (Invest, Not') is indeed a best response.

To summarize, the game in Fig. 7.4 has at least one equilibrium which satisfies consistency of beliefs and sequential rationality, namely ((Invest, Not'), (Trade, Not')), with the consistent beliefs $q = 1, r = 0$. (We will investigate if there are other equilibria of this game in the next section.) Recall that in Chap. 6, at such a high price of $P = 59$, there was no Bayesian Nash equilibrium in which trade occurs. However, when the seller can send a signal about the quality **in equilibrium**, then the good car can be traded, and adverse selection can be avoided.

7.2 Signaling Games

We define belief consistency and sequential rationality formally and define a perfect Bayesian equilibrium (Fudenberg and Tirole [5]) for **general signaling games** with finite types and actions. We use sequential equilibrium (Sect. 8.4) for general games with incomplete information, which is more widely used in applications other than signaling games.

Formally, a *signaling game* is a two-player game (except Nature) with three stages. The two players are often called the sender and the receiver. It is assumed that the sender's payoff function is not common knowledge. Let T_1 be the set of sender's types. In the first stage, Nature chooses a type $t_1 \in T_1$ according to the common prior probability distribution $p : T_1 \to [0, 1]$ (where $\sum_{t_1 \in T_1} p(t_1) = 1$), and only the sender learns the choice. In the second stage, the sender chooses an action from the set S_1, and the receiver observes only the actions by the sender, not the choices of Nature. In the third stage, the receiver chooses an action from S_2, and the game ends. A behavioral strategy of the sender is a function $\pi_1 : T_1 \to \Delta(S_1)$, while a behavioral strategy of the receiver is a function $\pi_2 : S_1 \to \Delta(S_2)$. Denote by $\pi_1(s_1 \mid t_1)$ the probability that the sender chooses action $s_1 \in S_1$ when her/his type is t_1 and the strategy is π_1. Similarly, let $\pi_2(s_2 \mid s_1)$ be the probability of the receiver's action $s_2 \in S_2$, after observing s_1 and when (s)he uses the strategy π_2. The payoff of a player depends on the action combination by all players and the sender's type, and the payoff function for player i is written as $u_i : S_1 \times S_2 \times T_1 \to \Re$ for $i = 1, 2$. For simplicity, assume that T_1, S_1, and S_2 are all finite.

The important information sets are the ones for the receiver. Each information set of the receiver corresponds to an action by the sender, and the decision nodes within an information set correspond to underlying types of the sender. The receiver must form a *belief* at each information set (corresponding to $s_1 \in S_1$), regarding the probabilities of the decision nodes (or the sender's types $t_1 \in T_1$). A belief function $\mu : T_1 \times S_1 \to [0, 1]$ assigns a probability distribution $\mu(\cdot \mid s_i)$ for each information set generated by $s_1 \in S_1$ such that $\sum_{t_1 \in T_1} \mu(t_1 \mid s_1) = 1$. For example, in Fig. 7.4, $q = \mu(G \mid \text{Invest})$ and $r = \mu(G \mid \text{Not})$.

Definition 7.1 A *perfect Bayesian equilibrium* (PBE) of a signaling game consists of a pair of behavioral strategies (π_1^*, π_2^*) and a belief function μ that satisfy the following properties.

Sequential Rationality: The sender's behavioral strategy π_1^* is optimal given the receiver's behavioral strategy π_2^*, and the receiver's behavioral strategy π_2^* is optimal given the sender's behavioral strategy π_1^* and the belief μ. That is, for any type[1] $t_1 \in T_1$ and any behavioral strategy $g_1 : T_1 \to \Delta(S_1)$,

$$\sum_{s_1 \in supp(\pi_1^*(t_1))} \pi_1^*(s_1 \mid t_1) \sum_{s_2 \in supp(\pi_2^*(s_1))} \pi_2^*(s_2 \mid s_1) u_1(s_1, s_2, t_1)$$

$$\geqq \sum_{s_1 \in supp(g_1(t_1))} g_1(s_1 \mid t_1) \sum_{s_2 \in supp(\pi_2^*(s_1))} \pi_2^*(s_2 \mid s_1) u_1(s_1, s_2, t_1),$$

and for any information set of the receiver, corresponding to $s_1 \in S_1$, and any behavioral strategy $g_2 : S_1 \to \Delta(S_2)$,

[1] Alternatively, the sender can maximize the ex-ante expected payoff, as with a Baysian Nash equilibrium.

$$\sum_{s_2 \in supp(\pi_2^*(s_1))} \sum_{t_1 \in T_1} \mu(t_1 \mid s_1)\pi_2^*(s_2 \mid s_1)u_2(s_1, s_2, t_1)$$
$$\geq \sum_{s_2 \in supp(g_2(s_1))} \sum_{t_1 \in T_1} \mu(t_1 \mid s_1)g_2(s_2 \mid s_1)u_2(s_1, s_2, t_1).$$

Weak Consistency: The belief function of the receiver is generated by Bayes' rule based on the sender's behavioral strategy π_1^* and the common prior p, as much as possible. That is, for any $s_1 \in S_1$, the belief (the ex-post probability) of each type $t_1 \in T_1$ is

$$\mu(t_1 \mid s_1) = \begin{cases} \frac{p(t_1)\pi_1^*(s_1|t_1)}{\sum_{t_1' \in T_1} p(t_1')\pi_1^*(s_1|t_1')} & \text{if} \sum_{t_1' \in T_1} p(t_1')\pi_1^*(s_1 \mid t_1') > 0 \\ \text{Arbitrary} & \text{if} \sum_{t_1' \in T_1} p(t_1')\pi_1^*(s_1 \mid t_1') = 0. \end{cases}$$

In other words, weak consistency requires that if some type of the sender (with a positive probability) chooses $s_1 \in S_1$ with a positive probability, then the belief probability distribution on the information set s_1 is the conditional probability distribution by Bayes' rule. If no type of the sender (with a positive probability) chooses s_1 with a positive probability, then the belief on the information set s_1 is arbitrary.

A careful reader may be concerned with the fact that weak consistency does not restrict the belief for the sender's *off-path* actions, which are not to be observed. However, it is a limitation of Bayes' rule that no conditional probability is defined for events with probability 0. Nonetheless, the off-path actions may be important, such as when we deal with *pooling equilibria* in the next section.

7.3 Pooling and Separating Equilibrium

In signaling games, it is crucial whether the sender chooses a strategy that plays the same action for all types or not. If such a strategy is chosen, the receiver gets no additional information by observing actions, and her/his posterior belief is the same as the prior distribution. If the sender chooses different actions for some of the types, the receiver can update the prior.

We classify the sender's strategies that choose the same action for all types as *pooling strategies*, and the others as *separating strategies*, and the equilibria are also classified accordingly as *pooling equilibria* and *separating equilibria*. In signaling games with only two types for the sender, like in the used car problem, the sender has only two kinds of pure strategies: to play the same action for both types or to play different actions. Therefore, the sender's pure strategy set consists of pooling strategies and separating strategies. For games with three or more types, sometimes the classification is more detailed and only strategies that choose different actions for every type are called separating strategies, while those that play the same action for some of the types are distinguished as *semi-separating strategies*.[2]

The used-car signaling game in Fig. 7.4 already has a separating equilibrium in pure strategies. Let us investigate if it has a pooling equilibrium. A pooling strategy chooses the same action for all types, and hence we look at (Invest, Invest') and (Not, Not'). It is easy to see that (Invest, Invest') is not a part of a perfect Bayesian equilibrium. The Lemon type seller S_L gets only a negative payoff if he invests in painting and gets at least 0 if he does not. Therefore, if there is a pooling equilibrium, it must involve (Not, Not').

Suppose that the seller uses the pooling strategy (Not, Not'). If the buyer B observes a used car without new paint (the on-path action), then by weak consistency, B should form a belief $r = 0.7$ that the car is Good, according to the prior. The optimal action in this case is Not' because $(0.7)(60 - 59) + (0.3)(30 - 59) < 0$. To check if the seller is choosing a best response, we need to determine what the buyer does when he observes the off-path action of Invest. (Of course, that must be specified in order to define a behavioral strategy.) When an off-path action is observed, weak consistency does not require a specific belief. We divide the beliefs into two cases, which differ in B's optimal action.

If B is at the left-side information set in Fig. 7.4, B's optimal action depends on whether the belief q satisfies $q(60 - 59) + (1 - q)(30 - 59) \geqq 0$ or not. The threshold is $q = 29/30$. If $q > 29/30$, then the buyer's optimal action at the left-side information set is Trade, and if $q < 29/30$, it is Not. When $q = 29/30$, any action is optimal. For all of these categories, we check if the seller's best response is (Not, Not').

Consider the case that the buyer has a belief $q \geqq 29/30$ in the left-side information set. Then the buyer's optimal behavioral strategy is (Trade, Not'). That is, the buyer will buy the car if and only if the car is re-painted. Knowing this, the Good seller S_G can earn $59 - 2$ instead of 55 by changing his action from Not to Invest. Therefore, (Not, Not') is not a best response, and this class of beliefs does not induce a perfect Bayesian equilibrium.

If the buyer has a belief $q \leqq 29/30$ in the left-side information set, then we can focus on his optimal behavioral strategy (Not, Not'). In this case, the car never gets sold, and thus it is optimal for both types of the seller Not to invest. Therefore, the pair of behavioral strategies $(\pi_S, \pi_B) =$ ((Not, Not'), (Not, Not')) and the belief function $(q, r = 0.7)$ for some $q \leqq 29/30$ are all perfect Bayesian equilibrium. (This class of pooling equilibria are illustrated as double arrows in Fig. 7.5.)

Overall, the used-car signaling game has one separating equilibrium ((Invest, Not'), (Trade, Not'), $q = 1, r = 0$) and a continuum of pooling equilibria ((Not, Not'), (Not, Not'), $q, r = 0.7$) where $q \leqq 29/39$. (The other separating strategy (Not, Invest') by the seller is not optimal for the type S_L.)

This example shows that when the receiver's action set is a discrete set, pooling equilibria can be obtained for a range of off-path beliefs, and thus the equilibria are

---

[2]There are also behavioral strategies that choose different mixed actions but with the same support for different types. These are called *hybrid strategies*.

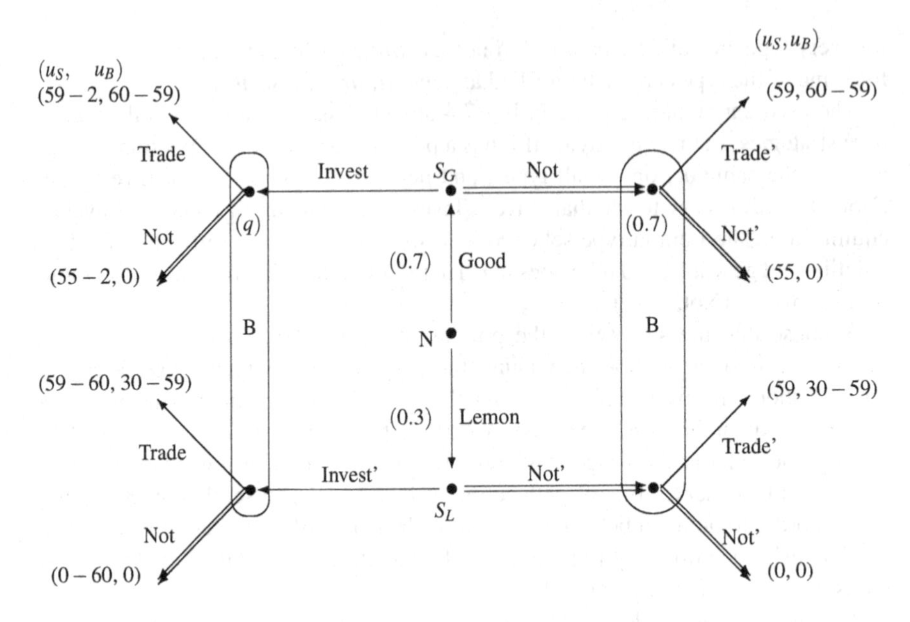

Fig. 7.5 Pooling equilibria ($q \leqq \frac{29}{30}$)

a continuum. The concept of perfect Bayesian equilibrium is unrestrictive in this sense. However, in signaling games, even the sequential equilibrium concept (see Chap. 8) does not restrict beliefs at off-path actions so much. Thus, there is some research on which class of beliefs are "more rational" (see the next section).

The above example also shows that a game can posses two kinds of fundamentally different perfect Bayesian equilibria: a separating equilibrium that reveals types, and pooling equilibria that do not. One reason for this is also the freedom of the off-path beliefs. Another reason is the payoff structure of this game, in particular that the cost of sending the signal (in this case to Invest in re-painting) differs across the types. If all types have the same cost for sending a signal, and if there is sufficient conflict of interest between the sender and the receiver, Crawford and Sobel [3] showed that there is no separating equilibrium in general signaling games. Their proof is quite technical, thus we explain the idea with a simple game below.

The sender has either type 0 or type 1. The receiver's payoff is maximized if he chooses the same number as the sender's true type. The sender's payoff is maximized if the receiver chooses a number which is the sender's type $+b$. This parameter b measures the degree of conflict of interest, and when $b = 0$, the two players have common interest. The sender can send a message, either 0 or 1. The receiver cannot observe the sender's type. Seeing only the message, the receiver chooses one number from $\{0, 1, 2\}$, and the game ends. When the type t-sender sends message m and the receiver chooses action a, the payoff of the sender is $u_S(m, a, t) = -\{a - (t + b)\}^2$, while that of the receiver is $u_R(m, a, t) = -(a - t)^2$. The game is illustrated in Fig. 7.6. (Notice that none of the players' payoff functions depend on m.)

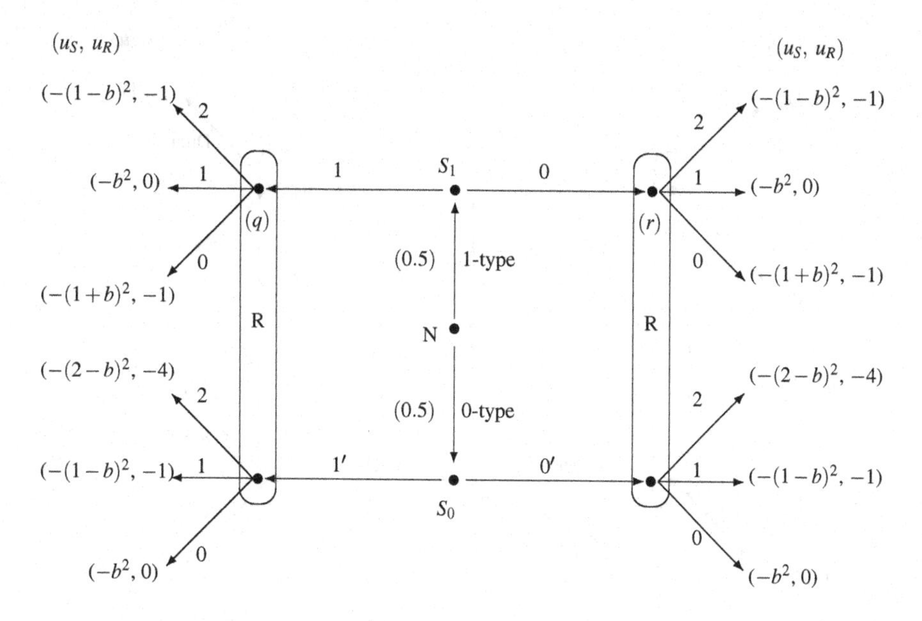

Fig. 7.6 A cheap-talk game

In this game, the sender's action is *cheap talk*, which does not affect any player's payoff. However, when $b = 0$ so that the players have common interest, the sender can reveal his type in equilibrium, and hence his message has a meaning.[3] That is, there is a perfect Bayesian equilibrium such that the sender uses a separating strategy $(1, 0')$ in which he sends the message 1 (resp. $0'$) when his type is 1 (resp. 0), and the receiver chooses action 1 (resp. 0) when the message is 1 (resp. 0). (Clearly the beliefs are $q = 1$ and $r = 0$.)

When $b = 1$, however, there is no separating equilibrium. Consider the separating strategy $(1, 0')$. The receiver can tell the sender's types, and chooses action 1 (resp. 0) when the message is 1 (resp. 0). This is not beneficial to the sender, and a 0-type sender would deviate to message 1 so that he can increase his payoff from -1 to 0. Analogously, if the sender uses the other separating strategy $(0, 1')$, the receiver chooses 1 (resp. 0) if the message is 0 (resp. 1). Then, again, the 0-type sender would deviate. Considering a continuum of types and all mixed strategies by the sender, including hybrid strategies, Crawford and Sobel [3] showed that when the two players' interests are sufficiently different, there is no perfect Bayesian equilibrium that separates the sender's types completely. This is an impossibility result of strategic information transmission.

[3] How much cheap talk can affect the outcomes of a game is itself an interesting research topic. Interested readers are referred to the survey paper by Farrell and Rabin [4] and references therein.

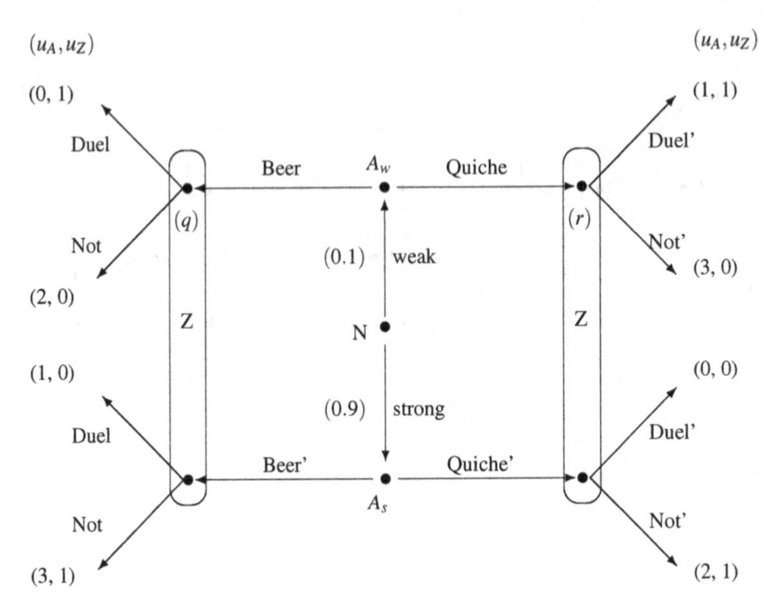

Fig. 7.7 Beer-Quiche game

7.4 Refinements of Equilibria*

Recall that the perfect Bayesian equilibrium concept often allows many equilibria, because it does not restrict beliefs on the information sets that are not reached. Equilibrium *refinements* are thus developed to narrow the set of equilibria by "rationally" restricting off-path beliefs. We introduce the *Intuitive Criterion* by Cho and Kreps [2] as a well-known refinement concept. Another refinement of equilibria in signaling games is the divine equilibrium by Banks and Sobel [1].[4]

Figure 7.7 is the famous Beer-Quiche game constructed by Cho and Kreps [2]. This is a game with two types and two actions for each player.

A possible underlying story is as follows.[5] The game takes place in the Wild West, and a gunman A has arrived at a town's (only) hotel late last night. A while ago, A had killed the brother of a cowboy Z who lives near the town, and Z quickly hears that A has come back to town. In the morning, as A comes down to the dining room of the hotel, A catches the glimpse of Z's face outside of the window. Z is peeping into the dining room, wondering whether to duel or not. Z is not sure how strong A is. The strength of A is his type and is private information. If A is weak, then Z prefers to Duel over Not to duel. If A is strong, it is better for Z Not to duel. The

[4]These refinement concepts require additional rationality properties on beliefs, given the payoff functions. There is a different approach such that the sender's rationality (i.e., the payoff function) is doubted, if an off-path action is observed. See Schulteis et al. [9].

[5]This story is totally made up by the author, who did not consult either In-Koo or David on this.

rumor says that A is strong, and the common prior is that A is weak with probability 0.1, while A is strong with probability 0.9.

A can choose either beer or quiche for his breakfast. The strong-type would prefer beer to quiche, and the weak-type would prefer quiche to beer. In either case, A would like to avoid having a duel. Thus, A gets a payoff of 2 if Z chooses Not to duel, and gets 0 if Z chooses to Duel. Moreover, A gets an additional payoff of 1 from the favorite breakfast, depending on his type. As for Z's payoff, if A is weak, Duel gives 1, while Not to duel gives him 0. If A is strong, the opposite payoff is obtained. These payoffs are depicted in Fig. 7.7. Z chooses whether to duel or not after observing A's breakfast, and the game ends.

Let us find all perfect Bayesian equilibria in pure strategies for this signaling game. The set of pure strategies of A is $S_A = \{(B, B'), (B, Q'), (Q, B'), (Q, Q')\}$, where the first coordinate is the action when A is weak. Z's set of pure strategies is $S_Z = \{(D, D'), (D, N'), (N, D'), (N, N')\}$, where the first coordinate is the action when he sees that A's breakfast is beer. Let q (resp. r) be the belief probability of the weak-type after Z observes action beer (resp. quiche).

First, consider the separating strategy (Q, B') with which A orders his favorite breakfast. Then Z's belief is $q = 0$ and $r = 1$ by Bayes' rule. This implies that Z's optimal behavioral strategy is (N, D'), that is, he believes that A is strong if and only if A orders beer, so that Z duels if and only if A orders quiche. Given this behavioral strategy, the weak-type A gets a payoff of 1 if he orders quiche, but he can get a payoff of 2 if he orders beer. Hence (Q, B') is not a best response.

Suppose that A uses the opposite separating strategy (B, Q'). Then the consistent beliefs of Z are $q = 1$ and $r = 0$. Z's optimal strategy is (D, N'). Given this, the weak-type A not only orders his un-favorite breakfast but also cannot avoid having a duel. By changing his action to quiche, the weak-type A can receive the best payoff of 3. Hence this type of equilibrium does not hold either. In sum, if A uses a separating strategy, Z duels when A is weak, and that is not optimal for A. Therefore, there is no separating perfect Bayesian equilibrium for this game.

Next, consider pooling strategies. Suppose that A chooses (B, B'). When observing beer, Z's belief is $q = 0.1$, because there is no additional information. In this case, the Duel action gives $(0.1)1 + (0.9)0 = 0.1$, while the Not action gives $(0.1)0 + (0.9)1 = 0.9$. Hence Z's optimal action after seeing beer is Not. The right-side information set of quiche corresponds to the off-path action. Depending on r, Z's optimal action differs. Duel' gives the expected payoff of $r \times 1 + (1 - r)0 = r$, and Not' gives $r \times 0 + (1 - r)1 = 1 - r$. The threshold is when $r = 1 - r$, or $r = 1/2$. If $r \geqq 1/2$, then Duel' is optimal, and if $r \leqq 1/2$, Not' is optimal. Consider the case of $r \geqq 1/2$, so that Z chooses Duel'. In this case, both types of A do not deviate from (B, B'). To see this, notice that if A orders quiche, Z will duel. If A is of weak type, beer gives a payoff of 2 and quiche gives only 1. If A is of strong type, beer gives 3 but quiche gives 0. Therefore, there is a continuum of pooling equilibria of the form $((B, B'), (N, D'), q = 0.1, r)$ such that $r \geqq 1/2$. (In Fig. 7.8, the equilibrium strategies are shown as double arrows.) When $r < 1/2$, Z will not duel after seeing quiche. Then A can avoid a duel for any action, and the weak type would want to deviate to quiche. Hence, such off-path beliefs do not make an equilibrium.

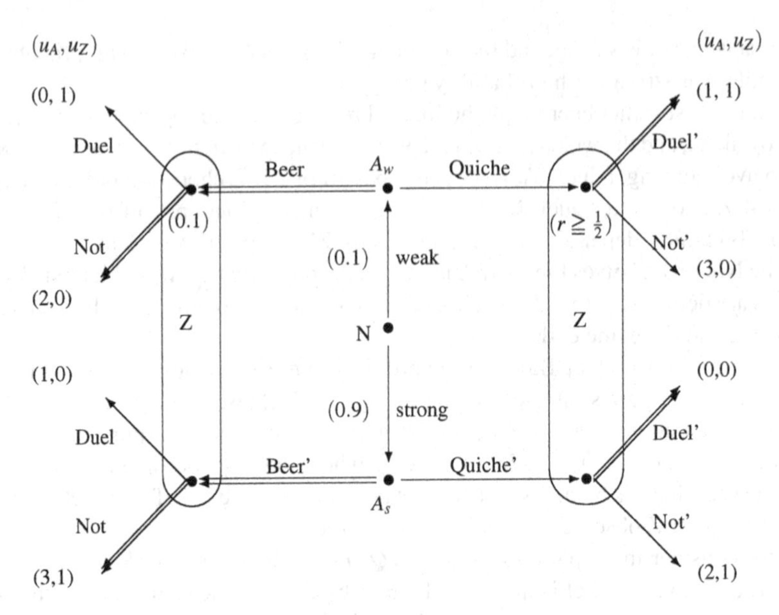

Fig. 7.8 Pooling equilibria such that both types of A order beer

Consider the other pooling strategy (Q, Q'), in which both types order quiche for breakfast. Bayes' rule implies that $r = 0.1$, and we must divide two classes of q, leading to different optimal actions. As in the case of (B, B'), when $q \geqq 1/2$, Duel is the optimal action for Z, and $q \leqq 1/2$ leads to Not. Again, when $q \geqq 1/2$, there are pooling equilibria of the form $((Q, Q'), (D, N'), q, r = 0.1)$ for any $q \geqq 1/2$. There is no equilibrium with $q < 1/2$.

To summarize, the Beer-Quiche game has two kinds of pure-strategy perfect Bayesian equilibria, both of which are a continuum of pooling equilibria, and there is no separating equilibrium. However, Cho and Kreps argued that the pooling equilibrium of the form $((Q, Q'), (D, N'), q \geqq 1/2, r = 0.1)$ is irrational. Notice that the weak type gets at most 2 by deviating from (Q, Q'), while quiche is giving him the best payoff of 3 on the play-path of (Q, Q'). Therefore, the weak type has no incentive to deviate to beer, and, if A deviates, he must be the strong type. Taking this into account, Z must believe that $q = 0$, and his optimal action after seeing beer must be Not to duel. Given this reasoning, the strong type would in fact deviate to beer, which gives him 3. Hence, the pooling equilibria in which both types order quiche are not rational, when we consider the sender's incentives to deviate.

The other class of pooling equilibria, in which both types order beer is still rational, under the above *Intuitive Criterion*. If the sender is to deviate from (B, B'), he must be the weak type. The adjusted belief is then $r = 1$, but $((B, B'), (N, D'), q = 0.1, r = 1)$ is still a perfect Bayesian equilibrium. Therefore only this pooling equilibrium is rational.

The idea of Intuitive Criterion is closely related to the equilibrium refinement notion of *strategic stability* advocated by Kohlberg and Mertens [6]. Strategic stability imposes extra-rationality conditions on Nash equilibria to refine among them. One of the conditions is that an equilibrium stays as an equilibrium after deleting weakly dominated strategies (this was already advocated at the early stages of game theory by Luce and Raiffa [8]). To interpret the Beer-Quiche game as a three-player game, the beer action by the weak-type player and the quiche action by the strong-type player are strictly dominated, because regardless of the receiver's action, the favorite breakfast gives a strictly greater payoff than the other breakfast. Therefore, an equilibrium which holds after removing these actions should be selected.

Problems

7.1 Consider a signaling game between a student and a potential employer R, as depicted in Fig. 7.9. By answering the following questions, find a separating equilibrium such that the High-type student S_H chooses Eduction and the Low-type student S_L chooses Not'. In the payoff vectors, the first coordinate is the student's payoff.
(a) When the student takes the separating strategy such that the High type chooses Education and the Low type chooses Not', compute the belief probabilities q and r which R assigns to the High type, after observing no education and education respectively.
(b) Derive an optimal behavioral strategy for player R to the beliefs computed in (a). Explain why it is optimal.

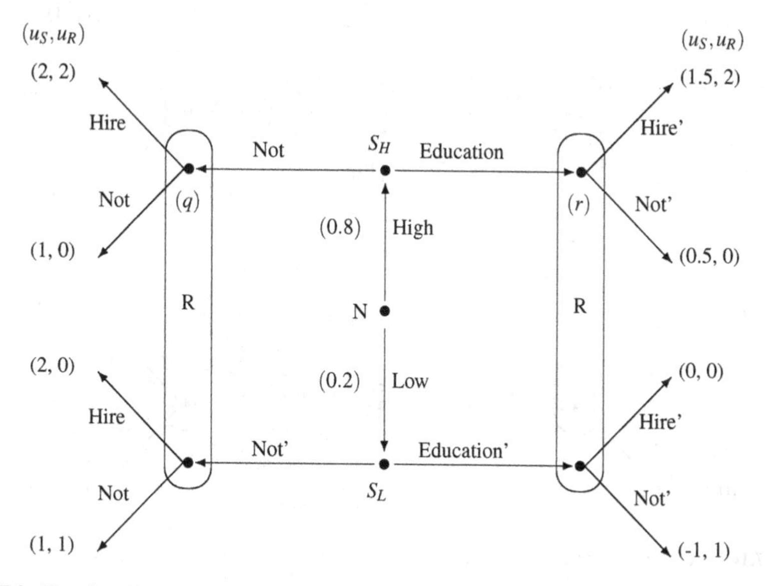

Fig. 7.9 The signaling game for Problem 7.1

(c) Given R's optimal behavioral strategy derived in (b), explain that the student's separating strategy is a best response.

7.2 Reconsider the melon game in Problem 6.3. The game is an incomplete information game between a fruit seller S and a customer C. The seller has a very expensive melon, but the customer does not know its quality. Nature chooses the quality, Ripe or Unripe, with probability 0.5 each, at the beginning of the game. The seller learns Nature's choice (i.e., the melon's quality) and decides what to do.

This time, we assume that the fruit seller S can offer a sample tasting of a similar melon. After S's decision of whether to offer a tasting or not, the customer decides whether to buy or not. The seller incurs some cost to offer the tasting because he must sacrifice a similar melon. Note also that the customer is not choosing the melon she tastes, and hence the tasting is only a signal.

The customer's payoff function is the same as in Problem 6.3, that is, it only depends on the quality of the expensive melon. The seller's payoff is the same as in Problem 6.3 if he does not offer a tasting. If the seller offers a tasting, the cost -1 is added to his payoff if his type is Ripe, and -2 is added if his type is Unripe (because he must find a better melon to sacrifice). The extensive-form game is illustrated in Fig. 7.10, where the first coordinate of each payoff vector is the seller's payoff.

(a) Is there a perfect Bayesian equilibrium in which Tasting is offered when the melon is Ripe and not offered when it is Unripe? If so, write down such an equilibrium (at least one), and explain why it is a perfect Bayesian equilibrium. If not, explain why there is no such equilibrium.

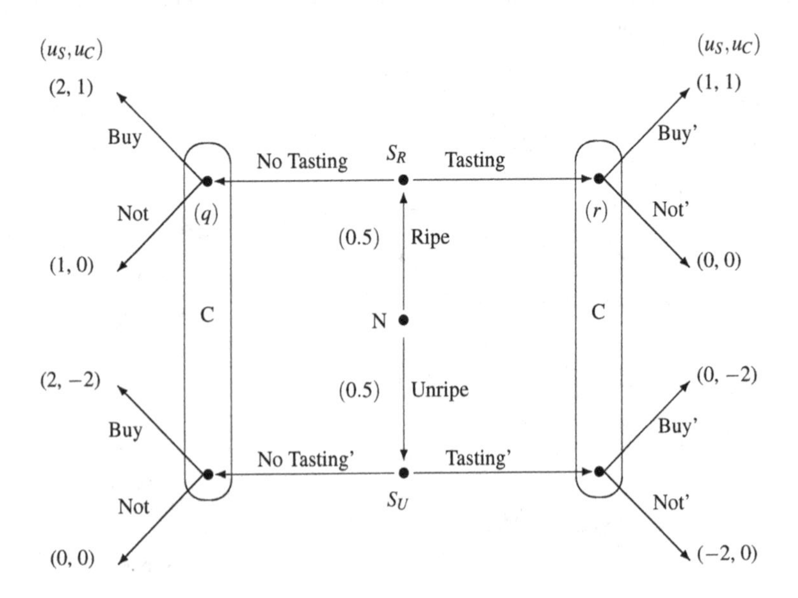

Fig. 7.10 Melon game for Problem 7.2

(b) Is there a perfect Bayesian equilibrium in which Tasting is not offered, regardless of the quality of the melon? If so, write down such an equilibrium (at least one), and explain why it is a perfect Bayesian equilibrium. If not, explain why there is no such equilibrium.
(c) From the above analysis and the analysis in Problem 6.3, derive some economic implications.

7.3 Consider a game between the Princess of the Sea and a fisherman called Urashima. When Urashima leaves the Princess' castle in the sea to go home, she is either Upset or still in Love, but Urashima does not know this. Hence, consider a Bayesian model such that Nature chooses her type, P_U or P_L. The common prior of the Upset type is 0.6. The princess learns her type and chooses whether to give a gift box (action Give) to Urashima or not (action Not Give).

The gift box from the upset princess is cursed with bad magic, and if Urashima opens it, his payoff is -5, but if he does not open it, his payoff is 0. The gift box from the princess who is still in love is blessed with good magic, and if Urashima opens it, his payoff is 10, but if he does not open it, his payoff is 0. Either type of princess gets a payoff of 5 if Urashima opens her gift box, and receives a payoff of -1 if he does not. If the princess decides not to give a gift, the game ends there, and regardless of the type of the princess, both players receive 0. This game is depicted in Fig. 7.11, where the first coordinate of each payoff vector is the princess' payoff.
(a) If both type of princesses give a gift box, should Urashima open it? Explain your answer.

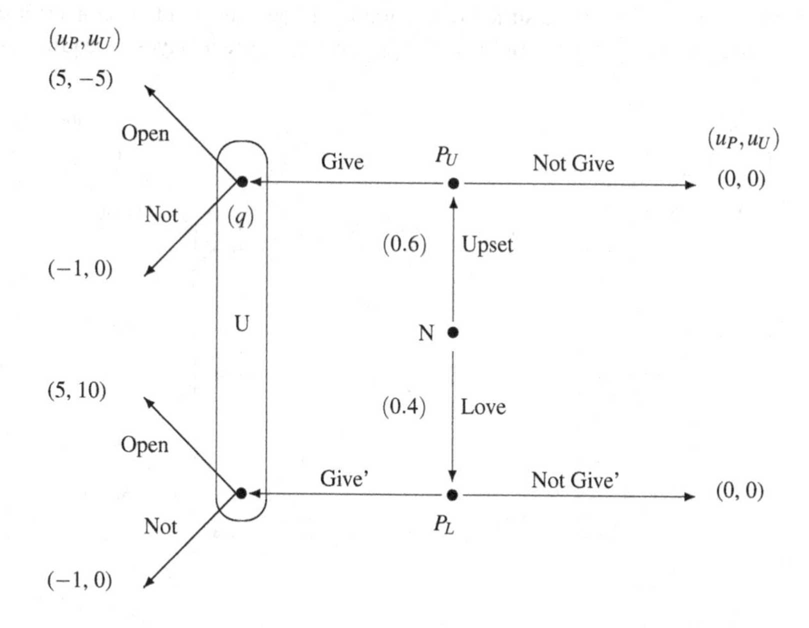

Fig. 7.11 Urashima game

(b) Given your answer in (a), is there a perfect Bayesian equilibrium in which both type of princesses give a gift box? If so, write down the (behavioral) strategy combination and Urashima's belief q, and explain why it is a perfect Bayesian equilibrium. If not, explain why there is no such equilibrium.

7.4 Soon it is time for a boss B to evaluate a worker W to decide whether to promote him or not. However, the boss does not know W's abilities well. The boss's payoff only depends on the worker's abilities (Good or Bad). If the boss promotes the worker, the boss's payoff is 4 if the worker is Good type, but it is -5 if the worker is Bad type. If the boss makes the worker stay in the current position (action Stay), the boss's payoff is 0 regardless of the worker's type. The common prior probability of Good type is 0.5.

(a) If the boss decides with only the prior information, would the boss promote this worker, to maximize his expected payoff? Explain your answer.

The worker W starts to think about whether or not to earn some certificate, before the boss decides on the promotion. This certificate itself does not show or improve the worker's abilities in the current job, and hence the boss's payoff is still dependent only on the worker's unobservable and fixed abilities, Good or Bad. However, to take the certificate (action Take), the worker W must study after work hours, and the cost to study may depend on his abilities. Specifically, let $c > 0$ be the cost to take the certificate if the worker is Good type, and it is $C > 0$ if he is Bad type.

The payoff from promotion is 5 for both types of worker, and if he is made to stay in the current position, the payoff is 0 for both types. The worker W maximizes the net payoff; the payoff from the promotion/staying minus the cost of study. The above incomplete information game structure is common knowledge, and we convert it to a Bayesian model in Fig. 7.12. Below, all equilibria are perfect Bayesian equilibria.

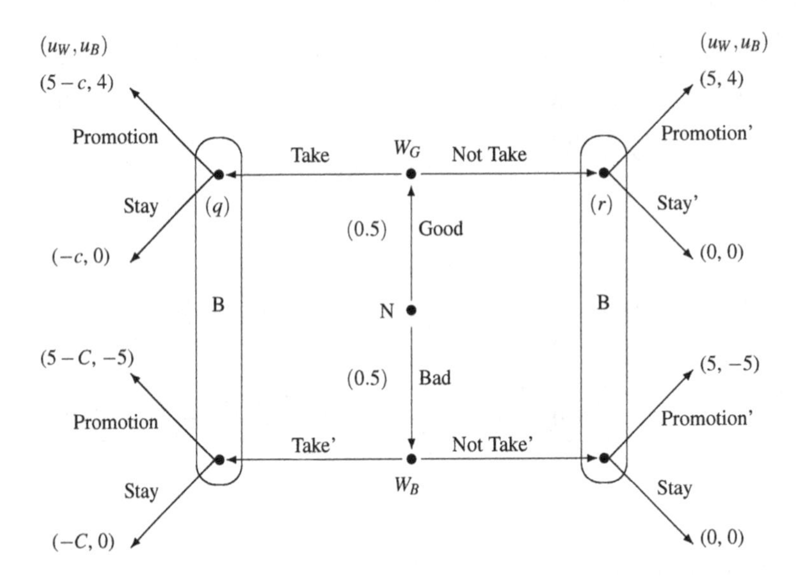

Fig. 7.12 Promotion game for Problem 7.4

Table 7.1 Simultaneous-move game for Problem 7.5

P1 \ P2	A	B
A	x, x	$x, 0$
B	$0, x$	$2, 2$

(b) Consider a separating equilibrium such that only the Good type worker takes the certificate. Find the ranges of c and C that support this equilibrium.
(c) Consider a class of pooling equilibria such that both types of worker do not take the certificate. Focus on the boss's beliefs such that $q > 5/9$, where q is the probability that the worker is Good if he takes the certificate. Find the ranges of c and C such that it is a best response for both types of W to not take the certificate.
(d) Using the above analysis, discuss the effect of the difficulty of getting a certificate on its screening function.

7.5 Consider a two-player incomplete information game as follows. The players are called P1 and P2. They play the simultaneous-move game in Table 7.1, but do not know the value of x.
(a) Assume that x is either -1 or 5. The common prior probability is 0.5 for each possible value. Write down the matrix representation of the Bayesian game in ex-ante expected payoffs. Note that both players do not know the payoff function.
(b) Find all Bayesian Nash equilibria of the game (a) in mixed strategies. Show that all other strategy combinations are not an equilibrium.

Now the situation changes so that P1 knows the value of x. P2 only knows that x takes a value of either -1 or 5 with the prior probability of 0.5 each. This new information structure for the two players is common knowledge.
(c) Draw the game tree for the Bayesian game of the incomplete information model as follows. Nature chooses the value of x (-1 or 5) at the beginning, with probability 0.5 each. P1 learns Nature's choice and chooses between actions A and B. P2 observes only P1's actions and chooses between actions A and B. Attach an appropriate payoff vector at each terminal node.
(d) Does the extensive-form game of (c) have a pooling (perfect Bayesian) equilibrium where P1 uses a pure strategy? If yes, write down at least one equilibrium, and explain why it is a perfect Bayesian equilibrium. If no, explain why there is no such equilibrium.

References

1. Banks J, Sobel J (1987) Equilibrium selection in signaling games. Econometrica 55(3):647–661
2. Cho I-K, Kreps D (1987) Signaling games and stable equilibria. Q J Econ 102(2):179–221
3. Crawford V, Sobel J (1982) Strategic information transmission. Econometrica 50(6):1431–1451
4. Farrell J, Rabin M (1996) Cheap talk. J Econ Perspect 10(3):103–118
5. Fudenberg D, Tirole J (1991) Perfect bayesian equilibrium and sequential equilibrium. J Econ Theor 53(2):236–260

6. Kohlberg E, Mertens J-F (1986) On the strategic stability of equilibria. Econometrica 54(5):1003–1037
7. Kreps D, Wilson R (1982) Reputation and imperfect information. J Econ Theor 27(2):253–279
8. Luce D, Raiffa H (1957) Games and decisions. Wiley, New York
9. Schulteis T, Perea A, Peters H, Vermeulen D (2007) Revision of conjectures about the opponent's utilities in signaling games. Econ Theor 30(2):373–384

Chapter 8
Equilibrium Refinements**

8.1 Further Stability

Non-cooperative game theory, with the fundamental equilibrium concept of Nash equilibrium, can give at least one equilibrium (a prediction) for any finite normal-form game and many infinite games as well. However, we are far from the completion of the theory, because not all games come with a unique Nash equilibrium. It is often the case that a game has many Nash equilibria or subgame perfect equilibria. This means that the prediction by non-cooperative game theory is too broad, and it would be better to refine the equilibria by requiring further stability.

In physics, there is a stability concept such that, even if an object moves slightly due to a small environmental change, it should return to the original point. For example, consider the two resting balls in Fig. 8.1. Right now they are both stable, because they are not in motion. What happens if a small earthquake occurs? The ball on top of the hill will fall, and it will not return to the top. By contrast, the ball resting in the valley will move around a little during the earthquake, but it will return to the same place. Therefore, the rest point at the bottom is more stable than the rest point at the top under small *perturbations*.

We can think of a similar stability in games. Players may **tremble** with a small probability, causing them to choose a strategy or an action that they are not supposed to choose. The cause may be the relevant player's computational mistakes or physical mistakes, as well as the other players' "worry". When considering opponents'

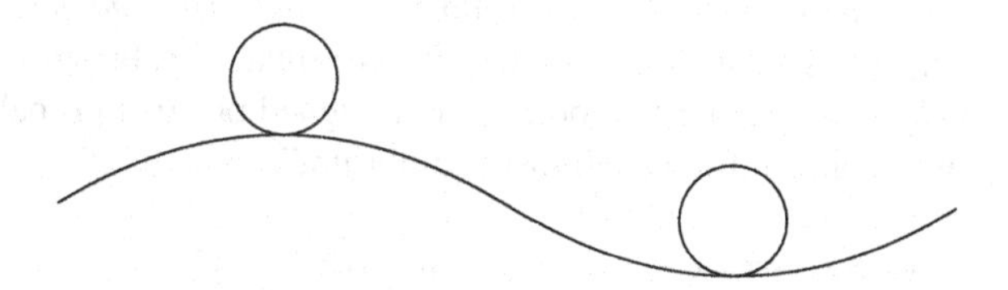

Fig. 8.1 Rest Points at a *Top* and at a *Bottom*

T. Fujiwara-Greve, *Non-Cooperative Game Theory*, Monographs in Mathematical Economics 1, DOI 10.1007/978-4-431-55645-9_8

trembling possibilities, a player must maximize the expected payoff of strategies, taking into account the opponents' possible strategy/action perturbations. Perturbations can be of various scales (a huge earthquake might move even the ball in a valley to a different valley), and depending on the scale we consider, the stability requirement differs. A weak requirement is to allow only very small trembles in opponents' strategies, which tests local stability. Such a weak requirement of further stability can preserve the existence of an equilibrium, but still select among Nash (or subgame perfect) equilibria. In this chapter, we introduce *equilibrium refinements* based on this stability under small perturbations. The basic idea is developed by Selten [42].

8.2 Trembling-Hand Perfect Equilibrium

Consider the normal-form game in Table 8.1. There are two Nash equilibria, (A, A) and (B, B).

The Nash equilibrium (B, B) is not robust against even a small mistake or *perturbation* in the opponents' strategy choice. If the opponent chooses strategy A with a positive probability, no matter how small it is, the pure-strategy B is no longer a best response to the mixed strategy. Generalizing this logic, Selten [42] defined an equilibrium under small perturbations of players' strategies.

Definition 8.1 A combination of mixed strategies $(\sigma_1^*, \ldots, \sigma_n^*) \in \Delta(S_1) \times \cdots \times \Delta(S_n)$ is a *trembling-hand perfect equilibrium* if, for any player $i \in \{1, 2, \ldots, n\}$, there exists a sequence of completely mixed strategies $\{\sigma_i^k\}_{k=1}^{\infty}$ such that:

(a) $\lim_{k \to \infty} \sigma_i^k = \sigma_i^*$; and
(b) for any $k = 1, 2, \ldots,$ σ_i^* is a best response to $\sigma_{-i}^k = (\sigma_1^k, \ldots, \sigma_{i-1}^k, \sigma_{i+1}^k, \ldots, \sigma_n^k)$.

The description "trembling-hand" is added to distinguish this concept from the "subgame" perfect equilibrium. This stability concept requires for an equilibrium to be robust against **at least one** diminishing sequence of strategy trembles. (It is too much to require stability against **all** sequences of diminishing trembles. Such a strategy combination may not exist. See Kohlberg and Mertens [29].) Note also that a trembling-hand perfect equilibrium need not be a completely mixed strategy combination. Only "supporting" perturbed strategy combinations must be. For the game of Table 8.1, the Nash equilibrium (B, B) is **not** trembling-hand perfect, because B is not a best response to **any** completely mixed strategy by the opponent.

There is an equivalent definition for the trembling-hand perfect equilibrium, as the limit of equilibria in a sequence of *perturbed games*, in which all player's pure strategies must have a (small) positive probability. In perturbed games, even strategies which are not best responses are assigned positive probabilities. Hence, we need a new concept for an "almost equilibrium".

Table 8.1 A game with multiple Nash equilibria

P1\P2	A	B
A	1, 1	0, 0
B	0, 0	0, 0

Definition 8.2 For any $\epsilon > 0$, a combination of mixed strategies $\sigma^{\epsilon} = (\sigma_1^{\epsilon}, \ldots, \sigma_n^{\epsilon}) \in \Delta(S_1) \times \cdots \times \Delta(S_n)$ is an "ϵ-constrained equilibrium" if there exists $\{\epsilon(s_i)\}_{s_i \in S_i, i \in \{1,2,\ldots,n\}}$ such that $0 < \epsilon(s_i) < \epsilon$ for all $s_i \in S_i$ and all $i \in \{1, 2, \ldots, n\}$, and σ_i^{ϵ} maximizes $Eu_i(x_i, \sigma_{-i}^{\epsilon})$ subject to $x_i(s_i) \geqq \epsilon(s_i)$ for all $s_i \in S_i$.
A combination of mixed strategies $(\sigma_1^*, \ldots, \sigma_n^*) \in \Delta(S_1) \times \cdots \times \Delta(S_n)$ is a *trembling-hand perfect equilibrium* if there exist a sequence of positive real numbers $\{\epsilon_k\}_{k=1}^{\infty}$ ($\epsilon_k > 0$ for all k) such that $\lim_{k\to\infty} \epsilon_k = 0$ and a corresponding sequence of ϵ_k-constrained equilibria $\{\sigma^{\epsilon_k}\}_{k=1}^{\infty}$ such that

$$\lim_{k\to\infty} \sigma^{\epsilon_k} = \sigma^*.$$

To see that Definitions 8.1 and 8.2 are equivalent, clearly a sequence of ϵ-constrained equilibria can be viewed as a "supporting" sequence in Definition 8.1. For the converse, let $\{\sigma_i^k\}_{k=1}^{\infty}$ be the "supporting" sequence of completely mixed strategies for each $i \in \{1, 2, \ldots, n\}$. For each $k = 1, 2, \ldots$, there exists $m_k > 0$ such that $\min_{i\in\{1,2,\ldots,n\}, s_i \in S_i} \sigma_i^k(s_i) > \frac{1}{k+m_k}$. This $\frac{1}{k+m_k}$ serves as $\epsilon(s_i)$ for all k and s_i. Define a sequence of positive real numbers $\{\epsilon_k\}_{k=1}^{\infty}$ by $\epsilon_k = \frac{1.1}{k+m_k}$ (so that $\epsilon_k > \epsilon(s_i)$) and a sequence of strategy combinations $\{\sigma^{\epsilon_k}\}_{k=1}^{\infty}$ by

$$\forall i \in \{1, 2, \ldots, n\},\ \forall s_i \in S_i,\ \ \sigma_i^{\epsilon_k}(s_i) = \begin{cases} \frac{1}{k+m_k} & \text{if } s_i \notin supp(\sigma_i^*) \\ \sigma_i^k(s_i) & \text{if } s_i \in supp(\sigma_i^*). \end{cases}$$

Consider the maximization problem $Eu_i(x_i, \sigma_{-i}^{\epsilon_k})$ subject to $x_i(s_i) \geqq \frac{1}{k+m_k}$. There exists $K < \infty$ such that for any $k \geqq K$ and any $i \in \{1, 2, \ldots, n\}$, σ_i^* is **a best response** to $\sigma_{-i}^{\epsilon_k}$, so that any pure strategy s_i not in the support of σ_i^* should have the minimum probability $\frac{1}{k+m_k}$ and any s_i in the support of σ_i^* can have the probability $\sigma_i^k(s_i)$. Over the subsequence $\{K, K+1, \ldots\}$, $\lim_{k'\to\infty} \epsilon_{k'} = 0$ and $\lim_{k'\to\infty} \sigma^{\epsilon_{k'}} = \sigma^*$ hold as well.

In Definition 8.2, the game itself is perturbed, while in Definition 8.1, the players' strategies are perturbed. Nonetheless, the two definitions are mathematically equivalent. The idea of the perturbed game is utilized in the proof of the existence Theorem 8.1 as well.

We have yet another equivalent definition for the trembling-hand perfect equilibrium. This version will be useful to compare the trembling-hand perfect equilibrium with the proper equilibrium in Sect. 8.3. It is based on another (and better known) "almost equilibrium" concept called ϵ-perfect equilibrium.

Definition 8.3 For any $\epsilon > 0$, a combination of mixed strategies $\sigma^{\epsilon} = (\sigma_1^{\epsilon}, \ldots, \sigma_n^{\epsilon}) \in \Delta(S_1) \times \cdots \times \Delta(S_n)$ is an *ϵ-perfect equilibrium* if, for any player $i \in \{1, 2, \ldots, n\}$, σ_i^{ϵ} is a completely mixed strategy and, for any $s_i \in S_i$,

$$s_i \notin BR_i(\sigma_{-i}^{\epsilon}) \Rightarrow \sigma_i^{\epsilon}(s_i) \leqq \epsilon.$$

Definition 8.4 A combination of mixed strategies $(\sigma_1^*, \ldots, \sigma_n^*) \in \Delta(S_1) \times \cdots \times \Delta(S_n)$ is a *trembling-hand perfect equilibrium* if there exist a sequence of positive real numbers $\{\epsilon_k\}_{k=1}^{\infty}$ ($\epsilon_k > 0$ for all k) such that $\lim_{k\to\infty} \epsilon_k = 0$ and a corresponding sequence of ϵ_k-perfect equilibria $\{\sigma^{\epsilon_k}\}_{k=1}^{\infty}$ such that

$$\lim_{k\to\infty} \sigma^{\epsilon_k} = \sigma^*.$$

For example, the Nash equilibrium (A, A) in Table 8.1 is a trembling-hand perfect equilibrium by the following sequence, where the first coordinate is the probability of strategy A for each player.

$$\sigma_1^{\epsilon_k} = (1 - \frac{1}{2}\epsilon_k, \frac{1}{2}\epsilon_k), \quad \sigma_2^{\epsilon_k} = (1 - \epsilon_k, \epsilon_k),$$

for some sequence of $\{\epsilon_k\}$ such that $\epsilon_k \in (0, 1)$ and $\lim_{k\to\infty} \epsilon_k = 0$. Given that the opponent uses a completely mixed strategy, strategy B is not a best response, and it has a probability of no more than ϵ_k for both players. Problem 8.1 asks the reader to formally prove that Definitions 8.1 and 8.4 are equivalent.

We now turn to the properties of trembling-hand perfect equilibria. Although the above definitions do not explicitly require that a trembling-hand perfect equilibrium be a Nash equilibrium, it is easily implied.

Proposition 8.1 *Any trembling-hand perfect equilibrium is a Nash equilibrium.*

Proof Let σ^* be a trembling-hand perfect equilibrium. Then there exists a sequence of completely mixed strategy combinations $\{\sigma^k\}_{k=1}^{\infty}$ such that for any $i \in \{1, 2, \ldots, n\}$ and any $k = 1, 2, \ldots$,

$$Eu_i(\sigma_i^*, \sigma_{-i}^k) \geqq Eu_i(\sigma_i, \sigma_{-i}^k), \quad \forall \sigma_i \in \Delta(S_i).$$

By the continuity of the expected payoff function Eu_i, at the limit as $k \to \infty$, we also have

$$Eu_i(\sigma_i^*, \sigma_{-i}^*) \geqq Eu_i(\sigma_i, \sigma_{-i}^*), \quad \forall \sigma_i \in \Delta(S_i).$$

This is precisely the definition of a Nash equilibrium. □

Clearly, the converse is not always true, as the example in Table 8.1 shows. That is, the stability concept of trembling-hand perfect equilibrium selects among Nash equilibria.

Trembling-hand perfection only requires a local stability, and thus it exists easily. In fact, its existence is guaranteed for any finite normal-form game. The proof utilizes the idea of the second definition of the equilibrium.

Theorem 8.1 *For any finite normal-form game, there is a trembling-hand perfect equilibrium.*

Proof For any $k = 1, 2, \ldots$ and any player i, consider a (minimum probability) function $\eta_i^k : S_i \to (0, 1)$ such that $\sum_{s_i \in S_i} \eta_i^k(s_i) < 1$ and, for any $s_i \in S_i$, $\lim_{k\to\infty} \eta_i^k(s_i) = 0$. For each $k = 1, 2, \ldots$, define the perturbed game $(G; \eta_1^k, \eta_2^k, \ldots, \eta_n^k)$ by restricting each player's set of strategies as the set of completely mixed strategies with each original pure-strategy s_i having at least probability $\eta_i^k(s_i)$. (The set of players and their payoff functions are the same as those of G.)

For each $k = 1, 2, \ldots$, the perturbed game possesses a Nash equilibrium $\sigma^{*(k)}$, and they constitute a sequence in a compact set Σ. Therefore, there is a convergent subsequence. Let the subsequence be $\{\sigma^{*(k')}\}_{k'=1}^{\infty}$, and let σ^* be the limit of the subsequence. To show that σ^* is a trembling-hand perfect equilibrium, it suffices to prove that for each player $i \in \{1, 2, \ldots, n\}$, and any $k' = 1, 2, \ldots$, σ_i^* is a best response to $\sigma_{-i}^{*(k')}$.

Notice that, for any $k' = 1, 2, \ldots$, if a pure-strategy y is not a best response to $\sigma_{-i}^{*(k')}$, then it should have the minimum probability, $\sigma_i^{*(k')}(y) = \eta_i^{k'}(y)$. If not, by slightly reducing the probability of the pure-strategy y from $\sigma_i^{*(k')}(y)$, player i can increase her/his payoff. This contradicts the definition of $\sigma^{*(k')}$ as a Nash equilibrium of the perturbed game.

Recall that as $k' \to \infty$, $\eta_i^{k'}(s_i) \to 0$ for all $s_i \in S_i$. Therefore, if we define

$$\epsilon(k', i) := \max_{s_i \in S_i} Eu_i(s_i, \sigma_{-i}^{*(k')}) - Eu_i(\sigma_i^{*(k')}, \sigma_{-i}^{*(k')}),$$

then

$$\lim_{k'\to\infty} \epsilon(k', i) = 0.$$

This is equivalent to

$$Eu_i(\sigma_i^*, \sigma_{-i}^{*(k')}) \geqq Eu_i(x, \sigma_{-i}^{*(k')}), \ \ \forall x \in S_i.$$

By the fact proved in Problem 3.1, σ_i^* is a best response in mixed-strategies to $\sigma_{-i}^{*(k')}$. □

As we discussed in Sect. 7.4, a desirable property of a stable strategy combination according to Kohlberg and Mertens [29] is to not include strategies which are weakly dominated. Trembling-hand perfection succeeds in this to some extent.

Definition 8.5 Player i's mixed strategy $\sigma_i \in \Delta(S_i)$ is *weakly dominated* if there exists another mixed strategy $\sigma_i' \in \Delta(S_i)$ such that

$$\forall\, s_{-i} \in S_{-i},\ Eu_i(\sigma_i, s_{-i}) \leqq Eu_i(\sigma_i', s_{-i}), \text{ and}$$
$$\exists\, s_{-i}' \in S_{-i};\ Eu_i(\sigma_i, s_{-i}') < Eu_i(\sigma_i', s_{-i}').$$

Proposition 8.2 *For two-person finite normal-form games, a Nash equilibrium* $\sigma^* \in \Delta(S_1) \times \Delta(S_2)$ *is a trembling-hand perfect equilibrium if and only if for each player* $i \in \{1, 2\}$, σ_i^* *is not a weakly dominated strategy.*

Some proofs of this Proposition use geometry (e.g., Osborne and Rubinstein [40]), but we construct a zero-sum game, like van Damme [9].

Proof of Proposition 8.2 Necessity. Let σ^* be a trembling-hand perfect equilibrium. Suppose that there is a player i such that σ_i^* is weakly dominated. We show that σ_i^* is never a best response to any completely mixed strategy combination by others, let alone to σ_{-i}^*. Since σ_i^* is weakly dominated, there exists $\sigma_i' \in \Delta(S_i)$ such that

$$\forall\, s_{-i} \in S_{-i},\ \ Eu_i(\sigma_i^*, s_{-i}) \leqq Eu_i(\sigma_i', s_{-i}),\ \text{ and}$$
$$\exists\, s_{-i}' \in S_{-i};\ \ Eu_i(\sigma_i^*, s_{-i}') < Eu_i(\sigma_i', s_{-i}').$$

For any completely mixed strategy combination σ_{-i} by the opponent(s), s_{-i}' has a positive probability. By decomposition,

$$\begin{aligned} Eu_i(\sigma_i^*, \sigma_{-i}) &= \sum_{s_{-i} \neq s_{-i}'} \sigma_{-i}(s_{-i}) Eu_i(\sigma_i^*, s_{-i}) + \sigma_{-i}(s_{-i}') Eu_i(\sigma_i^*, s_{-i}') \\ &< \sum_{s_{-i} \neq s_{-i}'} \sigma_{-i}(s_{-i}) Eu_i(\sigma_i', s_{-i}) + \sigma_{-i}(s_{-i}') Eu_i(\sigma_i', s_{-i}') \\ &= Eu_i(\sigma_i', \sigma_{-i}). \end{aligned}$$

Therefore σ_i^* cannot be a best response to the arbitrarily chosen σ_{-i} (at least σ_i' is better), let alone to σ_{-i}^*. This is a contradiction. (Note that we have not used the assumption of two-person games here.)

Sufficiency. Let $G = (\{1, 2\}, S_1, S_2, u_1, u_2)$ be a two-person finite normal-form game. Fix an arbitrary $i \in \{1, 2\}$. We will construct a convergent sequence of completely mixed strategies with the limit σ_i^*. In the same way as Lemma 3.1, define a new two-person zero-sum game $\overline{G} = (\{1, 2\}, S_1, S_2, \overline{u}_1, \overline{u}_2)$ as follows.

$$\forall (s_i, s_j) \in S_i \times S_j,\ \ \overline{u}_i(s_i, s_j) = u_i(s_i, s_j) - Eu_i(\sigma_i^*, s_j),$$
$$\forall (s_i, s_j) \in S_i \times S_j,\ \ \overline{u}_j(s_i, s_j) = -\overline{u}_i(s_i, s_j).$$

Take a Nash equilibrium $(\overline{\sigma}_i, \overline{\sigma}_j) \in \Delta(S_i) \times \Delta(S_j)$ of $\overline{G}$. (It exists by the Existence Theorem 3.1.) By the definition of a Nash equilibrium and $\overline{u}_i$, we have

$$E\overline{u}_i(\overline{\sigma}_i, \overline{\sigma}_j) \geqq E\overline{u}_i(\sigma_i^*, \overline{\sigma}_j) = 0.$$

We separate this into two cases.

Case 1: When $E\overline{u}_i(\overline{\sigma}_i, \overline{\sigma}_j) > 0$.

Because $(\overline{\sigma}_i, \overline{\sigma}_j)$ is a Nash equilibrium of the game $\overline{G}$ (in mixed strategies),

$$E\overline{u}_j(\overline{\sigma}_i, \overline{\sigma}_j) \geqq E\overline{u}_j(\overline{\sigma}_i, \sigma_j), \quad \forall \sigma_j \in \Delta(S_j).$$

Using the definition $\overline{u}_j = -\overline{u}_i$ and this case's assumption, we can rewrite the above inequality as

$$0 < E\overline{u}_i(\overline{\sigma}_i, \overline{\sigma}_j) \leqq E\overline{u}_i(\overline{\sigma}_i, \sigma_j), \quad \forall \sigma_j \in \Delta(S_j).$$

Converting $E\overline{u}_i(\overline{\sigma}_i, \sigma_j) > 0$ to a statement for u_i, we have

$$Eu_i(\overline{\sigma}_i, \sigma_j) > Eu_i(\sigma_i^*, \sigma_j), \quad \forall \sigma_j \in \Delta(S_j).$$

This means that σ_i^* is strictly dominated by $\overline{\sigma}_i$, so this case is impossible.

Case 2: When $E\overline{u}_i(\overline{\sigma}_i, \overline{\sigma}_j) = 0$.

Again, because $(\overline{\sigma}_i, \overline{\sigma}_j)$ is a Nash equilibrium of the game $\overline{G}$,

$$0 = E\overline{u}_i(\overline{\sigma}_i, \overline{\sigma}_j) \geqq E\overline{u}_i(\sigma_i, \overline{\sigma}_j), \quad \forall \sigma_i \in \Delta(S_i).$$

Converting $0 \geqq E\overline{u}_i(\sigma_i, \overline{\sigma}_j)$ into a statement for u_i, we have

$$Eu_i(\sigma_i^*, \overline{\sigma}_j) \geqq Eu_i(\sigma_i, \overline{\sigma}_j), \quad \forall \sigma_i \in \Delta(S_i).$$

Hence σ_i^* is a best response to $\overline{\sigma}_j$. We want to show that $\overline{\sigma}_j$ is a completely mixed strategy.

Suppose, to the contrary, that $\overline{\sigma}_j$ is not a completely mixed strategy. By the assumption of Case 2, $0 = E\overline{u}_i(\overline{\sigma}_i, \overline{\sigma}_j) = -E\overline{u}_j(\overline{\sigma}_i, \overline{\sigma}_j)$, therefore $E\overline{u}_j(\overline{\sigma}_i, \overline{\sigma}_j) = 0$. Since $(\overline{\sigma}_i, \overline{\sigma}_j)$ is a Nash equilibrium of $\overline{G}$, by Proposition 3.3,

$$\forall s_j \in supp(\overline{\sigma}_j), \quad E\overline{u}_j(\overline{\sigma}_i, s_j) = 0, \tag{8.1}$$

and

$$\forall s_j' \notin supp(\overline{\sigma}_j), \quad E\overline{u}_j(\overline{\sigma}_i, s_j') < 0. \tag{8.2}$$

By (8.1) and the definition of $\overline{u}_j = -\overline{u}_i$, we also have

$$\forall s_j \in supp(\overline{\sigma}_j), \quad E\overline{u}_i(\overline{\sigma}_i, s_j) = 0,$$

and this implies that

$$\forall s_j \in supp(\overline{\sigma}_j), \quad Eu_i(\overline{\sigma}_i, s_j) = Eu_i(\sigma_i^*, s_j).$$

Moreover, for any probability distribution over $supp(\overline{\sigma}_j)$, $\overline{\sigma}_i$ and σ_i^* give the same expected payoff.

Table 8.2 A game with a weakly dominated strategy

P1\P2	X	Y
A	0, 1	0, 1
B	−1, 2	1, 0
C	−1, 2	2, 3

From (8.2), we have

$$\forall s'_j \notin supp(\overline{\sigma}_j), \quad E\overline{u}_j(\overline{\sigma}_i, s'_j) < 0 = E\overline{u}_j(\overline{\sigma}_i, \overline{\sigma}_j).$$

Hence,

$$\forall s'_j \notin supp(\overline{\sigma}_j), \quad E\overline{u}_i(\overline{\sigma}_i, s'_j) > E\overline{u}_i(\overline{\sigma}_i, \overline{\sigma}_j) = 0.$$

In terms of Eu_i, this is

$$\forall s'_j \notin supp(\overline{\sigma}_j), \quad Eu_i(\overline{\sigma}_i, s'_j) > Eu_i(\sigma_i^*, s_j),$$

and by the assumption that $\overline{\sigma}_j$ is not a completely mixed strategy, there is a pure strategy $s'_j \notin supp(\overline{\sigma}_j)$. Combined with the fact that for any pure strategy in $supp(\overline{\sigma}_j)$, $\overline{\sigma}_i$ gives the same expected payoff as that of σ_i^*, we established that σ_i^* is weakly dominated by $\overline{\sigma}_i$. This is a contradiction.

Therefore, $\overline{\sigma}_j$ must be a completely mixed strategy, and $\sigma_i^* \in BR_i(\overline{\sigma}_j)$. Let us define a sequence of completely mixed strategies by $\sigma_j^k := (1 - \frac{1}{k+1})\sigma_j^* + \frac{1}{k+1}\overline{\sigma}_j$ $(k = 1, 2, \ldots)$. Then σ_i^* is a best response to σ_j^k for any k, and $\lim_{k\to\infty} \sigma_j^k = \sigma_j^*$. □

As we noted in the proof, the "only if" part holds for any finite normal-form game. That is, in any trembling-hand perfect equilibrium, no player uses a weakly dominated strategy. The "if" part does not hold for games with three or more players. A counter example is given in Problem 8.5. Moreover, even in two-person games, a strategy which is eliminated through a process of iterative elimination of weakly dominated strategies can be a part of a trembling-hand perfect equilibrium.

In the game in Table 8.2, if we eliminate the (weakly) dominated strategies in the order of B, X, and A, then the resulting "equilibrium" is (C, Y). However, the combination (A, X) is also a trembling-hand perfect equilibrium. (Problem 8.2.)

8.3 Proper Equilibrium

To establish a trembling-hand perfect equilibrium, we only need to find any sequence of perturbed strategy combinations that satisfies the two conditions of Definition 8.1. However, if we imagine that the perturbation is due to mistakes by rational players, some strategies are more likely to be chosen by mistake than others. Meyerson [38] used this idea to define a stronger stability concept by putting a restriction on a

Table 8.3 Different mistake probabilities are plausible

P1\P2	X	Y
A	2, 2	2, 2
B	4, 1	1, 0
C	0, 0	0, 1

sequence of perturbed strategies such that strategies with higher payoffs have higher probabilities to be trembled to. We explain how such a restriction makes a difference by the game in Table 8.3.

In Table 8.3, (A, Y) is a trembling-hand perfect equilibrium. For example, take any $\gamma \in (0, 1/3)$ and let $\epsilon_k = \gamma^k$ for each $k = 1, 2, \ldots$, so that $\lim_{k\to\infty} \epsilon_k = 0$. Take a sequence

$$\sigma_1^{\epsilon_k} = (1 - \frac{3}{2}\epsilon_k, \frac{1}{2}\epsilon_k, \epsilon_k),\ \sigma_2^{\epsilon_k} = (\epsilon_k, 1 - \epsilon_k).$$

Then $\gamma < 1/3$ implies that for any $k = 1, 2, \ldots$, $Eu_1(A, \sigma_2^{\epsilon_k}) = 2 > 4\gamma^k + 1 - \gamma^k = Eu_1(B, \sigma_2^{\epsilon_k})$. Notice that, to make Y a best response, we need to construct player 1's perturbed strategy sequence in which strategy C is chosen with a higher probability than that of strategy B. However, C is strictly dominated by B, and if player 1 is (almost) rational, it is more plausible that her mistakes should assign a higher probability to B than to C. The concept of the *proper equilibrium* formalizes the requirement that better strategies are more likely to be chosen.

Definition 8.6 For any $\epsilon > 0$, a combination of mixed strategies $\sigma^\epsilon = (\sigma_1^\epsilon, \ldots, \sigma_n^\epsilon) \in \Delta(S_1) \times \cdots \times \Delta(S_n)$ is an ϵ*-proper equilibrium* if, for any player $i \in \{1, 2, \ldots, n\}$ and any pure strategies $s_i, s_i' \in S_i$,

$$Eu_i(s_i, \sigma_{-i}^\epsilon) < Eu_i(s_i', \sigma_{-i}^\epsilon) \Rightarrow \sigma_i^\epsilon(s_i) \leqq \epsilon \cdot \sigma_i^\epsilon(s_i'). \tag{8.3}$$

Definition 8.7 A combination of mixed strategies $(\sigma_1^*, \ldots, \sigma_n^*) \in \Delta(S_1) \times \cdots \times \Delta(S_n)$ is a *proper equilibrium* if there exist a sequence of positive real numbers $\{\epsilon_k\}_{k=1}^\infty$ ($\epsilon_k > 0$ for all k) such that $\lim_{k\to\infty} \epsilon_k = 0$ and a corresponding sequence of ϵ_k-proper equilibria $\{\sigma^{\epsilon_k}\}_{k=1}^\infty$ such that

$$\lim_{k\to\infty} \sigma^{\epsilon_k} = \sigma^*.$$

For example, (B, X) of Table 8.3 is a proper equilibrium with the following sequence: Take a small $\gamma \in (0, 1)$ such that $1 - \gamma - \gamma^2 - \gamma^3 > 0$ (e.g., $\gamma = 0.1$). For each $k = 1, 2, \ldots$, define $\epsilon_k = \gamma^k$, so that $\lim_{k\to\infty} \epsilon_k = 0$. For each $k = 1, 2, \ldots$, define

$$\sigma_1^{\epsilon_k} = (\epsilon_k^2, 1 - \epsilon_k^2 - \epsilon_k^3, \epsilon_k^3),\ \sigma_2^{\epsilon_k} = (1 - \epsilon_k^2, \epsilon_k^2).$$

(The reader is asked to prove the rest in Problem 8.6.)

A proper equilibrium is defined in the same fashion as the second definition of the trembling-hand perfect equilibrium, Definition 8.4. Because the sequence of perturbed strategy combinations that supports the equilibrium must satisfy an additional property, the set of proper equilibria is a subset of trembling-hand perfect equilibria (which is already a subset of Nash equilibria). Indeed, the set of proper equilibria can be a proper subset of trembling-hand perfect equilibria. In the game of Table 8.3, (A, Y) is not a proper equilibrium. For any sequence such that $\sigma_1^{\epsilon_k}(C) \leqq \epsilon_k \cdot \sigma_1^{\epsilon_k}(B)$, the expected payoffs of the two strategies for player 2 satisfy

$$Eu_2(\sigma_1^{\epsilon_k}, X) = 2\sigma_1^{\epsilon_k}(A) + \sigma_1^{\epsilon_k}(B)$$
$$Eu_2(\sigma_1^{\epsilon_k}, Y) \leqq 2\sigma_1^{\epsilon_k}(A) + \epsilon_k \cdot \sigma_1^{\epsilon_k}(B),$$

so that there exists $K < \infty$ such that $Eu_2(\sigma_1^{\epsilon_k}, Y) < Eu_2(\sigma_1^{\epsilon_k}, X)$ for any $k \geqq K$. Then any sequence such that $\sigma_2^{\epsilon_k}(Y) \leqq \epsilon_k \cdot \sigma_2^{\epsilon_k}(X)$ does not converge to the pure strategy Y. For another example, see Problem 8.7. Formally, we can easily show that the concept of proper equilibrium is stronger than the concept of trembling-hand perfect equilibrium as follows.

Proposition 8.3 *A proper equilibrium is a trembling-hand perfect equilibrium.*

Proof It suffices to prove that, for any $\epsilon > 0$, an ϵ-proper equilibrium is an ϵ-perfect equilibrium. For any $s_i \notin BR_i(\sigma_{-i}^{\epsilon})$, there exists s_i' such that $Eu_i(s_i, \sigma_{-i}^{\epsilon}) < Eu_i(s_i', \sigma_{-i}^{\epsilon})$. By (8.3), we have $\sigma_i^{\epsilon}(s_i) \leqq \epsilon \cdot \sigma_i^{\epsilon}(s_i') \leqq \epsilon$. □

Interestingly, this stability concept appears again in evolutionary game theory (Sect. 10.1), which does not require rationality of players. The existence theorem of proper equilibria is proved in a very similar way to that of the existence theorem of trembling-hand perfect equilibria, and it is warranted for any finite normal-form game (see Myerson [38]).

There are other related stability concepts. For example, Kalai and Samet [23] advocate *persistent equilibrium*, which is based on local stability itself, but not selecting among Nash equilibria.

8.4 Sequential Equilibrium

From this section, we consider stability against perturbed strategies in extensive-form games. Selten [42] in fact started with a stability against trembles in extensive-form games, and applied it to normal-form games to define the trembling-hand perfect equilibrium in Sect. 8.2. Making Selten's idea more rigorous for extensive-form games, Kreps and Wilson [31] formulated *sequential equilibrium*. We will explain later how closely related the concepts of trembling-hand perfect equilibrium and sequential equilibrium are.

Let us consider the famous game called Selten's "horse" game in Fig. 8.2. (The game tree looks like a horse.) This is a three-person extensive-form game. The payoffs

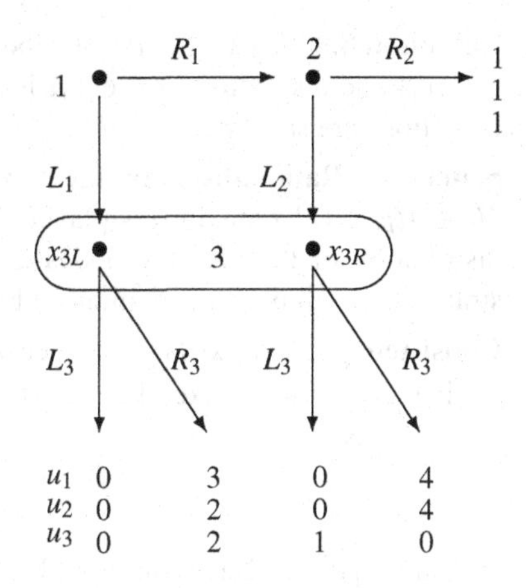

Fig. 8.2 Selten's horse game

are for players 1, 2, and 3 respectively from the top. This game does not have a proper subgame, and therefore the Nash equilibria and subgame perfect equilibria coincide. (See Problem 5.3 (b).)

There are two pure-strategy Nash equilibria for this game, namely (L_1, R_2, R_3) and (R_1, R_2, L_3). Consider the former. If player 1 trembles to R_1, then rational player 2 would not choose R_2, because player 3 is still expected to play R_3, independently from player 1's mistakes. Given R_3, strategy L_2 gives a higher payoff than R_2 to player 2. This argument suggests that, when trembles are introduced into extensive-form games, we should check the rational choices at *off-path* information sets, not reached by the original strategy combination, as well. Nash equilibria ignore optimization during the game, and subgame perfection has no bite when there is no proper subgame.

As we have seen in signaling game analysis (Chap. 7), when an information set contains multiple decision nodes, the optimal action choice depends on the *belief* (a probability distribution) over the possible decision nodes. Therefore, in order to formulate an equilibrium of an extensive-form game with non-singleton information sets, we need to specify not only a strategy combination, but also *the system of beliefs* for all players which makes the strategy combination optimal and is consistent with the strategy combination. Thus, Kreps and Wilson [31] defined a sequential equilibrium as a pair of a strategy combination and a system of beliefs which are mutually consistent (recall that a perfect Bayesian equilibrium as defined in Chap. 7 has the same structure).

Definition 8.8 A *system of beliefs* in an n-player extensive-form game is a function $\mu : \cup_{i=1}^{n} X_i \to [0, 1]$ that assigns non-negative real numbers to all decision nodes of all players such that each information set is endowed with a probability distribution over its decision nodes; for each $i \in \{1, 2, \ldots, n\}$ and each information set $H_i \in \mathcal{H}_i$,

$$\sum_{x \in H_i} \mu(x) = 1.$$

Definition 8.9 A pair (π, μ) of a behavioral strategy combination $\pi = (\pi_1, \pi_2, \ldots, \pi_n)$ and a system of beliefs μ is a *sequential equilibrium* if the following two conditions are satisfied.

Sequential Rationality: For any player $i \in \{1, 2, \ldots, n\}$ and any information set $H_i \in \mathcal{H}_i$, $\pi_i(H_i)$ maximizes player i's expected payoff, given that the probability distribution of the decision nodes in H_i is computed by μ and the continuation strategies by all players are induced by π.

Consistency: There exists a sequence of $\{(\pi^k, \mu^k)\}_{k=1}^{\infty}$ such that for any $k = 1, 2, \ldots$, π^k is a completely mixed behavioral strategy[1] combination, μ^k is induced by π^k using Bayes' rule, and

$$\lim_{k \to \infty} (\pi^k, \mu^k) = (\pi, \mu).$$

Unlike perfect Bayesian equilibrium, the consistency condition for sequential equilibrium requires a sequence of completely mixed behavioral strategies, so that all information sets are reached with a positive probability, and hence Bayes' rule is applicable at all information sets. Therefore, a sequential equilibrium can induce rational choices at information sets which are not originally reachable. For example, the (L_1, R_2, R_3) equilibrium in the horse game in Fig. 8.2 is not a sequential equilibrium. For any sequence of completely mixed (behavioral) strategy combinations that converges to (L_1, R_3), for sufficiently large k, player 3 chooses R_3 with a high probability, but player 2's information set is reached with a positive probability. Hence, R_2 is not a sequentially rational choice.

By contrast, (R_1, R_2, L_3) is a sequential equilibrium. We only need to construct a sequence of completely mixed behavioral strategy combinations and a corresponding sequence of systems of beliefs. Take a very small $\epsilon > 0$ and let $\pi^k = ((\epsilon^k, 1 - \epsilon^k), (3\epsilon^k, 1 - 3\epsilon^k), (1 - \epsilon^k, \epsilon^k))$, where the first number of each player is the probability on L_i. For each $k = 1, 2, \ldots$, μ^k is defined by Bayes' rule based on π^k. Then, for each $k = 1, 2, \ldots$, the probability of the left decision node of player 3 is

$$\mu^k(x_{3L}) = \frac{\epsilon^k}{\epsilon^k + (1 - \epsilon^k)3\epsilon^k} = \frac{1}{4 - 3\epsilon^k},$$

and this converges to $1/4$ as k goes to ∞. Under this system of beliefs, (R_1, R_2, L_3) satisfies sequential rationality. First, when x_{3L} has a probability of $1/4$, then player 3's optimal strategy is L_3. Second, if player 3 will choose L_3 in the future, player 2's optimal strategy is R_2. Finally, given the continuation strategies (R_2, L_3) by others, player 1's optimal strategy is R_1.

As is clear from the definition, sequential equilibrium is a generalization of subgame perfect equilibrium.

Proposition 8.4 *If (π, μ) is a sequential equilibrium, then π is a subgame perfect equilibrium.*

[1] A completely mixed behavioral strategy π_i satisfies $supp(\pi_i(H_i)) = A(H_i)$ for all i and $H_i \in \mathcal{H}_i$.

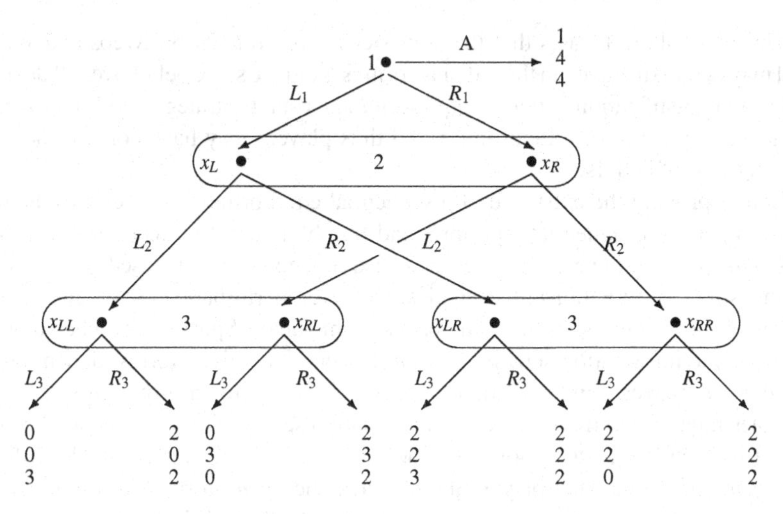

Fig. 8.3 Figure 8.5 of Fudenberg and Tirole

Subgame perfection is limited to proper subgames, and the sequential rationality reduces to a Nash equilibrium property in all proper subgames.

Although not clear in the example of the horse game, the consistency condition implies that all players have the same belief over all decision nodes of the game, including the off-path decision nodes. To see this clearly, consider the example by Fudenberg and Tirole [15] (their Fig. 8.5), illustrated in Fig. 8.3.

In the game in Fig. 8.3, player 2 and 3 do not know whether player 1 chose L_1 or R_1. Player 3 can observe player 2's actions. (The payoffs are for players 1, 2, and 3, from the top.) The induced normal-form of this three-player game has many pure-strategy Nash equilibria. Consider one of them, $(A, L_2, (L_3, L_3))$, where the first coordinate of player 3's strategy is the action at the left-side information set.

When player 2 gets to move, sequential rationality requires her/him to compare the expected payoffs of L_2 and R_2, based on the probability that player 1 chose L_1, and player 3's (continuation) strategy (L_3, L_3). Let p be the (deviation) probability that player 1 chose L_1, and $1 - p$ be the probability of R_1. The expected payoff of L_2 is $p \cdot 0 + (1 - p)3$, and that of R_2 is 2. Therefore, if the probability that player 1 chose L_1 is not more than 1/3, or if $\mu(x_L) \leqq \frac{1}{3}$, then L_2 is sequentially rational.

When player 3 gets to move, that is also when player 1 has deviated from A. Player 3 can observe player 2's actions so that her/his two information sets are distinguished. When player 3 is at the left information set, and if p is the probability that player 1 chose L_1, then the expected payoff of L_3 is $p \cdot 3 + (1 - p)0$, and that of R_3 is 2. Hence, L_3 is sequentially rational if and only if $p = \mu(x_{LL}) \geqq \frac{2}{3}$ (the same logic holds at the right information set as well). However, $\mu(x_{LL}) \geqq \frac{2}{3}$ is the same as $\mu(x_L) \geqq \frac{2}{3}$. This means that the system of beliefs that makes $(A, L_2, (L_3, L_3))$ sequentially rational differs for player 2 and player 3, so that $(A, L_2, (L_3, L_3))$ is not a sequential equilibrium.

This example illustrates that the consistency requirement by Kreps and Wilson [31] may be too stringent, in that all players must hold the same belief over all decision nodes. One justification is that an equilibrium is a stable strategy combination after the game is played for a long time,[2] and thus players may have come to hold the same system of beliefs.

Before proving the existence of a sequential equilibrium, let us clarify the relationship between sequential equilibria and trembling-hand perfect equilibria. This makes the proof of the existence easier. These concepts are both based on robustness against small perturbations of strategies. However, perturbations in extensive-form games are not as simple as those in normal-form games. Specifically, when a player has multiple information sets, perturbations can take various forms, depending on whether the player trembles independently across her/his information sets, or the trembles must be similar because the information sets belong to the same player.

Let us view the action choices at each information set as performed by different *agents* of the original player, and consider the *agent normal-form game* of an extensive-form game (this idea is in Kuhn [33] and Selten [42]). Each agent chooses an action only once, and the payoff function is the same as the one for the original player to whom the information set belongs. In this setup, it is natural that each agent trembles independently. By consolidating the action choices for the original players, the strategy combination in the agent normal-form game corresponds to the extensive-form game strategy under perfect recall condition. In this agent normal-form game, the set of trembling-hand perfect equilibria is a subset of strategy combinations induced by sequential equilibria.[3]

Proposition 8.5 *For any finite extensive-form game* Γ *with perfect recall, a trembling-hand perfect equilibrium of its agent normal-form game generates a sequential equilibrium of* Γ.

Proof Let σ be a trembling-hand perfect equilibrium of the agent normal-form game of Γ. There exists a sequence of completely mixed strategy combinations $\{\sigma^k\}_{k=1}^{\infty}$ such that $\lim_{k\to\infty} \sigma^k = \sigma$. From σ^k, it is straightforward to define the system of beliefs μ^k on Γ by Bayes' rule. The sequence $\{\mu^k\}_{k=1}^{\infty}$ is in a compact set, and hence there is a convergent subsequence $\{\mu^m\}_{m=1}^{\infty}$. Let μ be the limit. Then (σ, μ) satisfies the consistency.

The strategy combination σ is optimal for each agent, and thus by the one-step deviation principle of dynamic programming, the consolidated strategy for each original player is also optimal in the whole extensive-form game. Therefore σ satisfies sequential rationality. □

By Theorem 8.1, any finite normal-form game has a trembling-hand perfect equilibrium, and thus combining with Proposition 8.5, we have the existence of a sequential equilibrium. However, the converse of Proposition 8.5 is not true. That is, for some

[2] As we discussed in Sect. 3.8, how players learn/deduce to play an equilibrium is a different issue from what an equilibrium should be, and the former is also an ongoing research topic.

[3] Moreover, proper equilibria imply sequential rationality even in the induced normal form. See Problem 8.4 (b).

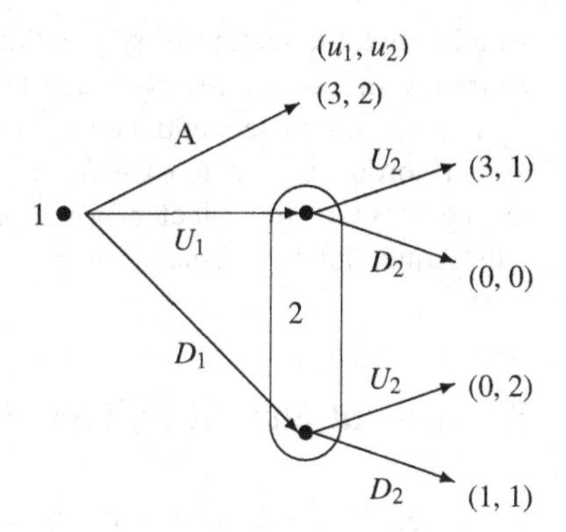

Fig. 8.4 A sequential equilibrium may not correspond to a trembling-hand perfect equilibrium

games, there is a sequential equilibrium which does not correspond to a termbling-hand perfect equilibrium in its agent normal-form game. Consider the game depicted in Fig. 8.4.

In the extensive-form game in Fig. 8.4, player 2's strategy D_2 is strictly dominated by strategy U_2, and hence, for any sequential equilibrium, U_2 must have a probability of 1. This implies that in any sequential equilibrium, the probabilities on A and U_1 can be arbitrary. However, there is a unique trembling-hand perfect equilibrium (A, U_2) in the (both induced and agent) normal-form of this game.

This game, however, is exceptional. If the play path A and the path $\{U_1, U_2\}$ do not give the same payoff to player 1, then the set of sequential equilibria coincides with the set of trembling-hand perfect equilibria of the normal-form game. In the space of payoff vectors (i.e., the space of games), those that give the same payoffs at multiple terminal nodes have a measure of 0. In other "generic" games, Kreps and Wilson [31] showed that almost all sequential equilibria are trembling-hand perfect.

As for the relationship between sequential equilibria and perfect Bayesian equilibria, Fudenberg and Tirole [16] proved that in a class of incomplete information extensive-form games, they coincide (we omit the proof).

Proposition 8.6 *Among extensive-form games with perfect recall and incomplete information, consider those in which the prior probabilities of types are independent across players. If either the number of types per player is no more than two, or the extensive-form game has two stages, then the set of perfect Bayesian equilibria coincides with the set of sequential equilibria.*

Therefore, for signaling games, the two equilibrium concepts predict the same strategy combinations. Sequential equilibria also have another similar feature to perfect Bayesian equilibria, that often there are many sequential equilibria for a game, corresponding to the freedom of choosing off-path beliefs or strategies. However, for generic extensive-form games, if we look at the induced probability distributions on the terminal nodes by sequential equilibria (possible ways of ending the game), the set of such probability distributions is finite (Theorem 2 of Kreps and Wilson [31]).

Another notable feature of sequential equilibria is that they may change if we add an irrelevant move to the extensive-form game. See Problem 8.8.

The concept of sequential equilibrium formalized **the stability during the game** quite rigorously, for extensive-form games with imperfect information. These days, most games with imperfect information (except probably signaling games) are solved with sequential equilibria.

8.5 A Solution to the Chain Store Paradox

Let us reconsider the *chain store paradox*, introduced in Sect. 4.4. We have shown that if there are finitely many markets, no matter how many they are, there is a unique solution by backward induction (hence, a subgame perfect equilibrium) in which the monopolist accommodates entry into the last market, and therefore it must accommodate entry into the second-to-last market, and so on, so that in all markets entry occurs. However, a more "realistic" story is that the monopolist fights with a very low price in early markets, even incurring deficits, to deter future entries. If entrants believe in such a threat, they can be in fact discouraged. Kreps and Wilson [32] succeeded in showing a rational equilibrium with this phenomenon. The story behind their game is a game with incomplete information, but using the Bayesian framework, they convert it to an imperfect information game with Nature's choice for the type of chain store at the beginning of the extensive-form game, and construct a sequential equilibrium of entry deterrence.

The players are a monopolist M which has branches in N markets, and N potential entrants for each market. For sequential rationality, it is easier to count how many markets are left until the end of the game, and therefore we name the potential entrant in the last market as the first firm, the potential entrant in the second-to-last market as the second firm, and so on. Thus, the nth firm is considering whether to enter the market when there are n markets left in the game, from the viewpoint of the monopolist.

Each potential entrant $n = 1, 2, \ldots, N$ does not know the monopolist's payoff function exactly. With the prior probability $(1 - \delta)$, the monopolist is believed to have the usual payoff function of the entry game in Fig. 7.1, reproduced in Fig. 8.5, and we call this monopolist the Rational type. With probability δ, the monopolist has the payoff function which makes Fight a dominant strategy, for example that of the Tough type in Fig. 7.2. This prior belief is common knowledge.

The game begins with Nature choosing one of the two types for the monopolist, and only the monopolist learns his type. Then the Nth firm and the monopolist play the entry game in Fig. 8.5, and the two firms receive the stage payoff (for M, depending on its type). After observing the outcome of the Nth market, the next $(N - 1)$th firm and the monopolist play the entry game and receive the stage payoff, and the game continues this way until the end. The outcomes of all past markets are observable to all players. (However, if a mixed action is played, only the realized

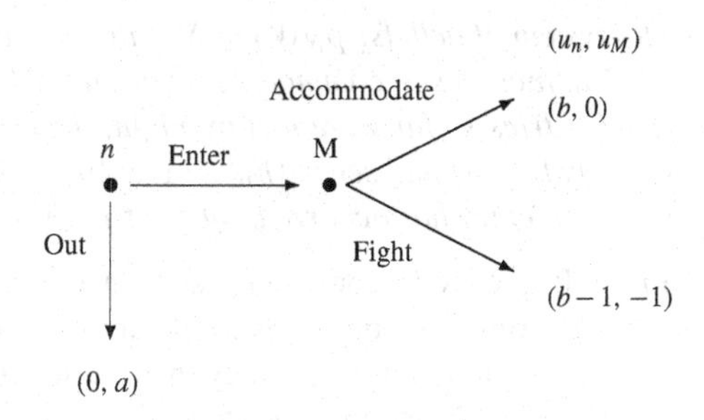

Fig. 8.5 Entry game between a rational monopolist and the nth potential entrant ($a > 1$, $1 > b > 0$)

actions are observable.) The monopolist's payoff for the whole game is the sum of the stage payoffs shown in Fig. 8.5 over the N markets, if he is the Rational type.

Let us find a sequential equilibrium of this finite-market entry game. Because the Tough type monopolist's optimal behavioral strategy is "fight after any history", we focus on the behavioral strategy by the Rational type, which must be optimal at each (singleton) information set, given the future potential entrants' strategies.

Consistency requires that each potential entrant updates the common prior probability δ of the monopolist being Tough, based on the observed history of actions and using Bayes' rule. For each potential entrant n, let h_n be the history of the actions before this market, and let $p_n(h_n)$ be the posterior probability that the monopolist is the Tough type. Each potential entrant $n = N, N-1, \ldots, 1$ chooses an optimal strategy against the belief $p_n(h_n)$.

We prove that when p_n is large, although the Rational type monopolist would choose Accommodate in every market under complete information, he chooses Fight in some initial markets, if entry occurs. By fighting initially, the monopolist can induce potential entrants to continue to believe in some probability of the Tough type. Knowing this, potential entrants in the first several markets choose Out.

Proposition 8.7 *(Kreps and Wilson [32]) The following pair of a behavioral strategy combination and a system of beliefs is a sequential equilibrium of the finite-market entry game with the Rational and Tough type.*
Rational Type monopolist: In stage $n = 1$ (the last market), Accommodate if entry occurs. In stage $n > 1$,

$$\pi_M(h_n) = \begin{cases} \text{Fight} & \text{if } p_n(h_n) \geqq b^{n-1} \\ \text{Fight with prob. } \frac{(1-b^{n-1})p_n}{(1-p_n)b^{n-1}} & \text{if } p_n(h_n) < b^{n-1}. \end{cases}$$

Potential entrant n:

$$\pi_n(h_n) = \begin{cases} \text{Out} & \text{if } p_n(h_n) > b^n \\ \text{Out with prob. } \frac{1}{a} & \text{if } p_n(h_n) = b^n \\ \text{Enter} & \text{if } p_n(h_n) < b^n. \end{cases}$$

The system of beliefs: $p_N(\emptyset) = \delta$ *at the beginning of the game. For each* $n < N$,
(i) if h_n *includes Accommodate, then set* $p_n(h_n) = 0$;
(ii) if entries so far were met by Fight, then let k *be the latest such market number (clearly* $k > n$*) and set* $p_n(h_n) = \max\{b^{k-1}, \delta\}$;
(iii) if no entry has occurred, then set $p_n(h_n) = \delta$.

Proof It is easy to construct a sequence of completely mixed behavioral strategy combinations that converges to the above strategy combination, and thus we show consistency at the limit strategy distribution. (Note that n is the number of markets left in the game, and thus the stage after n is $n-1$. We also abbreviate h_n to write p_n as the belief, because only the posterior probability of the Tough type is important.)

First, we derive the consistent relationship between p_n and p_{n-1}, for each $n > 1$. When $p_n \geqq b^{n-1} (> b^n)$, no entry occurs, and thus no new information is obtained. Therefore, it must be that $p_{n-1} = p_n$. Also, when $p_n = 0$, the type has been revealed so that $p_{n-1} = 0 (= p_n)$. When $0 < p_n < b^{n-1}$, the Rational type monopolist will mix actions. If Accommodate is realized, $p_{n-1} = 0$. If Fight is realized, by Bayes' rule,

$$
\begin{aligned}
p_{n-1} &= \frac{Pr(\text{Fight} \cap \text{Tough})}{Pr(\text{Fight} \cap \text{Tough}) + Pr(\text{Fight} \cap \text{Rational})} \\
&= \frac{p_n \cdot 1}{p_n \cdot 1 + (1 - p_n)[\frac{(1-b^{n-1})p_n}{(1-p_n)b^{n-1}}]} = b^{n-1}.
\end{aligned}
$$

Next, we extend the above argument to show that the specified system of beliefs is consistent across the whole game. The above "local" argument shows that, if Accommodate is observed, (i) is consistent. Recall that the initial belief is $p_N = \delta$. We divide the analysis into two cases, whether $\delta < b^N$ or not.

If $\delta < b^N (< b^{N-1})$, then entry occurs in the market N (the beginning of the game), and the Rational type randomizes his actions. If Accommodate is observed, it becomes the case of (i) and thus the system of beliefs is consistent afterwards. If Fight is observed, the above analysis implies that $p_{N-1} = b^{N-1}$, which is also $\max\{b^{N-1}, \delta\}$. In the next market $N-1$, $p_{N-1} = b^{N-1}$ implies that the $N-1$th firm randomizes between entry and staying out. Since $p_{N-1} = b^{N-1} < b^{N-2}$, the Rational monopolist also randomizes between Accommodate and Fight. Thus, if (Enter, Accommodate) is realized, (i) will be the case, so that the consistency holds afterwards. If Out is realized, then no new information is obtained, and (iii) is consistent. If Fight is observed, the "local" consistency implies that $p_{N-2} = b^{N-2} (= \max\{b^{N-2}, \delta\})$, and randomizations of both firms occur again. In this way, the system of beliefs is consistent throughout the game.

If $\delta \geqq b^N$, in the first market N there is no entry, and thus $p_{N-1} = \delta$ results. This is consistent. In the next market $N-1$, there are two possibilities: either $p_{N-1} = \delta < b^{N-1}$ or $\delta \geqq b^{N-1}$. The former case is analogous to the analysis of $\delta < b^N$ above, and consistency is warranted afterwards. The latter case brings us back to the same situation as $\delta \geqq b^N$. That is, no entry occurs in the market $N-1$, and again there are two possibilities: $p_{N-2} = \delta < b^{N-2}$ or $\delta \geqq b^{N-2}$. In this way, as long

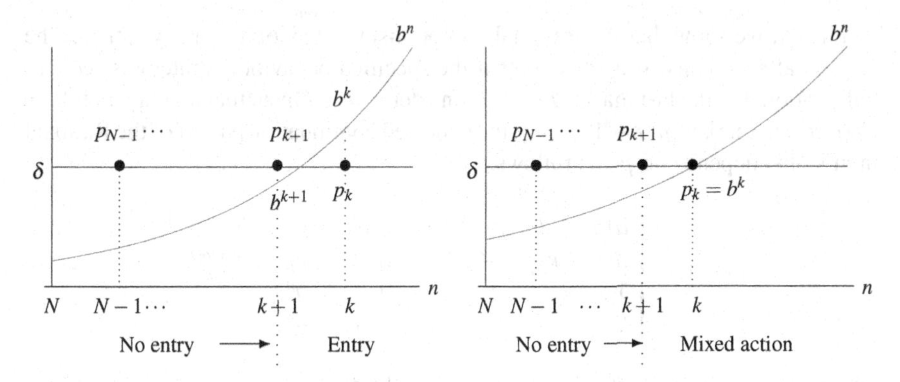

Fig. 8.6 Relationship between p_n and b^n ($\delta \geqq b^N$)

as $p_n > b^n$, no entry occurs, and the monopolist's type will not be revealed. If N is sufficiently large, however, at some point this inequality is reversed, as Fig. 8.6 illustrates.

The left-side figure in Fig. 8.6 depicts the case where $(\delta =)p_{k+1} > b^{k+1}$ and $(\delta =)p_k < b^k$ hold. In this case, no entry has occurred in markets $N, \ldots, k+1$, but in market k, entry occurs for sure. The right-side figure shows the other case where $p_{k+1} > b^{k+1}$ and $p_k = b^k$ hold. In this case, both the potential entrant and the Rational monopolist randomize in market k. The belief after market k depends on the realization in market k. If Out is realized, $p_{k-2} = p_{k-1} = \delta$, so (iii) is consistent. If (Enter, Fight) is realized (in either case of $p_k < b^k$ or $p_k = b^k$), then $p_{k-1} = b^{k-1}$ by Bayes' rule, and this corresponds to (ii), where k is the latest market number for which entry was met by Fight. In this way, the specified system of beliefs is consistent throughout the game.

Let us prove that each potential entrant firm n is choosing an optimal strategy. We have four cases.

Case 1: When $p_n \geqq b^{n-1} (> b^n)$.

In this case, for either type of monopolist, Fight is expected, and thus the optimal strategy is Out.

Case 2: When $b^n < p_n < b^{n-1}$.

Accommodate occurs with a positive probability but less than $1 - b$. (The proof of this statement is Problem 8.10. We have already shown that the threshold for the entrant is $1 - b$ in Sect. 7.1.) Therefore Out is optimal.

Case 3: When $p_n = b^n$.

In this case, Accommodate is chosen with probability $1 - b$. Then Out and Enter give the same expected payoff, so that any mixed strategy is optimal. (The mixed strategy $1/a$ is to make the monopolist's strategy sequentially rational.)

Case 4: When $p_n < b^n$.

The probability of Accommodate is greater than $1 - b$, so that Enter is optimal.

Finally, we show that the Rational monopolist's behavioral strategy satisfies the sequential rationality. It is obvious that the specified behavioral strategy is sequentially rational in the last market $n = 1$. Consider $n > 1$. Given that all players follow (π_M, π_n) in markets $n, n-1, \ldots, 1$, the expected continuation payoff of the Rational monopolist depends on p_n as follows:

$$v(p_n) = \begin{cases} a\{n - k(p_n) + 1\} + 1 & \text{if } b^n < p_n = b^{k(p_n)-1} \\ a\{n - k(p_n) + 1\} & \text{if } b^n < p_n < b^{k(p_n)-1} \\ 1 & \text{if } p_n = b^n \\ 0 & \text{if } p_n < b^n, \end{cases}$$

where $k(p_n) = \inf\{m \mid b^m < p_n\}$ if $p_n > 0$ and $k(0) = \infty$ if $p_n = 0$. This $k(p_n)$ is the number of the last market with no Entry.

Let us explain the derivation of $v(p_n)$. Consider the last case $p_n < b^n$, so that Entry occurs for sure. When $n = N - 1$, the Rational monopolist essentially compares the payoffs of this stage, because it will accommodate for sure in the next (last) market and get 0 after any outcome of this market. Therefore, in market $N - 1$, it should also accommodate, and the continuation payoff over the two markets is $0 + 0$. This argument can be repeated for any $n > 1$, and thus the continuation payoff is always 0.

Consider the case of $p_n = b^n$, so that both firm n and the Rational monopolist randomize. With probability $1/a$, the monopolist receives a in this market, and with probability $1 - (1/a)$, the monopolist gets to move. To simplify the notation, let $\gamma = \frac{(1-b^{n-1})p_n}{(1-p_n)b^{n-1}}$ be the probability of Fight. The continuation payoff can be formulated as

$$\frac{1}{a} \cdot a + (1 - \frac{1}{a})\{\gamma \cdot v_F + (1 - \gamma)0\}, \tag{8.4}$$

where v_F is the continuation payoff from $n - 1$ markets on, when Fight is realized in market n. This v_F itself has a recursive structure such as

$$v_F = -1 + \frac{1}{a} \cdot a + (1 - \frac{1}{a})\{\gamma \cdot v_F + (1 - \gamma)0\}.$$

To explain, in this market the monopolist receives -1 from the Fight action, which is the first term. The next belief is $p_{n-1} = b^{n-1}$, so that the monopolist faces the same situation as in this market, which are the second and the third terms. Note that the explicit solution is $v_F = 0$. Therefore, the continuation payoff (8.4) is $\frac{1}{a} \cdot a + (1 - \frac{1}{a})\{\gamma \cdot v_F + (1 - \gamma)0\} = 1$.

When $b_n < p_n$, no entry occurs in the markets $n, n - 1, \ldots, k(p_n)$, so that the monopolist receives $a\{n - k(p_n) + 1\}$ in these markets. In market $k(p_n) + 1$, entry occurs with a positive probability. If, in addition, $p_n < b^{k(p_n)-1}$ is the case, the entry in market $k(p_n) + 1$ is for certain, and the continuation payoff from then on is 0 (recall Fig. 8.6). If instead $p_n < b^{k(p_n)-1}$ is the case, then in market $k(p_n) + 1$, both players randomize, and by the above argument the continuation payoff from then on is 1. This completes the derivation of $v(p_n)$.

Now we prove that at each information set of the monopolist, his continuation strategy is sequentially rational. Consider the information set after Entry in market n. If the Rational monopolist accommodates to reveal his type, his continuation payoff is 0. If he fights, he gets -1 in this market but depending on p_n, his continuation payoff differs.

If $p_n = 0$, the type is already known and $v(p_{n-1}) = 0$. Hence Fight in market n gives the total of $-1+0$ and Accommodate is strictly optimal. If $0 < p_n < b^{n-1}$ (and the monopolist fights), from (ii), the next potential entrant's belief is $p_{n-1} = b^{n-1}$. Hence the monopolist's continuation payoff from $n-1$th market on is $v(p_{n-1}) = 1$, and the one in market n is $-1 + 1 = 0$. Therefore in this market n, any mixed action is optimal. When $p_n > b^{n-1}(> b^n)$, $v(p_n) > 1$ implies that following π_M is optimal. □

This result implies that, with a small amount of incomplete information, the chain store paradox resolves, in the sense that there is an equilibrium in which the monopolist fights in early markets to discourage entrants, and, knowing this, entrants in early markets do not enter.

Using a similar logic, the finitely repeated Prisoner's Dilemma also resolves. When there is a possibility that the opponent is committed to playing the Tit-for-Tat strategy (see Sect. 5.5), then a rational player who never cooperates in the finitely repeated Prisoner's Dilemma may cooperate for a while, to induce the opponent to cooperate. (See Kreps et al. [30].)

More generally, Fudenberg and Maskin [14] proved a Folk Theorem such that, for any sufficiently long but finitely repeated game, any feasible and individually rational payoff combination is (almost) sustainable by a sequential equilibrium, if there is adequate incomplete information.

8.6 Infinitely Repeated Games with Imperfect Monitoring

A repeated game of a normal-form game is already a game with imperfect information, because it contains a simultaneous-move game in each period. Moreover, in many realistic situations, players suffer from even less information. For example, competing firms in a market know their own production quantities and the market price, but may not know their rivals' production quantities, which are the rivals' actions. If one cannot detect a deviation from an agreement (e.g., a small production quantity to keep the price high), a non-myopic equilibrium seems to be difficult to sustain. Green and Porter [20] showed that in such repeated games with *imperfect monitoring*, a trigger strategy combination using observable signals may still constitute a cooperative equilibrium. Abreu et al. [1, 2] generalized this idea to prove a Folk Theorem for general infinitely repeated games with imperfect monitoring. In order to analyze games with imperfect monitoring, subgame perfect equilibrium is not useful, and we need the sequential equilibrium concept.

We consider an infinitely repeated Bertrand game (see Sect. 3.3) to illustrate the idea of a cooperative equilibrium under imperfect monitoring. It is based on Tirole [44] Sect. 6.7.1.

In the stage game, two firms $i = 1, 2$ choose a price p_i simultaneously. The production cost structure is symmetric, and each firm incurs the cost of $c \cdot q$ to produce q units. Their products are perfect substitutes, so that consumers buy only from the cheaper firm. If the prices are the same, the market demand is equally split between the firms. Specifically, for each firm $i = 1, 2$, (where $j \neq i$ is the rival firm) the demand is

$$D_i(p_i, p_j) = \begin{cases} A - p_i & \text{if } p_i < p_j \\ \frac{1}{2}(A - p_i) & \text{if } p_i = p_j \\ 0 & \text{if } p_i > p_j, \end{cases}$$

and assume that $A > c$ and $c \geqq 0$ hold.

As we have seen in Sect. 3.3.2, if this game is played only once, both firms choosing the marginal cost c as the price is the unique Nash equilibrium, and the resulting payoff (profit) of each firm is 0. If this game is infinitely repeated with perfect monitoring and with discounting, by the Folk Theorem, for sufficiently large discount factors, there are many cooperative (i.e., high price) equilibria. For example, there is a subgame perfect equilibrium in which both firms charge the monopoly price every period on the play path.

Let us consider an infinitely repeated Bertrand game such that a firm cannot observe its opponent's price. Even if the price is not observable, if a firm can observe the demand of its product and its own price, then it can infer whether the rival's price was lower than its own price. To make it impossible to infer the rival's action, we assume that there is a possibility that the market has no demand, with a small but positive probability ρ, in every period. With probability $1 - \rho$, the market demand of a period is as specified above. (Formally, we assume that the demand fluctuation is independently and identically distributed over time.) Then, even if a firm observes that its demand was 0 in some period, this can be due to either the rival's price-cut or the random demand shock.

We construct a sequential equilibrium in which both firms charge the same high price $\bar{p} > c$ as long as possible on the equilibrium path. To sustain this play path, deviations must be deterred. Notice that if a firm's demand becomes 0, both firms know that. This is because, if it is due to the demand shock, then both firms see no demand, and if it is due to the price cut of a firm i, then not only the betrayed firm j sees 0 demand, but also firm i knows that j must have no demand. Therefore, the event that "at least one firm's demand is 0" becomes common knowledge if this happens. Using this fact, we consider the following two-phase strategy b_T, where T is the parameter of the length of the punishment.

Cartel phase: In this phase, a firm charges $\bar{p}(> c)$. A firm stays in this phase if no firm has had zero demand from the beginning of the game (including the first period), or since the end of the punishment phase. If at least one firm has received zero demand, then move to the punishment phase.

Punishment phase: In this phase, a firm charges c for T periods, regardless of the demand observation. After that, shift to the cartel phase.

If both firms follow this strategy in a sequential equilibrium, the high price is continued until the demand shock hits, and, moreover, the price-fixing cartel revives even if it collapses from the demand fluctuation. Therefore, the long-run average payoff of the firms is quite high.

Because it is straightforward to construct a sequence of completely mixed behavioral strategy combinations with the limit being (b_T, b_T) and a corresponding sequence of systems of beliefs, we only show the sequential rationality of (b_T, b_T).

Proposition 8.8 *There exists a range of (δ, ρ, T) such that, for any $\bar{p} \in (c, A)$ and any $i = 1, 2$, given that the other firm follows b_T, b_T is optimal for firm i in any information set.*

Proof If the firms are in the punishment phase, then actions do not affect the continuation strategies of the players, and thus only the one-shot payoff matters for sequential rationality. Given the rival's price c, if firm i charges a price less than c and if there is a positive demand in the market, it receives a negative profit. For other combinations of firm i's price and demand shock, it gets 0 profit. Therefore following b_T is weakly optimal.

Next, we show that if the firms are in the cartel phase, a one-period deviation does not give a higher total expected payoff. If both firms are following b_T, then a firm's profit under the normal (positive) demand is $(\bar{p} - c)\frac{A-\bar{p}}{2}$. If firm i cuts the price slightly to $\bar{p} - \epsilon$, then it gets the one-shot deviation profit of $(\bar{p} - \epsilon - c)\{A - (\bar{p} - \epsilon)\}$, as we have seen in Sect. 3.3.2. For notational convenience, let us write $\overline{\Pi} = (\bar{p} - c)(A - \bar{p})$. Then the one-shot payoff (profit) in the cartel phase is $\overline{\Pi}/2$, and any one-shot deviation yields a one-shot payoff less than $\overline{\Pi}$.

Fix an arbitrary period in the cartel phase and consider the long-run payoff when both firms follow b_T. Denote by V the total expected payoff when both firms continue to follow b_T, and by W the total expected payoff of the punishment phase. Then V and W satisfy the following system of recursive equations:

$$\begin{cases} V = (1 - \rho)\{\frac{\overline{\Pi}}{2} + \delta V\} + \rho(0 + \delta W) \\ W = (1 + \delta + \cdots + \delta^{T-1})0 + \delta^T V. \end{cases}$$

To explain, V consists of two cases. The first term is the payoff when the "normal" demand occurs with probability $1 - \rho$. In this period, each firm receives $\frac{\overline{\Pi}}{2}$, and they stay in the cartel phase in the next period to return to the same situation as in this period. Hence, the discounted continuation payoff is δV. The second term is when the demand becomes 0 with probability ρ. In this case, although no firm has violated the cartel agreement, the profit in this period is 0, and they enter the punishment phase in the next period so that the continuation payoff is δW. The RHS of the second equation shows that in the punishment phase, the payoff of 0 continues for T periods for certain, and the firms return to the cartel phase.

We can solve the system of the simultaneous equations for V to get

$$V = \frac{(1-\rho)\overline{\Pi}/2}{1-(1-\rho)\delta - \delta^{T+1}\rho}. \tag{8.5}$$

By reducing the price for one period, a firm can get slightly less than

$$(1-\rho)[\overline{\Pi} + \delta W] + \rho(0 + \delta W).$$

The first term is when the demand is normal, and in this case the one-shot payoff is slightly less than $\overline{\Pi}$, but the rival will observe zero demand. Hence, they enter the punishment phase in the next period. The second term is when the demand is zero. Therefore, following b_T is optimal if

$$V \geqq (1-\rho)[\overline{\Pi} + \delta W] + \rho(0 + \delta W).$$

From Eq. (8.5) and $W = \delta^T V$, this condition is equivalent to

$$f(\delta) = \delta^{T+1}(2\rho - 1) + 2\delta(1-\rho) - 1 \geqq 0.$$

Because $f(1) = 0$, if the derivative of f with respect to δ is negative at $\delta = 1$, then there exists $\underline{\delta} \in (0, 1)$ such that for any $\delta \geqq \underline{\delta}$, $f(\delta) \geqq 0$ holds.

By differentiation, $f'(1) = T(2\rho - 1) + 1$. Therefore, there exist a range of $T \in \{1, 2, \ldots\}$ and $\rho \in (0, 1/2)$ that guarantee $T(2\rho - 1) + 1 \leqq 0$. □

The above result can be used to construct an "efficient" sequential equilibrium when $\rho \to 0$. Let both firms charge the monopoly price in the cartel phase, and for each $\rho < 1/2$, pick T which satisfies $T(2\rho - 1) + 1 \leqq 0$. The average equilibrium payoff $(1-\delta)V$ converges to half of the monopoly profit when $\rho \to 0$ and $\delta \to 1$.

The key to the above simple construction of the equilibrium was the availability of a *public signal* such that if at least one firm gets zero demand, this becomes common knowledge. Therefore the players can use the public signal to coordinate the punishment. The strategies which use only public signals are called *public strategies*, and the above equilibrium is sometimes called a *perfect public equilibrium* (see for example Fudenberg and Levine [13]). Moreover, in the Bertrand model, the value $\overline{\Pi}$ appears both in the on-path payoff and in the deviation payoff. Therefore, $\overline{\Pi}$ cancels out, and the incentive constraint $f(\delta) \geqq 0$ is independent from the target stage payoff $\overline{\Pi}/2$.

In general, even with public signals, the efficient outcome may not be achieved for two reasons. First, it is possible that the punishment phase needs to last for a very long period to wipe out any deviation gains. Since under imperfect monitoring, the punishment phase inevitably happens, this may reduce the total equilibrium payoff to an inefficient level. Second, when the public signals are not sufficiently informative, then an efficient equilibrium may be impossible to construct. In such a case, however,

Kandori and Obara [27] showed that using the player's own past actions in addition to the public signals may improve equilibrium payoffs.

In general, public signals may not be available. There are various possible signal structures with which non-myopic equilibria are constructed: *private signals* (e.g., Sekiguchi [41]) where each player receives noisy different signals, *conditionally independent signals* (e.g., Matsushima [36]) where players observe statistically independent signals, *almost public signals* (e.g., Mailath and Morris [34]) where players can believe that every player observes the same signal with a high probability, and *almost perfect signals* (e.g., Hörner and Olszewski [21]) where the probability that each player observes an erroneous signal is sufficiently small. A Folk Theorem in repeated games with discounting and a general monitoring structure is shown by Sugaya [43]. For a survey, see Kandori [25] and Mailath and Samuelson [35].

8.7 Random Matching Games

A *random matching game* is a dynamic game played by a population of homogeneous players. In every period, all players are randomly matched with each other to make pairs to play a two-person stage game. Even though the random matching process makes the opponents different every period, if the players can observe all other players' past actions, then it is easy to establish a Folk Theorem. However, a more natural model is that the players do not know the history of actions of their newly matched opponents. In this case, not only does the game have imperfect monitoring, but also the players cannot use a **personalized punishment** of trigger strategies. A player who deviates from an agreed action in a stage game will be matched with a different player afterwards, and thus future opponents of the deviator will not know of the deviation and cannot punish him.

In this section, following Kandori [24] and Ellison [10], we describe how a cooperative equilibrium can be constructed for a random matching game with the Prisoner's Dilemma as the stage game.

Consider a society of $2n$ players. Each period, n pairs are randomly formed (independently over time), and in each pair the normalized Prisoner's Dilemma in Table 8.4 is played, where $g > 0$ and $\ell \geqq 0$. Each player observes only the history of realized action profiles in the stage games which (s)he has played.

Consider a strategy which starts the game with playing C and continues to play C as long as D is not observed (this is called the C-phase). If D is observed, move to D-phase with some probability, where the player plays action D. Once in the D-phase, shift to C-phase with some probability each period. When all players use

Table 8.4 Normalized prisoner's dilemma

	C	D
C	$1, 1$	$-\ell, 1+g$
D	$1+g, -\ell$	$0, 0$

this strategy, a one-period deviation of playing D starts the "contagious process" of action D in the society. Because the society is finite, eventually the deviator faces a society with a lot of players playing D. However, if a player adheres to C, this contagion does not happen, and if the future is sufficiently important (i.e., if the discount factor is close to 1), no player deviates from the above strategy. To be precise, by setting the transition probabilities appropriately, deviations from both the D-phase and C-phase can be prevented. (For details, see Ellison [10].)

Proposition 8.9 *(Ellison [10]) Let $n \geqq 2$, and the society consists of $2n$ players. Let G be the normalized Prisoner's Dilemma in Table 8.4, where $g > 0$ and $\ell \geqq 0$.*[4] *In each period, pairs are randomly formed to play G, over the infinite horizon. Each player maximizes the expected total payoff with a common discount factor $\delta \in (0, 1)$. Then there exists $\underline{\delta} \in (0, 1)$ such that for any $\delta \geqq \underline{\delta}$, there is a sequential equilibrium in which (C,C) is played by all pairs in every period on the play path.*

The above equilibrium uses a private strategy because, in every period, a player knows nothing about her/his new opponent. If there is any information transmission among the players, cooperation, or "non-myopic" action profiles which are not stage game equilibria, would be easier to sustain. Okuno-Fujiwara and Postlewaite [39] added a social mechanism that assigns a "status" to each player after each period, depending on the action profile in that period. Then for any two-person stage game, a Folk Theorem-type result is obtained, with an appropriate social mechanism with two status levels. An example would be a credit-card system where the card company records each consumer's default history, and sellers can choose whether to allow a new customer to complete a credit-card transaction, depending on the record.

Another method to achieve non-myopic actions in equilibrium when players do not know each other's pasts when they first meet, is to add the possibility for the players to play the stage game again with the same opponent. This voluntary repetition option generates both personalized punishment if the players stay together and social punishment if newly-met players cannot earn a high stage payoff immediately, which makes betrayal and severance unattractive. Fujiwara-Greve and Okuno-Fujiwara [18] showed that this two-fold incentive structure is possible in equilibrium. Because their model is an evolutionary game model, we explain it in Sect. 10.6.

8.8 On the Informational Structure of Games

This chapter dedicates more pages to information than the other chapters of this book, and thus we would like to provide a (personal) perspective of the future of game theoretic research on information. Note, however, that this perspective depends on the author's (imperfect) information at this point of writing.

---

[4] In Kandori [24], not only δ but also ℓ is made sufficiently large.

In much of the research so far, the information structure has been assumed to be exogenous. We can relax this assumption in at least two directions. One direction is to allow players to strategically choose the quantity and/or the quality of information. In repeated games with imperfect monitoring, players may be able to acquire information regarding opponents' past actions with some cost. For example, one may hire an agent to spy on a rival company. Then there is a new trade-off between better information to sustain higher payoff strategies and the cost of the information. See for example, Miyagawa et al. [37] and Flesch and Perea [12]. Alternatively, players can give away costly information about opponents whom they may never see again, if everyone does so (e.g., Fujiwara-Greve et al. [19]). The latter line of research is related to repeated games with endogenous communications (e.g., Kandori and Matsushima [26], Compte [8], Ben-Porath and Kahneman [5], and Aoyagi [3]), where incentives to give information and incentives to play a non-myopic action should be jointly analyzed.

Information may be provided by "experts" who are also players in the game. Then, the equilibria must give the experts incentives to provide correct information, which may not be possible. See for example Ely and Välimäki [11]. Furthermore, the herding literature starting with Banerjee [4] and Bikhchandani et al. [6] investigates when players rationally ignore their (exogenously given) personal information.

Another direction is to introduce an interaction between playing the game (with a choice of opponents or an option to not play) and the arrival of information. Unlike the ordinary repeated games which assume that players play the game repeatedly, if the repetition itself is a choice and the relevant information to continue playing is endogenous, there is a trade-off between the possible low payoff by playing the game again and the additional information to play the game better later. A hungry fox need not to give up the grape at the top of a vine, if the cost to reach it is known to be less than the benefit, but the benefit may be revealed only after the fox pays the cost. Currently this type of learning by doing in games is not a fully developed area of research, but a good starting point can be the learning literature such as Bolton and Harris [7]. Kaneko and Matsui [28] consider a model in which players construct logically the game they are playing, as they play the game. This model can be also called an endogenous information game.

The endogeneity of information may generate dynamic asymmetries of actions, such as frequent transactions under good/precise information followed by infrequent transactions under imprecise/no information (Veldkamp [45]). A related model is analyzed by Fujiwara-Greve et al. [17], where players learn only the actions by their current partners, and the endogenous formation of partnerships leads to the fast loss of partners, when a bad signal occurs, and slow recovery.

Information and interpretation of the information may not be the same. For example, if a firm bribes an official of a foreign country, some people may interpret this news as damaging the reputation of the firm, but others may interpret that the bribery will improve the firm's business and is thus a necessary evil. When a big company has a scandal, people may over-interpret the scandal, so that its subsidiary firms may also lose their reputation (e.g., Jonsson et al. [22]). Therefore, what matters in a game is the interpretation of information by relevant players, which may be difficult to observe for researchers. This is more of an empirical problem, but an implication is that we must be careful in assuming an informational structure for a game. The model by Kaneko and Matsui [28] can be interpreted as saying that the players' actions depend on how the players interpret the game.

Problems

8.1 Prove that Definitions 8.1 and 8.4 are equivalent.

8.2 Prove that in the following game, (A, X) is a trembling-hand perfect equilibrium.

P1\P2	X	Y
A	0, 1	0, 1
B	−1, 2	1, 0
C	−1, 2	2, 3

8.3 Find all Nash equilibria of the following game and prove that there is a unique trembling-hand perfect equilibrium.

P1\P2	A	B	C
A	0, 0	0, 0	0, 0
B	0, 0	1, 1	3, 0
C	0, 0	0, 3	3, 3

8.4 We compare trembling-hand perfect equilibria and subgame perfect equilibria. Consider the following extensive form game.

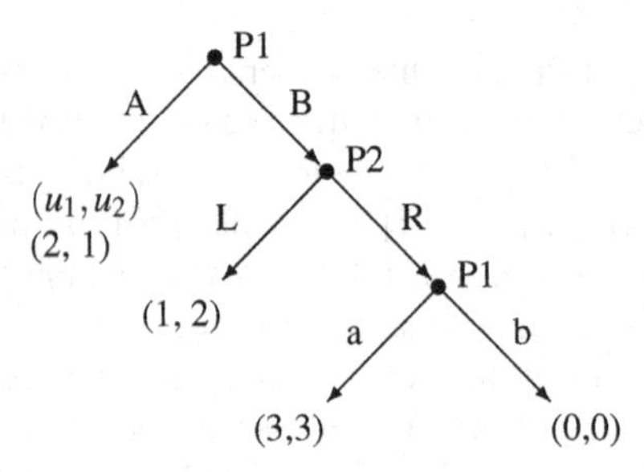

(a) What is the unique subgame perfect equilibrium of this game?

(b) Formulate the induced normal form game by the two players and show that $((A, a), L)$ is a trembling-hand perfect equilibrium.

(c) Separate player 1 into two agents, one that chooses actions in the first information set, and one who chooses actions in the last information set. Construct the corresponding normal form game by three agents, and find all pure-strategy trembling-hand perfect equilibria.

8.5 Consider the following three-player normal form game. Player 1 chooses rows x or y, player 2 chooses columns X or Y, and player 3 chooses matrices L or R. The payoffs are lined up in order of the player number.

P1 \ P2	X	Y
x	1, 1, 2	1, 0, 2
y	1, 1, 2	0, 0, 2

P3: L

P1 \ P2	X	Y
x	1, 2, 0	0, 0, 0
y	0, 2, 0	1, 0, 1

R

(a) Prove that (y, X, L) is a Nash equilibrium consisting of only strategies that are not weakly dominated.

(b) Prove that (y, X, L) is not a trembling-hand perfect equilibrium.

8.6 Prove that (B, X) in Table 8.3 is a proper equilibrium with the suggested sequence of mixed-strategy combinations in the text.

8.7 (Myerson) Find the Nash equilibria, trembling-hand perfect equilibria, and proper equilibria of the following game.

P1\P2	A	B	C
A	1, 1	0, 0	−9, −9
B	0, 0	0, 0	−7, −7
C	−9, −9	−7, −7	−7, −7

8.8 Prove that in the following two games, player 1's action A is an outcome of a sequential equilibrium of Game 1, but it is never played in any sequential equilibrium of Game 2.

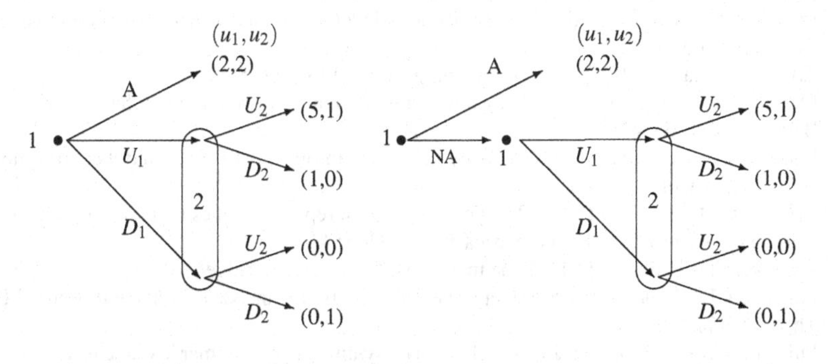

8.9 Compare sequential equilibria and trembling-hand perfect equilibria of the induced normal form game of the following game. Are they the same?

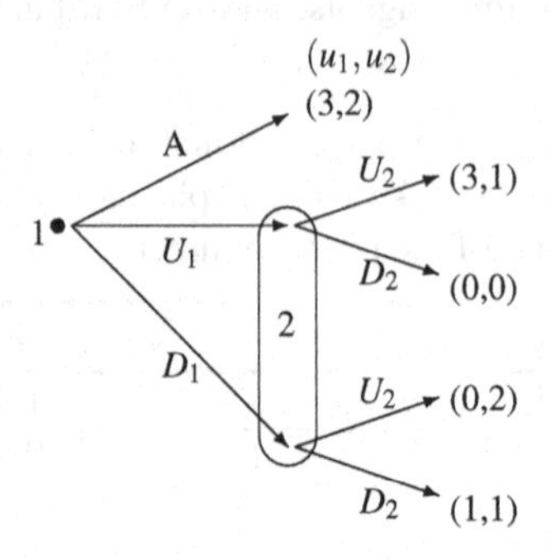

8.10 In the proof of Proposition 8.7, show that when $b^n < p_n < b^{n-1}$, Accommodate is taken with a positive probability but less than $1 - b$ by the monopolist.

References

1. Abreu D, Pearce D, Stacchetti E (1986) Optimal cartel equilibria with imperfect monitoring. J Econ Theory 39(1):251–269
2. Abreu D, Pearce D, Stacchetti E (1990) Toward a theory of discounted repeated games with imperfect monitoring. Econometrica 58(5):1041–1063
3. Aoyagai M (2002) Collusion in Bertrand oligopoly with correlated private signals and communication. J Econ Theory 102(1):229–248
4. Banerjee A (1992) A simple model of herd behavior. Q J Econ 107(3):797–818
5. Ben-Porath E, Kahneman M (2003) Communication in repeated games with costly monitoring. Games Econ Behav 44(2):227–250
6. Bikhchandani S, Hirshleifter D, Welch I (1992) A theory of fads, fashion, custom, and cultural change as informational cascades. J Polit Econ 100(5):992–1026
7. Bolton P, Harris C (1999) Strategic experimentation. Econometrica 67(2):349–374
8. Compte O (1998) Communication in repeated games with imperfect private monitoring. Econometrica 66(3):597–626
9. van Damme E (1987) Stability and perfection of Nash equilibria. Springer, New York
10. Ellison G (1994) Cooperation in the Prisoner's Dilemma with anonymous random matching. Rev Econ Stud 61(3):567–588
11. Ely J, Välimaäki J (2003) Bad reputation. Q J Econ 118(3):785–814
12. Flesch J, Perea A (2009) Repeated games with voluntary information purchase. Games Econ Behav 66(1):126–145
13. Fudenberg D, Levine D (1994) Efficiency and observability with long-run and short-run players. J Econ Theory 62(1):103–135
14. Fudenberg D, Maskin E (1986) The Folk theorem in repeated games with discounting or with incomplete information. Econometrica 54(3):533–554
15. Fudenberg D, Tirole J (1991) Game theory. MIT Press, Cambridge
16. Fudenberg D, Tirole J (1991) Perfect Bayesian equilibrium and sequential equilibrium. J Econ Theory 53(2):236–260
17. Fujiwara-Greve T, Greve H, Jonsson S (2014) Asymmetry of customer loss and recovery under endogenous partnerships: theory and evidence. Int Econ Rev (forthcoming)

18. Fujiwara-Greve T, Okuno-Fujiwara M (2009) Voluntarily separable repeated Prisoner's Dilemma. Rev Econ Stud 76(3):993–1021
19. Fujiwara-Greve T, Okuno-Fujiwara M, Suzuki N (2012) Voluntarily separable repeated Prisoner's Dilemma with reference letters. Games Econ Behav 74(2):504–516
20. Green E, Porter R (1984) Noncooperative collusion under imperfect price information. Econometrica 52(1):87–100
21. Hörner J, Olszewski W (2006) The Folk theorem for games with private almost-perfect monitoring. Econometrica 74(6):1499–1544
22. Jonsson S, Greve H, Fujiwara-Greve T (2009) Undeserved loss: the spread of legitimacy loss to innocent organizations in response to reported corporate deviance. Adm Sci Q 54(2):195–228
23. Kalai E, Samet D (1984) Persistent equilibria in strategic games. Int J Game Theory 13(3): 129–144
24. Kandori M (1992) Social norms and community enforcement. Rev Econ Stud 59(1):63–80
25. Kandori M (2002) Introduction to repeated games with private monitoring. J Econ Theory 102(1):1–15
26. Kandori M, Matsushima H (1998) Private observation. Communication and collusion. Econometrica 66(3):627–652
27. Kandori M, Obara I (2006) Efficiency in repeated games revisited: the role of private strategies. Econometrica 74(2):499–519
28. Kaneko M, Matsui A (1999) Inductive game theory. J Public Econ Theory 1(1):101–137
29. Kohlberg E, Mertens J (1986) On the strategic stability of equilibria. Econometrica 54(5): 1003–1037
30. Kreps D, Milgrom P, Roberts J, Wilson R (1982) Rational cooperation in the finitely repeated Prisoner's Dilemma. J Econ Theory 27(2):245–252
31. Kreps D, Wilson R (1982) Sequential equilibrium. Econometrica 50(4):863–894
32. Kreps D, Wilson R (1982) Reputation and imperfect information. J Econ Theory 27(2):253–279
33. Kuhn H (1953) Extensive games and the problem of information. In: Kuhn HW, Tucker AW (eds) Contributions to the theory of games, vol II. Princeton University Press, Princeton, pp 193–216
34. Mailath G, Morris S (2002) Repeated games with almost-public monitoring. J Econ Theory 102(1):189–228
35. Mailath G, Samuelson L (2006) Repeated games and reputations: long-run relationships. Oxford University Press, Oxford
36. Matsushima H (2004) Repeated games with private monitoring: two players. Econometrica 72(3):823–852
37. Miyagawa E, Miyahara Y, Sekiguchi T (2008) The Folk theorem for repeated games with observation costs. J Econ Theory 139(1):192–221
38. Myerson R (1978) Refinements of the Nash equilibrium. Int J Game Theory 7(2):73–80
39. Okuno-Fujiwara M, Postlewaite A (1995) Social norms and random matching games. Games Econ Behav 9(1):79–109
40. Osborne M, Rubinstein A (1994) A course in game theory. MIT Press, Cambridge
41. Sekiguchi T (1997) Efficiency in the prisoner's dilemma with private monitoring. J Econ Theory 76(2):345–361
42. Selten R (1975) Reexamination of the perfectness concept for equilibrium points in extensive games. Int J Game Theory 4(1):25–55
43. Sugaya T (2013) Folk theorem in repeated games with private monitoring. Stanford Graduate School of Business, Mimeo
44. Tirole J (1988) The theory of industrial organization. MIT Press, Cambridge
45. Veldkamp L (2005) Slow boom, sudden crash. J Econ Theory 124(2):230–257

Chapter 9
Equilibrium Selection*

9.1 Payoff Dominance and Risk Dominance

Because many games have multiple Nash equilibria, *equilibrium refinement* research adds extra rationality/stability requirements to narrow down the set of equilibria. By contrast, *equilibrium selection* research keeps the concept of Nash equilibrium intact, but adds some external structure, such as incomplete information or a long-run process of learning or evolution, to see which Nash equilibrium is more likely to be played. The latter idea originates from the book by Harsanyi and Selten [4], and there are two ways to select among equilibria, based on efficiency comparison of the payoff combinations (*payoff dominance*) and comparison of the product of "risk" that arises by unitary deviation (*risk dominance*). To this day, there is no agreement among researchers which approach is better. Depending on the underlying model, either selection can become more plausible than the other.

In this chapter, we introduce the concepts of payoff dominance and risk dominance, explain their meanings, and also extend the risk dominance concept to p-dominance. In addition, we illustrate how equilibrium selection is performed using the example of Global Games, which is relevant to many economic situations.

Reconsider the game represented by Table 8.1 with multiple Nash equilibria (Table 9.1).

The idea of the trembling-hand perfect equilibrium was that the Nash equilibrium (B, B) is unstable under small perturbations of the opponent's strategic choice. Another idea is to consider the two players' reasoning: among the two Nash equilibria, only (A, A) is beneficial for the two players, and thus can be a "focal point" to play. Alternatively, (A, A) is the easiest strategy combination that the two players

Table 9.1 Game of Table 8.1 (reproduced)

P1\P2	A	B
A	1, 1	0, 0
B	0, 0	0, 0

T. Fujiwara-Greve, *Non-Cooperative Game Theory*, Monographs in Mathematical Economics 1, DOI 10.1007/978-4-431-55645-9_9

Table 9.2 Stag hunt game

P1\P2	Stag	Hare
Stag	3, 3	0, 2
Hare	2, 0	2, 2

can agree to play, because its payoff is better for both players than the other Nash equilibrium (B, B). (Of course, the players cannot agree to play a strategy combination that is not a Nash equilibrium.) The payoff dominance idea formalizes this argument.

Definition 9.1 A Nash equilibrium $\sigma^* \in \Delta(S_1) \times \cdots \times \Delta(S_n)$ *payoff dominates* another Nash equilibrium $\sigma' \in \Delta(S_1) \times \cdots \times \Delta(S_n)$ if

$$Eu_i(\sigma^*) > Eu_i(\sigma'), \ \ \forall i \in \{1, 2, \ldots, n\}.$$

If a payoff-dominance relationship holds between two Nash equilibria, then the players would not agree to play the dominated Nash equilibrium. If there is a Nash equilibrium that payoff-dominates all other Nash equilibria, which is called a *payoff-dominant equilibrium*, then it can be a focal point to be played. Note, however, that the binary relation based on payoff dominance does not satisfy completeness, and thus an arbitrary pair of Nash equilibria may not be comparable. For example, none of the two pure-strategy Nash equilibria of the Meeting Game in Table 3.1 is payoff dominated by the other.

It is, nonetheless, not easy to construct a formal model in which rational players come to agree on a payoff-dominant Nash equilibrium among multiple Nash equilibria. This is because all Nash equilibria are self-enforcing. One approach is to add an external structure such as pre-play communication to a game. Even if communication is *cheap talk* which does not affect payoffs, players can send signals to each other to deviate from a payoff-dominated Nash equilibrium and move to a dominating one, but they cannot move from a payoff-dominant Nash equilibrium. (For a rigorous analysis, we give an evolutionary game model. See Sect. 10.3 and Robson [10].)

Let us consider a slightly different game called the *Stag Hunt game*.[1] The story behind this game is as follows. Two hunters simultaneously choose whether to hunt Hare or Stag. Hare hunting can be done by a single hunter, and regardless of the other hunter's strategy, one can get a payoff of 2. Stag hunting requires two men, but the reward is larger than a hare hunt. Specifically, if both hunters choose Stag, they get a total payoff of 6, which will be divided into 3 each. If a hunter chooses Stag alone, he gets 0. The matrix representation of the game is shown in Table 9.2.

The Stag Hunt game has two pure-strategy Nash equilibria with a payoff-dominance relationship. Namely, the equilibrium (Stag, Stag) payoff dominates the

---

[1]The origin of this game is said to be Rousseau's book, *Discourse on the Origin and Basis of Inequality among Men* (1755).

equilibrium (Hare, Hare). However, this is not the end of the comparisons. If a player changes strategies unilaterally from (Stag, Stag), he loses only $3 - 2 = 1$ unit of the payoff, while the loss of unitary deviation from (Hare, Hare) is $2 - 0 = 2$. This means that the (Hare, Hare) combination has less risk of the opponent changing his mind than (Stag, Stag) has. Hence, (Hare, Hare) can be chosen over (Stag, Stag) under this risk consideration.

Looking at the equilibria of the game in Table 8.1, the (A, A) equilibrium is less risky than the (B, B) equilibrium as well, because players can change strategies unilaterally from (B, B) without loss of payoffs, while changing from (A, A) gives a loss to both players. This comparison is generalized below as *risk dominance*. (For simplicity, we only consider two-person finite pure-strategy games.)

To prepare, let us define the *unitary deviation loss*[2] from a strategy combination (s_i, s_j) when player i changes to strategy s'_i:

$$u_i(s_i, s_j) - u_i(s'_i, s_j).$$

The product of the two players' loss is the total loss by moving from (s_1, s_2) to (s'_1, s'_2). Then a strategy combination with a larger total loss of unitary deviation is safer than the one with a smaller total loss.

Definition 9.2 A Nash equilibrium $(s^*_1, s^*_2) \in S_1 \times S_2$ *risk dominates* a Nash equilibrium $(s'_1, s'_2) \in S_1 \times S_2$ if

$$\begin{aligned}&\{u_1(s^*_1, s^*_2) - u_1(s'_1, s^*_2)\}\{u_2(s^*_1, s^*_2) - u_2(s^*_1, s'_2)\}\\ &> \{u_1(s'_1, s'_2) - u_1(s^*_1, s'_2)\}\{u_2(s'_1, s'_2) - u_2(s'_1, s^*_2)\}.\end{aligned}$$

With this criterion, the total risk of unitary deviation from (Stag, Stag) is $(3-2)^2 = 1$, while that from (Hare, Hare) is $(2-0)^2 = 4$. Therefore, the latter risk-dominates the former. The notion of risk dominance is also introduced in the book by Harsanyi and Selten [4] along with payoff-dominance, and they provide an axiomatization for it. That is, a selection which satisfies a set of certain axioms necessarily becomes the risk dominance concept. Also, the selections by risk dominance and payoff dominance are independent.

Another way of measuring the risk to play an equilibrium is how much a player must "trust" the opponent to play the equilibrium strategy. A totally uncertain situation is often formulated as a uniform belief, i.e., in this case $1/2$ probability on each strategy by the opponent. With this belief, the expected payoff from Hare is 2, and that from Stag is 1.5, which implies that Hare is optimal. In order for Stag to be optimal, the player must have a belief that puts at least $2/3$ on the opponent's strategy Stag. Or, alternatively, in order to play Hare, a player only needs to believe that the opponent plays Hare with a probability of at least $1/3$. We conclude that (Hare, Hare)

[2] We follow the definition by Carlsson and van Damme [1] for later references. There is another definition using the opposite subtraction, but for the product of the unitary deviation losses, the signs are irrelevant.

requires a lower probability to guarantee that the same strategy is optimal than (Stag, Stag) requires, and thus (Hare, Hare) is more robust in the uncertainty about the opponent's choice. From this argument, we can extend the notion of risk dominance.

Definition 9.3 For any $p \in [0, 1]$, a Nash equilibrium $s^* \in S_1 \times S_2$ is *p-dominant* if, for each $i = 1, 2$, and any probability distribution λ over the set of the opponent's pure strategies S_j such that $\lambda(s_j^*) \geqq p$,

$$\sum_{s_j \in S_j} \lambda(s_j) u_i(s_i^*, s_j) > \sum_{s_j \in S_j} \lambda(s_j) u_i(s_i, s_j), \ \ \forall s_i \in S_i \setminus \{s_i^*\}.$$

That is, if the belief that the opponent plays s_j^* with at least probability p implies that s_i^* is the unique optimal strategy for both players, then (s_1^*, s_2^*) is p-dominant.

For symmetric 2×2 games, risk dominance is equivalent to $\frac{1}{2}$-dominance. To see this, note that a Nash equilibrium (s^*, s^*) risk dominates another Nash equilibrium (s', s') if $u_i(s^*, s^*) - u_i(s', s^*) > u_i(s', s') - u_i(s^*, s')$ for both $i = 1, 2$. By moving terms and multiplying with $1/2$,

$$\frac{1}{2}\{u_i(s^*, s^*) + u_i(s^*, s')\} > \frac{1}{2}\{u_i(s', s^*) + u_i(s', s')\},$$

and by the definition of a Nash equilibrium, for any λ such that $\lambda(s^*) \geqq 1/2$,

$$\begin{aligned}\lambda(s^*)u_i(s^*, s^*) + \lambda(s')u_i(s^*, s') &\geqq \frac{1}{2}\{u_i(s^*, s^*) + u_i(s^*, s')\}\\ > \frac{1}{2}\{u_i(s', s^*) + u_i(s', s')\} &\geqq \lambda(s^*)u_i(s', s^*) + \lambda(s')u_i(s', s').\end{aligned}$$

Hence, for symmetric 2×2 games, risk dominance is equivalent to the fact that a dominating equilibrium (s^*, s^*) is $\frac{1}{2}$-dominant. We can derive general properties for p-dominance easily. (The reader is asked to prove these in Problem 9.1.)

Remark 9.1 (1) (s_1^*, s_2^*) is a strict Nash equilibrium if it is 1-dominant.
(2) If (s_1^*, s_2^*) is p-dominant, then it is q-dominant for any $q \in [p, 1]$.
(3) If (s_1^*, s_2^*) is 0-dominant, then each s_i^* $(i = 1, 2)$ strictly dominates all other pure strategies of player i.

Notice that the concept of p-dominance is not a binary relation, and it can be used to rank Nash equilibria. Let us explain this by the example of Table 9.3. This game has three pure-strategy Nash equilibria. (T, L) risk dominates (M, C), and (M, C) risk dominates (D, R), but (D, R) risk dominates (T, L). By the binary relation of risk dominance, we cannot make a ranking among the Nash equilibria.

If we use the notion of p-dominance, (T, L) is p-dominant for any $p \geqq \frac{8}{15}$, (D, R) is p-dominant for any $p \geqq \frac{2}{3}$, and (M, C) is p-dominant for any $p \geqq \frac{7}{9}$. For any p such that $\frac{8}{15} \leqq p < \frac{2}{3} < \frac{7}{9}$, only (T, L) is p-dominant. The smaller the required

Table 9.3 Game by Morris et al. [7]

P1\P2	L	C	R
T	7, 7	0, 0	0, 0
M	0, 0	2, 2	7, 0
D	0, 0	0, 7	8, 8

p is, the safer the strategy combination is, and therefore (T, L) is the most robust. This way of comparing strict Nash equilibria by the level of p in p-dominance is valid for almost all games.

Proposition 9.1 *(i) If $p \leqq 1/2$, then there is at most one p-dominant strategy combination.*
(ii) For generic payoff functions, every normal-form game has either completely mixed Nash equilibria only, or exactly one p-dominant Nash equilibrium for some p.

Proof (i) Let $p \leqq 1/2$ and $(a_1, a_2) \neq (b_1, b_2)$ be both p-dominant strategy combinations. Then, for each $i = 1, 2$, we have

$$\frac{1}{2}\{u_i(a_i, a_j) + u_i(a_i, b_j)\} > \frac{1}{2}\{u_i(b_i, a_j) + u_i(b_i, b_j)\},$$

and

$$\frac{1}{2}\{u_i(b_i, b_j) + u_i(b_i, a_j)\} > \frac{1}{2}\{u_i(a_i, a_j) + u_i(a_i, b_j)\}.$$

This is a contradiction.
(ii) For generic payoff functions, there is no pair of strategy combinations with the same payoff for any of the players. Thus, if there is a pure-strategy Nash equilibrium, there is a strict Nash equilibrium. Even if there are multiple strict Nash equilibria, they are p-dominant for different ranges of p. Hence, there is a unique lower bound to p at which the statement holds. □

Payoff dominance, risk dominance, and p-dominance are all plausible selection criteria. In the following sections, we show normal-form games with multiple Nash equilibria with an external structure which always selects a risk-dominant Nash equilibrium of the component game. A similar result for p-dominance has been obtained, for example, in Oyama [9] by an extension of Matsui and Matsuyama [6]. He showed that with a dynamic process called perfect foresight dynamics, a strict Nash equilibrium is linearly stable if and only if it is p-dominant with $p < 1/2$.

9.2 Global Games**

Let us introduce another formulation of incomplete information other than the Bayesian game model by Harsanyi. It is the *global game* framework by Carlsson and van Damme [1]. A rigorous analysis of a global game is mathematically involved,

Table 9.4 Game $G(x)$

P1\P2	α_2	β_2
α_1	x, x	$x, 0$
β_1	$0, x$	$4, 4$

and hence in this section we show only the idea of how a risk-dominant equilibrium emerges over a payoff-dominant equilibrium in global games, with an example in Carlsson and van Damme [1].

Two players are facing the symmetric game $G(x)$ in Table 9.4. If $x > 4$, then β_i is strictly dominated by α_i so that $\alpha = (\alpha_1, \alpha_2)$ is the unique dominant-strategy equilibrium. (Clearly it is also efficient, i.e., payoff dominant.) If $4 > x > 2$, then $x^2 - (4-x)^2 > 0$ implies that the Nash equilibrium α risk dominates the Nash equilibrium $\beta = (\beta_1, \beta_2)$, but β is the unique efficient Nash equilibrium. If $2 > x > 0$, then β not only risk dominates α but also payoff dominates it. Finally, if $0 > x$, then β is the unique dominant-strategy equilibrium.

Suppose that x is random and follows the uniform distribution over an interval $[\underline{x}, \overline{x}]$, where $\underline{x} < 0$ and $\overline{x} > 4$. Assume that player i can only observe a noisy signal of x, which is a random variable X_i that follows the uniform distribution over the interval $[x - \epsilon, x + \epsilon]$ for some $\epsilon > 0$. We also assume that the two players' observation errors $X_1 - x$ and $X_2 - x$ are independent.

The game is as follows. At the beginning, Nature chooses a value of x. The two players make noisy observations of x simultaneously, choose α_i or β_i simultaneously, and the game ends. The payoff for each player is given by the true $G(x)$. The structure of this incomplete information game is common knowledge.

We use the notion of iterative elimination of strictly dominated strategies to find a stable behavioral strategy combination. Because the game is symmetric, we focus on player 1 below.

First, when player 1 observes a value x_1, the ex-post probability distribution of the true x is the uniform distribution over $[x_1 - \epsilon, x_1 + \epsilon]$. Hence, if x_1 belongs to the interval $[\underline{x} + \epsilon, \overline{x} - \epsilon]$, then the expected payoff from the strategy α_1 is x_1, because the possible true values of x are contained in $[\underline{x}, \overline{x}]$.

Next, we consider the ex-post probability distribution of player 2's observation x_2, given player 1's observation value x_1. The opponent's observation x_2 is at most $x_1 + 2\epsilon$ and at least $x_1 - 2\epsilon$. Therefore x_2 distributes over the interval $[x_1 - 2\epsilon, x_1 + 2\epsilon]$. In addition it has a symmetric distribution with the center x_1, as shown below.

Step 1: The ex-post probability distribution for $x_2 \in [x_1 - 2\epsilon, x_1]$.

Let Δ be the distance between $x_1 - 2\epsilon$ and x_2, then $x_2 = \Delta + x_1 - 2\epsilon$. From Fig. 9.1, we can see that the range of x which can give rise to the two players' observations x_1 and x_2 respectively is $[x_1 - \epsilon, x_2 + \epsilon] = [x_1 - \epsilon, x_1 - \epsilon + \Delta]$. Hence, the length of this interval is Δ. (Figure 9.1 shows the case where $x_2 < x_1 - \epsilon$, but as long as $x_2 < x_1$, the range of x is the same.)

The ex-post probability that x falls in this interval $[x_1 - \epsilon, x_2 + \epsilon]$ is the probability of falling in an interval of length Δ in $[x_1 - \epsilon, x_1 + \epsilon]$, which is $\Delta/(2\epsilon)$. The ex-post probability of x_2 is proportional to this. Hence, as x_2 increases from $x_1 - 2\epsilon$ towards x_1, its ex-post probability increases.

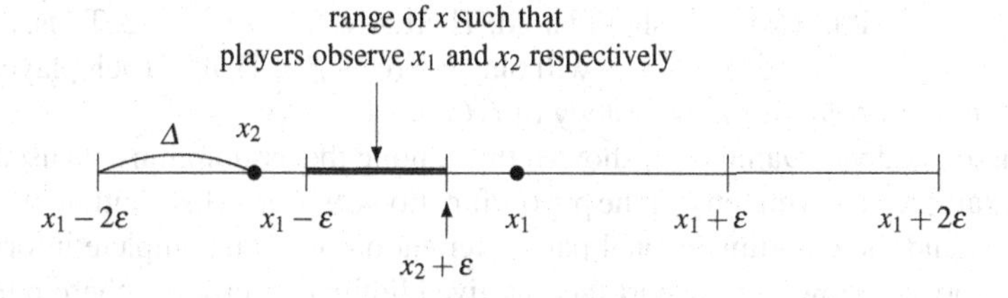

Fig. 9.1 When $x_2 \in [x_1 - 2\epsilon, x_1]$

Step 2: The ex-post probability distribution for $x_2 \in [x_1, x_1 + 2\epsilon]$.
This case is the opposite of Step 1, so that as x_2 decreases from $x_1 + 2\epsilon$ towards x_1, its ex-post probability increases.

Therefore, for any x_1,

$$Pr[X_2 < x_1 \mid x_1] = Pr[X_2 > x_1 \mid x_1] = \frac{1}{2}.$$

(Only this symmetry is used in the following analysis.)

Assume that $\epsilon < -\underline{x}/2$, and let us eliminate strictly dominated strategies iteratively, depending on the observation value x_1. Consider the case where player 1 observes $x_1 < 0$. Then the expected payoff of strategy α_1 is negative. To see this, if $x_1 + \epsilon \leqq 0$, then the whole range of possible x is in the negative area. If $x_1 + \epsilon > 0$, still the range $[x_1 - \epsilon, x_1 + \epsilon]$ is contained in $[\underline{x}, \overline{x}]$ so that the expected value is x_1 itself. If player 1 chooses β_1, the payoff is at least 0, and thus α_1 is strictly dominated by β_1. Notice that the same argument holds for player 2.

When player 1 observes $x_1 = 0$, then $Pr[X_2 < 0 \mid x_1 = 0] = 1/2$ implies that player 2 will not play α_2 half of the time, by the above argument. This in turn implies that the ex-post expected payoff of strategy β_1 is at least 2, and that of α_1 is 0. Hence, α_1 is iteratively eliminated.

Let x_1^* be the lowest observation value at which α_1 is **not** iteratively eliminated. By symmetry, $x_1^* = x_2^*$, and let the common value be x^*. By iterative elimination, for any $x_i < x^*$, player i chooses β_i. Suppose that player 1 observes $x_1 = x^*$. Then $Pr[X_2 < x^* \mid x_1 = x^*] = 1/2$ implies that β_2 has at least 1/2 probability, so that the ex-post expected payoff of β_1 is at least 2, while that of α_1 is x^*. By the definition of x^*, it must be that $x^* \geqq 2$.

Now we do an analogous analysis from the above. Assume $\epsilon < (\overline{x} - 4)/2$ and consider the case where player 1 observes $x_1 > 4$. The ex-post expected payoff of α_1 is more than 4, while β_1 gives at most 4. Hence, β_1 is strictly dominated by α_1. The same argument holds for player 2. When the observation is $x_1 = 4$, still $Pr[X_2 > 4 \mid x_1 = 4] = 1/2$ implies that β_1 is iteratively eliminated. Let x^{**} be the largest observation value at which β_1 is **not** iteratively eliminated. By the same argument as above, if player 1 observes x^{**}, then the ex-post expected payoff of β_1 is at least 2, while that of α_1 is x^{**}. Therefore $x^{**} \leqq 2$.

By the definition, $x^* \leqq x^{**}$ should hold. Therefore, $x^* = x^{**} = 2$. Thus, a player who observes $x_i > 2$ (resp. $x_i < 2$) will play α_i (resp. β_i). That is, each player plays the risk dominant equilibrium strategy of $G(x_i)$.

Carlsson and van Damme [1] showed that among the assumptions we used in the above example, the symmetry of the payoff functions, uniform distribution of random variables, and the one-dimensional parameterization of the incomplete information can be dropped. However, to start the iterative elimination process, there must be a game with a dominant strategy in the set of possible games. Frankel et al. [3] generalized the analysis even further to consider games with strategic complementarities by any number of players, including a continuum of players.

The advantage of the framework of global games is that it uses the most uncontroversial equilibrium concept of iterative elimination of strictly dominated strategies and produces a unique equilibrium (the risk dominant one) under a small degree of incomplete information for a game with multiple Nash equilibria. Thus, it is useful for applications. Morris and Shin [8] gave a rationale for monetary crises, which can be formulated as the payoff-dominated but risk-dominant equilibrium of a coordination game.

9.3 The Kandori-Mailath-Rob Model**

Kandori et al. [5] formulated a dynamic model in which the stage game is a symmetric 2×2 game with multiple strict Nash equilibria. Over time, players adjust strategies towards a best response to the previous period strategy combination. Given this myopia and a finite population, the dynamical system is a finite Markov chain defined on the number of players who adopt the first strategy. They add "mutation" or experimentation to the process so that the transition probability from any state to any state is positive. Then we can investigate the proportion of time that the society spends in each state, at the limit stationary distribution of the Markov chain.

We consider a simple example to explain their idea. There are 10 students and two kinds of computers, s_1 and s_2 (e.g., Mac and Windows), as possible strategies to use. In each period, the students randomly meet in pairs, and if both use the same s_1 computer, they can share the same good software to study together and get a payoff of 3 each. If students using different strategies meet, the one with s_2 can study without modifying any files and gets 2, while the one with s_1 must create a new file and gets 0. However, s_2 computers are in fact not as efficient as s_1 computers, so if two students with s_2 meet, they get only 2 each. In effect, the game within a pair of students is a Stag Hunt game (Table 9.5).

Table 9.5 Stag hunt game (reproduced)

P1\P2	s_1	s_2
s_1	3, 3	0, 2
s_2	2, 0	2, 2

The Stag Hunt game has two strict Nash equilibria (s_1, s_1) and (s_2, s_2), and one completely mixed Nash equilibrium, in which both players choose s_1 with probability $2/3$. This probability is the threshold for choosing one of the pure strategies before meeting an opponent. That is, the (ex-ante) best response is s_1 if at least $2/3$ of the students (in this case at least 7 students) use s_1, and it is s_2 otherwise. Students repeatedly play the random matching game, and occasionally get a chance to change their computers. When the occasion comes (say, when they have saved enough money or the computer brakes down), a student chooses a best response to the previous-period strategy combination. (Therefore, unlike the players we considered in Chaps. 5 and 8, these players do not maximize their long-run payoff.) In the long run, depending on the initial strategy distribution, there are two possible absorbing states: if at least 7 students started with s_1, then all players eventually choose s_1, and otherwise all students end up with s_2. However, such a path dependence is not useful in predicting which equilibrium is more stable in the long run. We want a prediction that does not depend on the exogenous initial state.

Let us introduce a small degree of randomness in strategy choices. In each period, with a small probability $\epsilon > 0$, each student is replaced by a new student, who has the opposite strategy than the predecessor. With this random change of strategies, the dynamic process of strategy distribution is no longer convergent nor dependent on the initial state. This is because if all players use strategy s_2 initially, it is possible that 7 or more students change to strategy s_1 simultaneously, and after that the strategy distribution moves towards the distribution that all players use s_1. Similarly, even if all players use s_1 initially, if at least 4 students change to s_2 simultaneously at some point, then the distribution moves to s_2 by all players. The number of "mutants" required to upset the s_1-equilibrium is 4, while 7 are needed to upset the s_2-equilibrium. Hence, we expect that the symmetric s_2-equilibrium is more likely to be played for a longer time.

Let us show this with a simple two-state Markov process. The two possible states are when all players choose s_1 (state 1) and when all choose s_2 (state 2). If the strategy adjustment rule is the *best response dynamic* such that all players play a best response to the previous period strategy distribution, then only these states occur. Let the transition probability from state 1 to state 2 be p, and from state 2 to state 1 be p'. To move from state 1 to state 2, we only need four or more players to "mutate" to strategy s_2 so that

$$p = \sum_{n=4}^{10} \binom{10}{n} \epsilon^n (1-\epsilon)^{10-n}.$$

Note that the terms in the summation have ϵ^4 as the common coefficient. Similarly, the transition from state 2 to state 1 needs at least 7 students to be replaced:

$$p' = \sum_{k=7}^{10} \binom{10}{k} \epsilon^k (1-\epsilon)^{10-k}.$$

In this, ϵ^7 is the common coefficient of the terms. By computation,

$$\frac{p'}{p} = \epsilon^3 \frac{120 - 315\epsilon + 280\epsilon^2 - 84\epsilon^3}{210 - 1008\epsilon + 2100\epsilon^2 - 2400\epsilon^3 + 1575\epsilon^4 - 560\epsilon^5 + 84\epsilon^6}$$

so that as ϵ converges to 0, p'/p converges to 0.

For each state $i = 1, 2$, let $x_i(t)$ be the probability that the dynamical system is in the state i in period t (where $x_1(t) + x_2(t) = 1$). In the next period $t + 1$, the probability that the state is 1 is the sum of the probability that the current state is 1 and the process does not move to state 2, and the probability that the current state is 2 and the process moves to state 1. The transition probability to state 2 is similarly computed, and the transition of the distribution $(x_1(t), x_2(t))$ to $(x_1(t+1), x_2(t+1))$ is formulated as follows:

$$\begin{pmatrix} x_1(t+1) \\ x_2(t+1) \end{pmatrix} = \begin{pmatrix} 1-p & p' \\ p & 1-p' \end{pmatrix} \begin{pmatrix} x_1(t) \\ x_2(t) \end{pmatrix}$$

A stationary distribution (x_1^*, x_2^*) of the dynamic process is a fixed point of this difference equation. Hence,

$$(x_1^*, x_2^*) = (\frac{p'}{p+p'}, \frac{p}{p+p'}).$$

(The reader is asked to prove this in Problem 9.2.) As ϵ converges to 0,

$$x_2^* = \frac{1}{1 + p'/p} \to 1,$$

so that the probability becomes 1 for the system to be in state 2 of the risk-dominant Nash equilibrium.

Kandori et al. [5] restricted attention to 2×2 stage games, while Young [12] derived a similar conclusion for a larger class of "weakly acyclic" stage games. He showed that with a different (but finite Markov) dynamic process with random strategy changes, a risk-dominant Nash equilibrium is more robust than a payoff-dominant Nash equilibrium. Intuitively, a dynamic process with random strategy changes (interpreted as experimentations or mistakes) has a stationary distribution that is independent from the initial distribution. The states which remain should be robust against random strategy changes, and what matters is not the payoff values but the set of states from which the system moves towards the limit distribution. Such a set of states is called the *basin of attraction*, and the risk-dominant Nash equilibrium has a larger basin of attraction than the payoff-dominant equilibrium (see Problem 9.3).

In general, however, it is difficult for a general dynamic process to even converge to a Nash equilibrium, let alone an efficient strategy combination which is not a Nash equilibrium. For example, a Prisoner's Dilemma has an efficient strategy combination

Table 9.6 A cyclic game

1\2	s_1	s_2	s_3
s_1	2, −2	−2, 2	−3, −3
s_2	−2, 2	2, −2	−3, −3
s_3	−3, −3	−3, −3	1, 1

that is not a Nash equilibrium, but it is very difficult even for a dynamic process with a payoff improving adjustment rule to converge to the efficient strategy combination.

Moreover, even if a pure-strategy Nash equilibrium exists, the stage game may have a cycle of best responses which does not include a pure-strategy Nash equilibrium. Consider the game in Table 9.6.

This game has a pure Nash equilibrium (s_3, s_3), but the "monotone" dynamic processes which adjust the strategy distribution towards a payoff-improving one, such as the Cournot dynamic $s_i(t+1) = BR_i(s_j(t))$ for $i = 1, 2$ and all t, may only cycle among (s_1, s_1), (s_1, s_2), (s_2, s_2), and (s_2, s_1). Some random change of strategies is needed to force the process to explore enough strategy distributions to find the pure Nash equilibrium. Therefore, perturbed dynamic processes are useful, but favor a Nash equilibrium which is stable against random strategy changes, instead of a payoff-dominant one. (Of course, risk-dominance and payoff-dominance are not mutually exclusive, so that they sometimes give the same prediction. See Problem 9.4.)

Problems

9.1 Prove the three statements in Remark 9.1.

9.2 Prove that the stationary distribution of the differential equation

$$\begin{pmatrix} x_1(t+1) \\ x_2(t+1) \end{pmatrix} = \begin{pmatrix} 1-p & p' \\ p & 1-p' \end{pmatrix} \begin{pmatrix} x_1(t) \\ x_2(t) \end{pmatrix}$$

is

$$(x_1^*, x_2^*) = (\frac{p'}{p+p'}, \frac{p}{p+p'}).$$

9.3 For symmetric 2×2 games, we investigate the equivalence of risk dominance and the largest area of the basin of attraction. Let a_1 and a_2 be positive real numbers. Consider a very large society, and two players are randomly chosen from the society to play the game in Table 9.7. Each player is endowed with one of the pure strategies, 1 or 2. Let $p \in [0, 1]$ be the fraction of the players with strategy 1 in the society. Assume that this is also the probability that a randomly met opponent chooses strategy 1.[3]

[3]The Law of Large Numbers (LLN) guarantees that the fraction of the strategy in the population is the meeting probability. For a continuum population, see Duffie and Sun [2] and Sun [11].

Table 9.7 A symmetric 2×2 coordination game

	1	2
1	a_1, a_1	$0, 0$
2	$0, 0$	a_2, a_2

(a) Prove that, if $p > \frac{a_2}{a_1+a_2}$, then the expected payoff of the players with strategy 1 is greater than the expected payoff of players with strategy 2, and, if $p < \frac{a_2}{a_1+a_2}$, then the opposite holds.

Therefore, if there is a chance to revise one's strategy, a player changes to strategy 1 if $p > \frac{a_2}{a_1+a_2}$ and changes to strategy 2 if $p < \frac{a_2}{a_1+a_2}$. In this case, the basin of attraction of strategy 1 is the interval $(\frac{a_2}{a_1+a_2}, 1]$, and that of strategy 2 is $[0, \frac{a_2}{a_1+a_2})$.

(b) Prove that the strategy combination (1, 1) is a risk-dominant Nash equilibrium if and only if the measure (the length) of the basin of attraction of strategy 1 is greater than that of strategy 2.

9.4 Prove that in the above symmetric 2×2 coordination game, the risk-dominant Nash equilibrium coincides with the payoff-dominant Nash equilibrium.

References

1. Carlsson H, van Damme E (1993) Global games and equilibrium selection. Econometrica 61(5):989–1018
2. Duffie D, Sun Y (2012) The exact law of large numbers for independent random matching. J Econ Theory 147(3):1105–1139
3. Frankel D, Morris S, Pauzner A (2003) Equilibrium selection in global games with strategic complementarities. J Econ Theory 108(1):1–44
4. Harsanyi J, Selten R (1988) A general theory of equilibrium selection in games. MIT Press, Cambridge, MA
5. Kandori M, Mailath G, Rob R (1993) Learning, mutation, and long run equilibria in games. Econometrica 61(1):29–56
6. Matsui A, Matsuyama K (1995) An approach to equilibrium selection. J Econ Theory 65(2):415–434
7. Morris S, Rob R, Shin H (1995) p-Dominance and belief potential. Econometrica 63(1):145–157
8. Morris S, Shin H (1998) Unique equilibrium in a model of self-fulfilling currency attacks. Am Econ Rev 88(3):587–597
9. Oyama D (2002) p-Dominance and equilibrium selection under perfect foresight dynamics. J Econ Theory 107(2):288–310
10. Robson A (1990) Efficiency in evolutionary games: darwin, nash and the secret handshake. J Theor Biol 144(3):379–396
11. Sun Y (2006) The exact law of large numbers via fubini extension and characterization of insurable risks. J Econ Theory 126(1):31–69
12. Young P (1983) The evolution of conventions. Econometrica 61(1):57–84

Chapter 10
Evolutionary Stability*

10.1 Evolutionarily Stable Strategy

So far in this book, we have assumed that strategies are chosen by individual players, and the choice criterion is payoff maximization (rationality hypothesis). However, it is possible to talk about the stability of a strategy distribution without relying on an individual player's choice. This is evolutionary stability. Suppose that each player is endowed with a strategy from birth until death, but the number of offspring who inherit the same strategy from a player depends on the strategy combination of the player and a randomly assigned opponent. In the long run, a strategy (or a distribution of strategies) that generates the greatest number of offspring dominates the society by natural selection, and thus it is a stable strategy (distribution).

There are two ways to formulate evolutionary stability of a strategy distribution of a society. One way is to compare the payoffs (viewed as *fitness*) of strategy distributions under mutations, introduced by Maynard-Smith and Price [11] and the book by Maynard-Smith [10].[1] The other way is to explicitly formulate a dynamic selection process of strategy distributions, and define the limit as the stable distribution. In this chapter we give some "static" stability notions based on the payoff comparison, and one dynamic stability based on the *replicator dynamic*, which is most widely used.

We focus on the simplest model of a single population society. That is, the society consists of a single species (of an animal, if you want to have a concrete idea), and players meet randomly at feeding places. Each player has a strategy (a behavior pattern) from a set Σ, and it behaves according to its strategy. (This set Σ can be the set of pure strategies or mixed strategies.) When a player endowed with strategy $s \in \Sigma$ meets another player with strategy $t \in \Sigma$, the number of offspring of the s-player is determined by a *fitness function* $u(s, t)$. For simplicity, we only consider

[1] Before Maynard-Smith and Price [11], Hamilton [7] and others already formulated a mathematical model of natural selection.

T. Fujiwara-Greve, *Non-Cooperative Game Theory*, Monographs in Mathematical Economics 1, DOI 10.1007/978-4-431-55645-9_10

asexual reproduction to ignore non-essential details such as the probability of meeting with the opposite sex. The fitness function is symmetric so that when s-player meets t-player, the latter's number of offspring is governed by $u(t, s)$.

Maynard-Smith [10] defined an *evolutionarily stable strategy* as a stability notion of a strategy (used by all players in the society) with the property that, against any possible mutant strategy, if the mutants are a sufficiently small minority, then the original strategy has greater post-entry fitness than the mutants' so that eventually all mutants will be driven out by natural selection.

Definition 10.1 A strategy $s^* \in \Sigma$ is an *evolutionarily stable strategy* (ESS) if, for any other strategy $s \in \Sigma \setminus \{s^*\}$, there exists $\epsilon_s > 0$ such that for any $\epsilon \in (0, \epsilon_s)$,

$$(1-\epsilon)u(s^*, s^*) + \epsilon\, u(s^*, s) > (1-\epsilon)u(s, s^*) + \epsilon\, u(s, s). \tag{10.1}$$

The LHS of the inequality (10.1) is the expected fitness value of players endowed with the s^*-strategy, when a small fraction ϵ of mutants with a strategy s enters the society, while the RHS is the mutant's expected fitness value. (We define the ESS by the expected value of the fitness. In some literature, the expected fitness function notation, $u(s^*; (1-\epsilon)s^* + \epsilon s)$, is used instead of $(1-\epsilon)u(s^*, s^*) + \epsilon u(s^*, s)$. We use this notation also to conserve space, when no confusion arises.)

Moreover, the upper bound to the mutants' measure (*invasion barrier*) can depend on the mutants' strategy. Thus, the notion of ESS is weak, not requiring a *uniform invasion barrier* against all possible mutants. Nonetheless, Vickers and Cannings [17] showed that, for finite pure-strategy games, the requirement by ESS is equivalent to the existence of a uniform invasion barrier.

Lemma 10.1 *(Vickers and Cannings [17]) Let Σ be the set of mixed strategies over a finite set of pure strategies. Then $s^* \in \Sigma$ is an ESS if and only if there exists $\overline{\epsilon} > 0$ such that for any $\epsilon \in (0, \overline{\epsilon})$ and any $s \in \Sigma \setminus \{s^*\}$,*

$$(1-\epsilon)u(s^*, s^*) + \epsilon\, u(s^*, s) > (1-\epsilon)u(s, s^*) + \epsilon\, u(s, s).$$

The proof of this Lemma (i.e., the only-if part) is rather complex, so interested readers should refer to [17] or Proposition 2.5 of the book by Weibull [18].

Let us investigate the properties of evolutionarily stable strategies. In the Eq. (10.1), when we take the limit as $\epsilon \to 0$,

$$u(s^*, s^*) \geqq u(s, s^*)\ \forall s \neq s^*$$

holds, so that the combination (s^*, s^*) of an ESS is a symmetric Nash equilibrium of a symmetric two-person game $G = (\{1, 2\}, \Sigma, \Sigma, u, u)$. Although we had not assumed any rational choice, robustness against mutants induces a Nash equilibrium! In addition, ESS is a strictly stronger equilibrium concept than Nash equilibrium. To see this, let us give an equivalent definition to Definition 10.1.

Table 10.1 A Hawk-Dove game

	H	D
H	$-10, -10$	$10, 0$
D	$0, \ 10$	$5, 5$

Definition 10.2 A strategy $s^* \in \Sigma$ is an *evolutionarily stable strategy* if (s^*, s^*) is a symmetric Nash equilibrium of $G = (\{1, 2\}, \Sigma, \Sigma, u, u)$, and moreover, for any $s \in \Sigma \setminus \{s^*\}$,
(1) $u(s^*, s^*) > u(s, s^*)$, or
(2) $u(s^*, s^*) = u(s, s^*)$ and $u(s^*, s) > u(s, s)$ must hold.

Problem 10.1 asks the reader to prove the equivalence of the two definitions. It is now clear that the concept of ESS leads to a subset of symmetric Nash equilibria. In fact, symmetric coordination games have a symmetric mixed strategy which constitutes a Nash equilibrium but is not an ESS. See the discussion before Proposition 10.1. Another immediate consequence of Definition 10.2 is as follows.

Lemma 10.2 *If (s^*, s^*) is a strict Nash equilibrium of $G = (\{1, 2\}, \Sigma, \Sigma, u, u)$, then s^* is an ESS.*

Proof Since it is a strict Nash equilibrium, for any $s \neq s^*$, $u(s^*, s^*) > u(s, s^*)$ holds. Thus (1) of Definition 10.2 is satisfied. □

A strict symmetric Nash equilibrium strategy would satisfy all of the evolutionary stabilities we are going to introduce. This is because the fundamental idea of evolutionary stability is the superiority against a very small fraction of other strategies (mutants), and if a strategy is a best response to itself with a strict inequality, a very small perturbation does not upset the inequality. However, the problem is that there are not many symmetric games which possess a strict symmetric Nash equilibrium.

An interpretation of $s \in \Sigma$ is a strategy distribution in a society. If s is a pure strategy, it can be interpreted as a symmetric strategy distribution such that all players in a society use the same pattern of behavior. If s is a mixed strategy, there are two interpretations. One is that it is still a symmetric strategy distribution where all players use the same mixed-strategy. The other is that it is an asymmetric strategy distribution such that the fraction of players using a pure strategy in the support of the mixed strategy corresponds to the mixture probability of the pure strategy. With the latter interpretation, an ESS can be formulated as a combination of pure strategies which equates the expected fitness of strategies in the support with an additional stability property. To see this, consider the Hawk-Dove game in Table 10.1.

When two players meet, if they are both endowed with the Hawk strategy (H), then they attack each other and get injured badly, so that their fitness is negative. If one player has the Hawk strategy and the other has the Dove strategy (D), then the Hawk player can bully the Dove player and monopolizes the food, resulting in a fitness of 10 for the Hawk and 0 for the Dove. If two Dove players meet, they share the food peacefully, and each get a fitness of 5. If Σ is the set of pure strategies, there is no symmetric Nash equilibrium. Hence, there is no ESS.

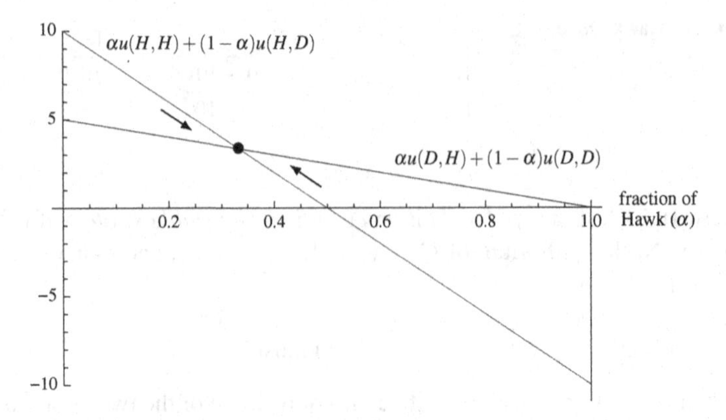

Fig. 10.1 Expected fitness of Hawk and Dove strategies

Let Σ be the set of mixed strategies. Then there is a unique symmetric Nash equilibrium, which uses Hawk with probability $\frac{1}{3}$ and Dove with probability $\frac{2}{3}$. It is easy to check that the mixed strategy $(\frac{1}{3}, \frac{2}{3})$ satisfies the ESS condition, but we can also analyze the property graphically using the expected fitness figure. Following the second interpretation for the mixed strategy in the society, assume that all players have one of the pure strategies. Let $\alpha \in [0, 1]$ be the fraction of players using the Hawk strategy, and the rest are using the Dove strategy. The expected fitness of each pure strategy is

$$\alpha u(H, H) + (1-\alpha)u(H, D) = \alpha(-10) + (1-\alpha)10 = 10 - 20\alpha$$
$$\alpha u(D, H) + (1-\alpha)u(D, D) = \alpha \cdot 0 + (1-\alpha)5 = 5 - 5\alpha.$$

Their graphs are displayed in Fig. 10.1. The expected fitnesses of the two strategies coincide if and only if $\alpha = \frac{1}{3}$. If the fraction of the Hawk strategy increases from $\alpha = \frac{1}{3}$, the expected fitness of the Dove strategy exceeds that of the Hawk strategy. Hence, natural selection implies that the Hawk fraction should decline. Alternatively, if the fraction of players using the Hawk strategy decreases from $\alpha = \frac{1}{3}$, then the expected fitness of Hawk is greater than that of Dove, and thus the Hawk fraction should increase. Therefore, when $\alpha = \frac{1}{3}$, not only do the two pure strategies have the same expected fitness, but also if one of the pure strategies increases, natural selection restores the balance. Thus, an ESS can be interpreted as a stable strategy distribution under natural selection via the relative expected fitness.

Next, consider a symmetric *coordination game*, where two players receive a positive payoff if and only if they use the same strategy.

The game in Table 10.2 has three Nash equilibria: all players using s_0-strategy, all players using s_1-strategy, and a mixed-strategy equilibrium such that s_0 is used with probability $\frac{2}{5}$ and s_1 is used with $\frac{3}{5}$. Let $\alpha \in [0, 1]$ be the fraction of s_0-players in the society. The expected fitness of the two pure strategies is illustrated in Fig. 10.2.

Table 10.2 A symmetric coordination game

	s_0	s_1
s_0	6, 6	0, 0
s_1	0, 0	4, 4

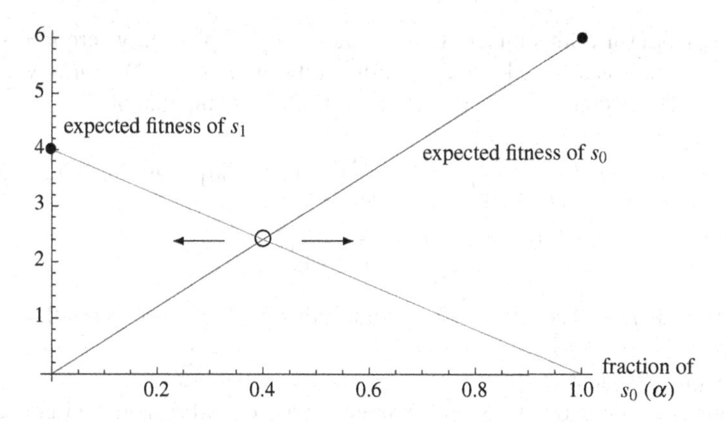

Fig. 10.2 Expected fitness graphs of the symmetric coordination game (Table 10.2)

Unlike the Hawk-Dove game, the mixed-strategy equilibrium strategy is not an ESS in the coordination game. If the s_0-strategy increases from $\frac{2}{5}$, then its expected fitness is greater than that of the s_1-strategy. Conversely, if the s_0-strategy decreases from $\frac{2}{5}$, then its expected fitness becomes less than that of the s_1-strategy. Hence, the symmetric distributions such that all players use s_0 or s_1 are the only stable distributions. At these extreme distributions, the incumbent strategy has a strictly greater fitness than any mutant for sufficiently small fractions of mutants.

Following Weibull [18], we can prove the existence of an ESS for any symmetric 2×2 games. A generic normalized symmetric 2×2 game can be formulated as Table 10.3, where $a_0 \cdot a_1 \neq 0$.

Proposition 10.1 For any normalized symmetric 2×2 game with $a_0 \cdot a_1 \neq 0$, let Σ be the set of mixed strategies. Then there is an ESS in Σ.

Proof We divide generic symmetric games into three classes depending on the signs of a_0 and a_1.

Class 1: $a_0 \cdot a_1 < 0$.

In this class of 2×2 games, there is a unique strict symmetric Nash equilibrium in a pure strategy, and Lemma 10.2 implies that the Nash equilibrium strategy is an ESS.

Class 2: $a_0 < 0$ and $a_1 < 0$.

In this class, the game is essentially a chicken game (Table 3.3) or a Hawk-Dove game (Table 10.1). There is a unique symmetric Nash equilibrium, which consists of a mixed strategy $s^* = (\frac{a_1}{a_0+a_1}, \frac{a_0}{a_0+a_1})$. (The first coordinate is the probability of the

Table 10.3 Normalized symmetric 2 × 2 games

	s_0	s_1
s_0	a_0, a_0	0, 0
s_1	0, 0	a_1, a_1

s_0-strategy.) Let any other mixed strategy be $s = (p, 1-p) \in \Sigma$ (where $p \neq \frac{a_1}{a_0+a_1}$). Because (s^*, s^*) is a mixed-strategy Nash equilibrium, $u(s^*, s^*) = u(s, s^*)$ holds. Hence, we check condition (2) in Definition 10.2. By computation,

$$u(s^*, s) - u(s, s) = \frac{a_1}{a_0 + a_1} p \cdot a_0 + \frac{a_0}{a_0 + a_1}(1-p)a_1 - p^2 \cdot a_0 - (1-p)^2 a_1$$
$$= -(a_0 + a_1)\{p - \frac{a_1}{a_0 + a_1}\}^2.$$

Because $a_0 + a_1 < 0$ and $p \neq \frac{a_1}{a_0+a_1}$, it holds that $u(s^*, s) - u(s, s) > 0$, i.e., (2) is satisfied.
Class 3: $a_0 > 0$ and $a_1 > 0$.

In this class, there are two strict symmetric Nash equilibria and, by Lemma 10.2, the pure strategies are both an ESS. □

Note that, as we saw in the example of the coordination game in Table 10.2, when $a_0 > 0$ and $a_1 > 0$, the completely mixed strategy which constitutes a Nash equilibrium is not an ESS. (Problem 10.2 asks the reader to prove this with Table 10.3.) This shows that the set of ESS is a proper subset of the symmetric Nash equilibrium strategies.

Unfortunately, the existence of an ESS is warranted up to symmetric 2 × 2 games. Among symmetric two-person normal-form games, as soon as the number of pure strategies becomes three, an ESS may not exist even in mixed strategies. In general, the main problem of evolutionary stability concepts is the existence. Let us use the example of the rock-paper-scissors game in Table 3.5 (reproduced in Table 10.4) to show that an ESS may not exist for a symmetric game with three pure strategies.

The rock-paper-scissors game has a unique Nash equilibrium, which consists of $s^* = (\frac{1}{3}, \frac{1}{3}, \frac{1}{3})$ by both players, i.e., a symmetric equilibrium. Because all pure

Table 10.4 Rock-paper-scissors game (reproduced)

1 \ 2	R	P	S
R	0, 0	−1, 1	1, −1
P	1, −1	0, 0	−1, 1
S	−1, 1	1, −1	0, 0

strategies in the support of s^* earn the same expected payoff against s^*, the condition (1) in Definition 10.2 is violated. To see that (2) is also violated, consider a mutant using the pure strategy R. Because

$$u(s^*, R) = 0 = u(R, R)$$

holds, (2) is violated. In fact, any pure strategy can invade the symmetric strategy distribution of s^*. Thus the rock-paper-scissors game has no ESS.

The problem of non-existence of an ESS can be attributed to its "strong" stability requirement. The strength of the requirement, however, gives an ESS nice properties as well. For example, as Kohlberg and Mertens [9] advocated, a good equilibrium should not involve a weakly dominated strategy, and ESS satisfies this property.

Proposition 10.2 No weakly dominated strategy can be an ESS.

Proof Let $s \in \Sigma$ be a weakly dominated strategy. Then, there exists $s' \in \Sigma$ such that

$$\begin{aligned} u(s,t) &\leqq u(s',t) \;\; \forall t \in \Sigma, \\ u(s,t') &< u(s',t') \;\; \exists t' \in \Sigma. \end{aligned} \tag{10.2}$$

We show that this strategy s' can invade the strategy distribution s played by all players. If (s, s) is not a Nash equilibrium, s cannot be an ESS. Hence, assume that (s, s) is a Nash equilibrium. Then, for any $t \in \Sigma$, $u(s, s) \geqq u(t, s)$. In particular, $u(s, s) \geqq u(s', s)$ holds. However, by letting $t = s$ in (10.2), the opposite weak inequality also holds. Thus $u(s, s) = u(s', s)$ must be the case. This violates (1) of Definition 10.2.

Next, letting $t = s'$ in (10.2), we have $u(s', s') \geqq u(s, s')$, violating (2) of Definition 10.2 as well. □

Moreover, an ESS not only constitutes a symmetric Nash equilibrium, but also a proper equilibrium. Recall that proper equilibrium is a refinement of Nash equilibria based on additional rationality in perturbations. Thus, it is quite interesting that an evolutionary stability concept without any rationality assumption implies a rational refinement of Nash equilibria.

Proposition 10.3 (van Damme [1]) If s^* is an ESS, then (s^*, s^*) is a proper equilibrium of $G = (\{1, 2\}, \Sigma, \Sigma, u, u)$.

The proof of this Proposition is in the book by van Damme [1], and it can be also indirectly derived by Proposition 10.4 in the next section. The intuition is that for some perturbed sequence of strategies that converges to (s^*, s^*), natural selection implies that the perturbed strategy does not put much probability on strategies which are not best responses to itself.

10.2 Weaker Stability Concepts

Maynard-Smith [10] was aware of the existence problem of an ESS, and also defined a weaker stability concept called *neutrally stable strategy* (NSS). This simply replaces the strict inequality of (10.1) with a weak inequality. The motivation of this stability concept is that, although a NSS does not have a strictly greater fitness than mutants, at least mutants do not have a strictly greater fitness than the incumbents, either. Thus, all mutants are "neutral" to the incumbent strategy distribution in the society.

Definition 10.3 A strategy $s^* \in \Sigma$ is a *neutrally stable strategy* (NSS) if, for any other strategy $s \in \Sigma \setminus \{s^*\}$, there exists $\epsilon_s > 0$ such that for any $\epsilon \in (0, \epsilon_s)$,

$$(1-\epsilon)u(s^*, s^*) + \epsilon\, u(s^*, s) \geqq (1-\epsilon)u(s, s^*) + \epsilon\, u(s, s). \tag{10.3}$$

The symmetric Nash equilibrium strategy of the rock-paper-scissors game in Table 10.4 was not an ESS, but a NSS. However, if we modify the payoffs of the rock-paper-scissors game slightly, the NSS ceases to exist. Consider the game represented by Table 10.5 due to Weibull [18].

In this game, again $s^* = (\frac{1}{3}, \frac{1}{3}, \frac{1}{3})$ constitutes a unique symmetric Nash equilibrium. When $a = 0$, s^* is a NSS. To see this, the mixed-strategy equilibrium implies that the first terms of both sides of (10.3) are the same. Thus, we only check the second terms. For any $s \neq s^*$, we have that $u(s^*, s) = \frac{a}{3} = 0 = u(s, s)$. Therefore if $a = 0$, then the second terms are also the same, and therefore s^* is a NSS. However, if $a < 0$, then $u(s^*, s) = \frac{a}{3} < 0 = u(s, s)$, so that s^* is not a NSS. Because there is no other symmetric Nash equilibrium, this game with $a < 0$ does not have a NSS.

There are other ways to weaken the notion of ESS. Swinkels [16] restricted the set of possible mutants with an idea similar to proper equilibrium, but the restriction works in the opposite way. For equilibrium refinements, restricting the set of possible perturbations strengthens the equilibrium, while for evolutionary stability, restricting the set of possible mutant strategies makes the robustness requirement weaker.

By Swinkels [16], mutations are interpreted as experimentations of new behavior patterns by some players. The players do not experiment with all possible strategies, but only "rational" ones that yield a reasonable fitness. To be precise, when the incumbent strategy (distribution) is s^*, a different strategy s enters the population with the fraction ϵ, if it is a best response to the post-entry distribution $(1-\epsilon)s^* + \epsilon\, s$. Otherwise the strategy s will not arise in the population, and thus we do not need to worry about it. The new stability notion is defined as follows. Let Σ be the set of mixed strategies, and, against a (mixed) strategy distribution σ in the population, let

Table 10.5 A modified rock-paper-scissors game ($a \leqq 0$)

1 \ 2	1	2	3
1	0, 0	−1, 1 + a	1 + a, −1
2	1 + a, −1	0, 0	−1, 1 + a
3	−1, 1	1 + a, −1	0, 0

the set of best responses in expected fitness be

$$BR(\sigma) = \{s \in \Sigma \mid u(s;\sigma) \geqq u(s';\sigma)\ \forall s' \in \Sigma\}.$$

Definition 10.4 The strategy distribution $s^* \in \Sigma$ is *robust against equilibrium entrants* (REE) if there exists $\overline{\epsilon} > 0$ such that for any $\epsilon \in (0, \overline{\epsilon})$ and any $s \in \Sigma \setminus \{s^*\}$,

$$s \notin BR((1-\epsilon)s^* + \epsilon s).$$

The concept of REE is defined by the uniform invasion barrier, which deters all possible mutants. As we have seen in the previous subsection, when Σ is the set of mixed strategies over a finite set of pure strategies, by Lemma 10.1, ESS also essentially requires a uniform invasion barrier. Therefore, if s^* is an ESS, then there exists $\overline{\epsilon} > 0$ such that for any $\epsilon \in (0, \overline{\epsilon})$ and any $s \in \Sigma \setminus \{s^*\}$, $s \notin BR((1-\epsilon)s^* + \epsilon s)$ holds. That is, an ESS strategy is REE, so that REE is a weaker stability concept than ESS. Furthermore, Swinkels [16] proved that REE is stronger than Nash equilibrium or even proper equilibrium.

Proposition 10.4 (Swinkels [16]) If $s^* \in \Sigma$ is REE, then (s^*, s^*) is a symmetric Nash equilibrium of $G = (\{1,2\}, \Sigma, \Sigma, u, u)$. Moreover, it is a symmetric proper equilibrium.

Proof We only prove the first statement. Let s^* be REE. By the definition, there exists $\overline{\epsilon} > 0$ such that, for any $\epsilon \in (0, \overline{\epsilon})$ and any $s \in \Sigma \setminus \{s^*\}$, $s \notin BR((1-\epsilon)s^* + \epsilon s)$. Fix an arbitrary $\epsilon \in (0, \overline{\epsilon})$. For any $s \in \Sigma$, consider a correspondence $B(s) = BR((1-\epsilon)s^* + \epsilon s)$. This correspondence $B : \Sigma \rightarrow\rightarrow \Sigma$ has a fixed point by Kakutani's Fixed Point Theorem, as in the proof of Theorem 3.1. Thus, there exists $x \in \Sigma$ such that $x \in B(x)$. By the definition of REE, it must be that $x = s^*$. Therefore, $s^* \in B(s^*)$, which means that (s^*, s^*) is a symmetric Nash equilibrium. For the proof of the second statement, see Swinkels [16]. □

Hence, if a strategy is an ESS, then it is REE, and moreover, it is a symmetric proper (Nash) equilibrium strategy. To show that REE is a strictly weaker stability concept than ESS, reconsider the rock-paper-scissors game in Table 10.4. This game does not have an ESS, but the strategy $s^* = (\frac{1}{3}, \frac{1}{3}, \frac{1}{3})$, which constitutes the unique and symmetric Nash equilibrium, is REE. For any $s \neq s^*$, the post-entry distribution $(1-\epsilon)s^* + \epsilon s$ does not assign the same probability to all pure strategies. With such a skewed distribution, there is a unique best response in a pure strategy. Moreover, the best response is not in the direction towards which s shifts s^*. For example, when the post-entry distribution has a higher probability on R than the two others, then P is the best response, and so on. Thus, s will not be a best response to $(1-\epsilon)s^* + \epsilon s$.

Nonetheless, the REE and NSS concepts are independent. As we have seen, when an ESS exists, it is a NSS and REE. However, there is a game with a NSS which is not REE. For example, consider a (trivial) game in which all strategy combinations have the same fitness. Then all strategies are NSSs, but no strategy is REE. This example also shows that REE does not always exist. Swinkels [16] also defined a set-equilibrium concept by extending the idea of REE, but even the set-equilibrium may not exist for all symmetric games.

Gilboa and Matsui [6] also considered a set-equilibrium concept called a *cyclically stable set*. This is a limit of a dynamic process, and its existence is guaranteed.

In summary, there are many concepts of evolutionary stability with both advantages and disadvantages. This area of research is still evolving.

10.3 Cheap Talk

As we discussed in Sect. 9.1, it may be possible to select the payoff-dominant Nash equilibrium among others, if players can communicate before playing a game. Although the stage game is a normal-form game, adding the pre-play communication stage makes it an extensive-form game. To analyze the communication effect in terms of evolutionary stability, we consider the induced normal-form game of the two-stage game. Note, however, that strategies which behave the same way on the play path receive the same payoff in the induced normal-form game (this can be seen in the example in Sect. 10.5 as well), and thus there is no Evolutionarily Stable Strategy. Hence, we adopt Neutrally Stable Strategy as the stability concept. The following analysis is inspired by Robson [12].

Consider the Stag Hunt Game in Sect. 9.1 again. (Reproduced below in Table 10.6.)

The Stag Hunt Game has two strict Nash equilibria, and (Stag, Stag) is efficient (payoff-dominant), while (Hare, Hare) risk dominates (Stag, Stag). In a large population of homogeneous players, randomly matched pairs of players play this game. We add a communication stage so that, before playing the game, each player can send a message from a set $M = \{a, b\}$ to the other player. The messages do not affect any player's fitness, and thus are cheap talk. A player's strategy is now a behavioral strategy to choose a message and an action from $\{S, H\}$ after observing the opponent's message. (We only consider pure strategies. It is possible to make a player's action in the Stag Hunt Game dependent on her/his own message as well, but this additional freedom does not alter the analysis below, and thus is omitted.)

To count the pure strategies, there are two possible messages and two possible ways to react to the two possible messages from the randomly matched opponent. Hence there are eight pure strategies, which are listed in Table 10.7. The first column is the player's own message, and the second column is the conditional action in the Stag Hunt Game depending on the opponent's message. To name a pure strategy, we line up the actions in this order (for example, the first strategy will be denoted as aSS).

Table 10.6 Stag Hunt game (reproduced)

P1 \ P2	Stag	Hare
Stag	3, 3	0, 2
Hare	2, 0	2, 2

Table 10.7 Strategies for Stag Hunt game with cheap talk

Message to send	Action after seeing a	Action after seeing b
a	Stag	Stag
a	Stag	Hare
a	Hare	Stag
a	Hare	Hare
b	Stag	Stag
b	Stag	Hare
b	Hare	Stag
b	Hare	Hare

We now show that, given an arbitrary message $m \in \{a, b\}$, the symmetric strategy mHH played by all players is not a NSS. For example, consider $s^* = aHH$. As a mutant, suppose that $s = bHS$ enters the population. A mutant player using s observes message a when it meets an incumbent player, and plays Hare, so that the outcome will be (Hare, Hare). When mutants meet each other, they observe message b and play Stag, hence the outcome will be (Stag, Stag). Therefore, when $\epsilon > 0$ is the fraction of s in the population, the mutant strategy's expected fitness is

$$(1-\epsilon)u(s, s^*) + \epsilon\, u(s, s) = (1-\epsilon)2 + \epsilon \cdot 3,$$

while the incumbent players play Hare regardless of the opponent's message, so that their expected fitness is

$$(1-\epsilon)u(s^*, s^*) + \epsilon\, u(s^*, s) = 2.$$

Hence, the incumbents' fitness is smaller than the mutants'. The reason is that mutants can recognize each other by sending a different message from the incumbents' message, and they can play a more efficient action combination among mutants. In addition, mutants can also recognize when they meet an incumbent and coordinate with incumbents as well.

By contrast, for any message $m \in \{a, b\}$, the strategy mSS is a NSS, when Σ is the pure strategies in Table 10.7 (the proof is Problem 10.5). The idea is that incumbents can earn the efficient payoff 3 among themselves, and for sufficiently small ϵ, even if they earn 0 against a mutant, their expected fitness will not be lower than the mutants'. This is due to the payoff-dominance of the (Stag, Stag) equilibrium.

Unfortunately, when the set of messages is fixed, and mixed messages (a probability distribution over messages) can be used, it is possible that an inefficient Nash equilibrium strategy becomes a part of a NSS. The intuition is that an incumbent strategy can randomize all possible messages with a positive probability to "jam the signal", and after that can play the inefficient action regardless of the observation. Then no mutants can recognize each other to play an efficient action combination among themselves. For example, let mHH be the incumbent strategy where $m = \alpha a + (1-\alpha)b$ (with $\alpha \in (0, 1)$) is a completely mixed message. Then both messages are observed with a positive probability, and the mutants cannot play (Stag, Stag) among themselves. In summary, an inefficient Nash equilibrium strategy becomes unstable when there is a message that is not used by incumbents (see Farrell [4]).

10.4 Asymmetric Games

So far, we have dealt with only symmetric games. It is desirable to also analyze asymmetric games in the evolutionary setting. For example, even animals of the same species behave differently depending on which one arrives first to a feeding place, as if the game is an asymmetric game. We keep the assumption that players come from a single population and play a two-person game with a randomly-matched opponent. To incorporate an asymmetric stage game, it is assumed that each player is randomly assigned to one of the "roles" of the stage game.

Let the stage game be $G = (\{1, 2\}, \Sigma_1, \Sigma_2, u_1, u_2)$, to be played by randomly paired players. Each player is endowed with a role-contingent strategy, that is, a behavior pattern $\sigma_1 \in \Sigma_1$ when it becomes player 1 in the stage game, and another behavior pattern $\sigma_2 \in \Sigma_2$ when it is assigned the role of player 2. The probability of becoming one of the players is $1/2$. Then, the game can be converted to an ex-ante symmetric game as follows. The set of strategies is the set of role-contingent strategies (σ_1, σ_2), and we denote it as $\Sigma = \Sigma_1 \times \Sigma_2$. This is a common strategy set for all players. The (expected) fitness of a player endowed with (σ_1, σ_2) when it meets a player with (τ_1, τ_2) is defined as

$$U((\sigma_1, \sigma_2), (\tau_1, \tau_2)) = \frac{1}{2}u_1(\sigma_1, \tau_2) + \frac{1}{2}u_2(\tau_1, \sigma_2).$$

Since this fitness function is symmetric, the game is symmetric, and we can use the same stability concepts defined thus far to asymmetric game situations.

10.5 Stability for Extensive Form Games

When randomly-met players engage in an extensive-form game, what strategy (combination) should be considered as evolutionarily stable? There is in fact no clear agreement among researchers on this problem yet. An extensive-form game can be converted into the induced normal-form game, and hence one might think that we can simply use the stability concepts developed for normal-form games. But this method is not desirable. To see this, consider the following example provided in Selten [15].

This is an asymmetric game, and thus using the framework in Sect. 10.4, we assume that when two players meet, each player has $1/2$ probability to become each of the roles 1 and 2. A strategy of a player is a role-contingent action plan.

Fig. 10.3 Selten's example

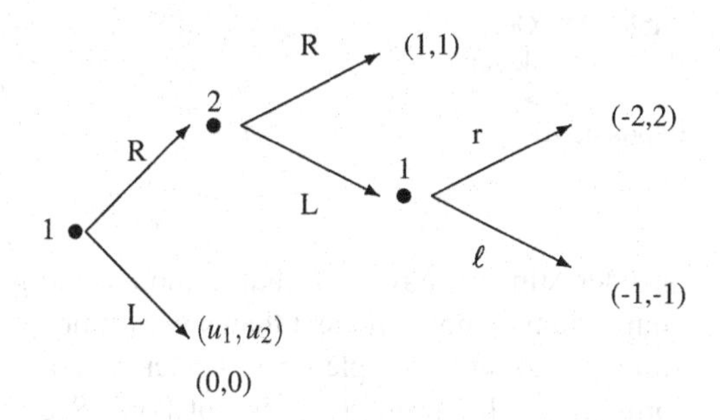

Then the converted normal-form game of Fig. 10.3 has many Nash equilibria. The pure strategies which constitute a symmetric Nash equilibrium are $(L\ell, L)$, (Lr, L), and $(R\ell, R)$, where the first coordinate is the actions for two information sets of role 1 and the second coordinate is the action of role 2. Among them, the one that corresponds to the subgame perfect equilibrium of the extensive form is $(R\ell, R)$. However, this strategy is not an ESS. Suppose that all players use the role-contingent strategy $(R\ell, R)$, that is, when the player is assigned role 1, then she plays $R\ell$ and, in role 2, she plays R. The expected fitness of this strategy against the same strategy is $U((R\ell, R), (R\ell, R)) = \frac{1}{2}u_1(R\ell, R) + \frac{1}{2}u_2(R\ell, R) = 1$. Consider a mutant strategy (Rr, R). This strategy earns the same expected fitness against the incumbents, $U((Rr, R), (R\ell, R)) = 1$, and, moreover, $U((Rr, R), (Rr, R)) = 1 = U((R\ell, R), (Rr, R))$ holds. Therefore, neither of the two conditions of Definition 10.2 is satisfied.

The problem is, as we discussed in Sect. 10.3, that strategies which play differently at off-path information sets have the same fitness as the incumbents in extensive-form games. Therefore an ESS may not exist. The NSS concept can admit mutants which behave differently off the play path, but then a lot of (even mixed-strategy) Nash equilibrium strategies become neutrally stable.

Selten proposed to convert the extensive-form game into a sequence of perturbed *agent normal-form game* $\{G(\epsilon_k) : k = 1, 2, \ldots\}$ such that each information set belongs to a different agent of the original role, and all agents choose all actions with at least probability $\epsilon_k > 0$, where $\lim_{k\to\infty} \epsilon_k = 0$. If each $G(\epsilon_k)$ possesses an ESS, the limit as $k \to \infty$ is a stable strategy called *limit evolutionarily stable strategy*. This idea is similar to sequential equilibrium and induces sequential rationality. In fact, for the example in Fig. 10.3, only $(R\ell, R)$ is a limit evolutionarily stable strategy. (The reader is asked to prove this in Problem 10.6.)

10.6 Voluntarily Separable Repeated Prisoner's Dilemma**

We introduce the Voluntarily Separable Repeated Prisoner's Dilemma (VSRPD) by Fujiwara-Greve and Okuno-Fujiwara [5] to show that the concept of NSS is also effective as a stability concept for extensive-form games. Moreover, this game has

Table 10.8 General Prisoner's Dilemma ($g > c > d > \ell$) (reproduced)

	C	D
C	c, c	ℓ, g
D	g, ℓ	d, d

an interesting feature such that, although the game is symmetric, an asymmetric equilibrium is more efficient than any symmetric equilibrium, if we restrict the pure strategies to some simple ones. Under random matching with a one-shot normal-form game, this property does not hold. See for example Figs. 10.1 and 10.2 in Section 10.1. Among Nash equilibria, the asymmetric mixed-strategy equilibrium is not more efficient than some of the symmetric ones.

In ordinary repeated games, the same set of players play the same stage game for an exogenously given number of periods (including infinitely many periods). However, in reality, it should be possible to exit a repeated interaction when a player decides to. By contrast, the random matching game we have been implicitly assuming in this chapter changes the opponents every time. This is another extreme situation, and if players agree to play again, they should be able to repeat the same interaction.

The VSRPD model encompasses these two extreme cases. At the beginning of the dynamic game, pairs are randomly formed from a homogeneous population, and matched players engage in a general Prisoner's Dilemma shown in Table 10.8. After observing the action combination, the matched players simultaneously decide whether to keep the partnership or to end it. If at least one player chooses to end the interaction, the pair is dissolved and each player gets a random new opponent in the next period.[2] If both players choose to keep the partnership, they can play the Prisoner's Dilemma game with each other, skipping the random matching process. (To make the long-run total fitness well-defined, we assume that each player lives to the next period only with probability δ. Hence, if one of the partners dies, even if they have agreed to play again, the surviving player must get a random new opponent in the next period. When a player dies, a new player enters the dynamic game to keep the population size constant.) Thus, if matched players choose to stay together, the game becomes a repeated game, and if at least one player ends the partnership every period, it becomes a random matching game. Each player is assumed to be endowed with a strategy (an action plan) of the VSRPD when they enter the dynamic game.

The informational structure of the dynamic game is as follows. Each player perfectly observes her/his current opponent's action, but none of the actions in the other pairs. The population is assumed to be a continuum, so that a player never meets the same opponent again. In addition, due to the player turnover by the death and entry of a new player (who is assumed to have no information about any other player's past action history), players who have no partner at the beginning of a period have

[2]This assumption is plausible in many realistic cases, such as animal encounters. If both players need to agree to end the partnership, like in a divorce process, then the same set of outcomes as that of an infinitely repeated game arises, because players can force each other to stay in the game forever.

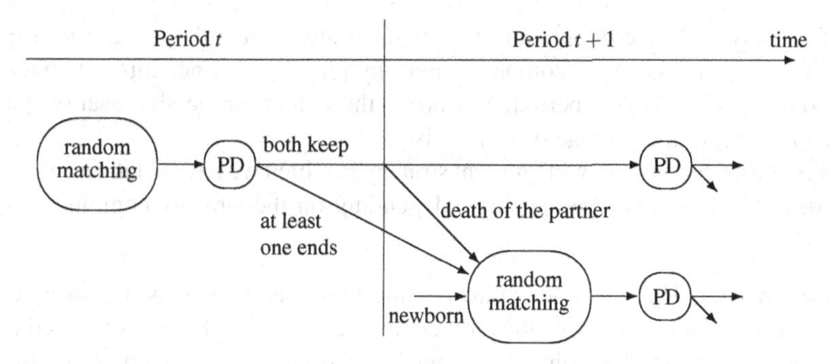

Fig. 10.4 Voluntarily separable repeated Prisoner's Dilemma

various backgrounds, and thus one cannot find out whether a player looking for a new partner has done something bad in the past, and so on. Figure 10.4 shows the outline of the game (abbreviating the random matching details and Prisoner's Dilemma).

Let us consider the relevant "fitness" in this game. So far, the fitness comparison has been based on one-shot payoffs of the stage game, because each player is endowed with a strategy for the one-shot game as well. In the VSRPD, a pure strategy specifies a Prisoner's Dilemma action to the current partnership history[3] and the keep/end decision to the partnership history including this period's action combination. This is a long-run action plan over an endogenously determined horizon for each partnership. Therefore, we consider the (long-run) average payoff from the entire dynamic game as fitness. In ordinary infinitely repeated game with discount factor δ, the average payoff is defined by the total payoff multiplied with $(1-\delta)$. However, in the VSRPD, the average payoff computation is more complex, because if there are multiple strategies in the population, depending on which one the randomly met opponent has, the (expected) length of the partnership varies. Therefore, we have to take the endogenous length of a partnership with another strategy into account to take the average. For evolutionary stability, even if the incumbent strategy is unique, we must consider another mutant strategy, and thus the average payoff computation with various lengths of partnerships is important.

Let us give an example to show how the (long-run) average fitness is computed. First, consider a myopic strategy, which is optimal if the game is a one-shot Prisoner's Dilemma.

Definition 10.5 The d_0-strategy plays action D for any partnership history, including the empty history (the first period of a newly-formed partnership), and ends the partnership for any partnership history.

[3]Other possible information that a player may base her/his actions upon is her/his own past histories (including actions by past opponents, e.g., Kandori [8] and Ellison [3]) and the time count since (s)he entered the game. For simplicity, we only consider partnership-independent strategies, ignoring these kinds of information.

If all players in the population use the d_0-strategy, then the (long-run) average fitness of a player is easily computed. For any player, a partnership lasts only one period and gives d in any period. Therefore, the within-partnership average payoff is d, and the lifetime average payoff is also d.

Next, consider the following mutant strategy in which we have to take into account two different lengths of partnerships, depending on the strategy combination in a match.

Definition 10.6 The c_1-strategy plays action D if it is the first period of a newly-formed partnership, and keeps the partnership if and only if (D,D) is observed in the first period. If the partnership is continued, the strategy plays action C in the t-th period (where $t \geqq 2$) of a partnership, and keeps the partnership if and only if (C,C) is observed in that period. (Hence, if D is observed in the t-th period with $t \geqq 2$, then this strategy ends the partnership.[4])

The c_1-strategy is called the *one-period trust-building* strategy. A c_1-player does not cooperate in the first period of any new match, but does not end the partnership even if the partner also played D. The first period can be called the "trust-building" phase to see if the opponent is willing to continue the partnership or not. If the two players agree to continue the partnership, then "trust" is built, and from then on a c_1-player cooperates as long as the partner also cooperates. The part from the second period on is similar to the grim-trigger strategy in the repeated Prisoner's Dilemma, except that the punishment is to end the partnership. (The trust-building phase need not be one period. We also consider a two-period trust-building strategy below.)

Suppose that in the symmetric population of d_0-strategy players, $\epsilon > 0$ of the players mutates to the c_1-strategy, in the matching pool. The incumbent d_0-player earns d from both a d_0-strategy opponent and a c_1-strategy opponent, and the partnership always ends in one period. Therefore, the post-entry average fitness of the d_0-strategy is d. By contrast, a c_1-player has different lengths of partnerships and within-partnership total payoffs, depending on the randomly-matched partner's strategy. If a c_1-player meets a d_0-player, (s)he gets d and the partnership ends immediately. (Recall that the partnership dissolves if at least one player chooses to end it.) If a c_1-player meets another c_1-player, then they continue the partnership with probability δ^2, with which both partners survive, and the one-shot payoff is d in the first period and c after that. Hence, the long-run total payoff in this partnership is

$$V(c_1, c_1) = d + \delta^2 c + \delta^4 c + \cdots = d + \delta^2 \frac{c}{1-\delta^2}. \tag{10.4}$$

The expected length of this partnership is

$$1 + \delta^2 + \delta^4 + \cdots = \frac{1}{1-\delta^2}.$$

[4] To be precise, this definition does not specify actions in off-path information sets, because they do not affect the NSS analysis.

Hence, we define the long-run average fitness of a c_1-strategy in the population of $(1-\epsilon)d_0 + \epsilon\, c_1$ as

$$v(c_1; (1-\epsilon)d_0 + \epsilon\, c_1) = \frac{(1-\epsilon)d + \epsilon(d + \delta^2 \frac{c}{1-\delta^2})}{(1-\epsilon)\cdot 1 + \epsilon \frac{1}{1-\delta^2}}.$$

The numerator is the expected (long-run) payoff and the denominator is the expected length of the two possible partnerships. The average fitness of the d_0-strategy can also be formulated in the same way:

$$v(d_0; (1-\epsilon)d_0 + \epsilon c_1) = \frac{(1-\epsilon)d + \epsilon \cdot d}{(1-\epsilon)\cdot 1 + \epsilon \cdot 1} = d.$$

Let us analyze a stable strategy distribution, using the above setup. First, we show that the d_0-strategy played by all players is a Nash equilibrium in the following sense.

Proposition 10.5 For any individual survival probability $\delta \in (0, 1)$, the symmetric distribution of the d_0-strategy in the population is a Nash equilibrium of the VSRPD; namely, for any other strategy s,

$$v(d_0; d_0) \geqq v(s; d_0).$$

Proof Classify the other strategies into two families: ones which play C in the first period of a new match and others. When all players use the d_0-strategy, any strategy in the latter family has the (long-run) average fitness $v(s; d_0) = d$, which does not exceed $v(d_0; d_0) = d$. Any strategy in the former family gets $\ell < d$ in the first period of a new partnership, and then the partnership is ended by the d_0-player. Hence, $v(s; d_0) < d$. □

The definition of a Nash equilibrium in an evolutionary game means that only a single player mutates to a different strategy, and this player must meet incumbents all the time. In the above Proposition, $s \neq d_0$ is always matched with a d_0-player and cannot have a higher average fitness than the d_0-players. By contrast, evolutionary stability considers mutations by a positive fraction of players, and mutants can form a partnership among themselves (although with a small probability). In this case, the fitness of a mutant pair becomes important, just like in the case of cheap talk in Sect. 10.3. Moreover, mutants may be able to form a long-term cooperative relationship in the VSRPD, and we can show that the d_0-strategy is in fact not a NSS.

Let us define a NSS by the average fitness. A strategy s^* is a Neutrally Stable Strategy if, for any other strategy s, there exists $\epsilon_s > 0$ such that for any $\epsilon \in (0, \epsilon_s)$,

$$v(s^*; (1-\epsilon)s^* + \epsilon s) \geqq v(s; (1-\epsilon)s^* + \epsilon s).$$

Proposition 10.6 For any $\delta \in (0, 1)$, d_0-strategy is not a NSS.

We leave the proof for the readers, in Problem 10.7. The idea is that if mutants using the c_1-strategy enter the population with a positive fraction $\epsilon > 0$, they earn the same average fitness against the incumbents (d_0-strategy), and earn a strictly higher average fitness when they meet each other. Therefore in the long-run, c_1-players are not neutral to the population of d_0-players. This logic is similar to the equilibrium selection by cheap talk. Mutants can distinguish each other by agreeing to continue the partnership after (D, D) in the first period.

The above analysis shows that the neutral stability concept yields a proper subset of Nash equilibria in the VSRPD as well. Let us turn to constructing a NSS. Since the VSRPD game includes an infinitely repeated game, it is easy to guess that for large δ, there should be many NSSs, by a similar logic to the Folk Theorem. For details, the reader should consult Fujiwara-Greve and Okuno-Fujiwara [5], but here we at least explain how the c_1-strategy may become a NSS. The following strategy features strong mutants against c_1-strategy (called a two-period trust-building strategy and denoted as c_2).

Definition 10.7 A strategy c_2 specifies actions as follows. In the first two periods of a new partnership, play D regardless of the partnership history, and keep the partnership if and only if (D, D) is observed in that period. If the partnership continues to the t-th period (with $t \geqq 3$), play C, and keep if and only if (C, C) is observed in that period. Otherwise, end the partnership.[5]

A mutant using the c_2-strategy can earn a large payoff g in the second period if it meets a c_1-player, and $c, c, \ldots$ if it meets a mutant. It can be shown (see Fujiwara-Greve and Okuno-Fujiwara [5]) that it is sufficient to deter this strategy for the c_1-strategy to be a NSS, that is, the c_2-strategy is the relevant one-step deviation from the c_1-strategy.

Suppose that $(1 - \epsilon)$ of the population in the matching pool consists of the c_1-strategy and ϵ of the population consists of the c_2-strategy. We investigate when the c_1-strategy gets a higher average fitness than the c_2-strategy does. When a c_1-player meets another c_1-strategy, their long-run in-match payoff is $V(c_1, c_1) = d + \frac{\delta^2 c}{1-\delta^2}$ from (10.4). When a c_1-player meets a mutant, the pair lasts at most two periods, and the long-run in-match payoff of the c_1-player is

$$V(c_1, c_2) = d + \delta^2 \ell.$$

The expected length of this type of partnership is $1 + \delta^2$. Next, if a mutant c_2-player meets an incumbent c_1-player, the long-run in-match payoff is

$$V(c_2, c_1) = d + \delta^2 g,$$

[5] Again, this specification is not complete for off-path information sets. Any strategy with this on-path action plan works.

and the expected length of this match is $1 + \delta^2$. If two mutants meet each other, the long-run in-match payoff is

$$V(c_2, c_2) = d + \delta^2 d + \delta^4 c + \delta^6 c + \cdots = d + \delta^2 d + \delta^4 \frac{c}{1-\delta^2},$$

and the expected length of this type of match is $1/(1-\delta^2)$. Therefore, the average fitness of c_1-players and c_2-players are as follows:

$$v(c_1; (1-\epsilon)c_1 + \epsilon c_2) = \frac{(1-\epsilon)(d + \delta^2 \frac{c}{1-\delta^2}) + \epsilon(d + \delta^2 \ell)}{(1-\epsilon)\frac{1}{1-\delta^2} + \epsilon(1+\delta^2)},$$

$$v(c_2; (1-\epsilon)c_1 + \epsilon c_2) = \frac{(1-\epsilon)(d + \delta^2 g) + \epsilon(d + \delta^2 d + \delta^4 \frac{c}{1-\delta^2})}{(1-\epsilon)(1+\delta^2) + \epsilon\frac{1}{1-\delta^2}}.$$

To guarantee that a c_1-player earns no less than a c_2-player does, when ϵ is very small, consider the limit case:

$$\lim_{\epsilon \to 0} v(c_1; (1-\epsilon)c_1 + \epsilon c_2) = (1-\delta^2)d + \delta^2 c$$

$$\lim_{\epsilon \to 0} v(c_2; (1-\epsilon)c_1 + \epsilon c_2) = \frac{1}{1+\delta^2}(d + \delta^2 g).$$

Hence, if

$$(1-\delta^2)d + \delta^2 c > \frac{1}{1+\delta^2}(d + \delta^2 g) \iff \delta > \sqrt{\frac{g-c}{c-d}},$$

then by the continuity of the average fitness with respect to ϵ, there exists $\bar{\epsilon} > 0$ such that for any $\epsilon \in (0, \bar{\epsilon})$, the average fitness of c_1 is not less than that of c_2. This condition holds when $g + d < 2c$ and δ is sufficiently large.

Finally, we also consider whether players can play C in the Prisoner's Dilemma from the beginning of any new partnership (i.e., with a total stranger). For example, consider the following trigger-strategy-like behavior rule.

Definition 10.8 A strategy c_0 specifies actions as follows. In any period of a partnership (including the first period), play C, and keep the partnership if and only if (C, C) is observed in that period. Otherwise, end the partnership.

Note that it is not even a Nash equilibrium when all players use the c_0-strategy. This is because, if a single player mutates to the d_0-strategy, then (s)he earns g in every partnership (which ends in one period), and thus $v(d_0; c_0) = g$, which is clearly greater than the average fitness $v(c_0; c_0) = c$ of the c_0-players. The problem is that in the VSRPD, a player can unilaterally end a partnership and cannot be recognized by future opponents. Hence, hit and run by the d_0-strategy is advantageous.

Interestingly, a (mixed-strategy) combination of c_0-players and c_1-players (who have an advantage against d_0-players) may become stable. We explain the stability logic, and compare the fitness with that of the symmetric c_1-strategy equilibrium.

Let $\alpha \in [0, 1]$ be the fraction of the population who have the c_0-strategy and the rest be the c_1-strategy. If the same strategy players meet, they continue the partnership as long as they live, but if different strategy players meet, in the first period the c_0-player plays C and the c_1-player plays D, so that the partnership is ended immediately by the c_0-player.

The long-run in-match payoff of a c_0-player when meeting another c_0-player is

$$V(c_0, c_0) = c + \delta^2 c + \delta^4 c + \cdots = \frac{c}{1-\delta^2}.$$

Similarly, the long-run in-match payoff of a c_1-player against another c_1-player is $V(c_1, c_1) = d + \frac{\delta^2 c}{1-\delta^2}$ by (10.4). Notice that $V(c_0, c_0) > V(c_1, c_1)$. The long-run payoff of the non-Nash equilibrium pair is greater. (This is because the stage game is a Prisoner's Dilemma.)

When a c_0-player meets a c_1-player, the former gets ℓ and the latter receives g, and the match is ended by the c_0-player. Thus, the long-run in-match payoffs are respectively,

$$V(c_0, c_1) = \ell,$$
$$V(c_1, c_0) = g.$$

Clearly, $V(c_1, c_0) > V(c_0, c_1)$, but for large δ, a long-term partnership is beneficial so that $V(c_1, c_1) > V(c_1, c_0)$ should hold.

Overall, the average fitness of a c_0-player is

$$v(c_0; \alpha\, c_0 + (1-\alpha)c_1) = \frac{\alpha V(c_0, c_0) + (1-\alpha)V(c_0, c_1)}{\alpha \frac{1}{1-\delta^2} + (1-\alpha)\cdot 1},$$

and the average fitness of a c_1-player is

$$v(c_1; \alpha\, c_0 + (1-\alpha)c_1) = \frac{\alpha V(c_1, c_0) + (1-\alpha)V(c_1, c_1)}{\alpha \cdot 1 + (1-\alpha)\frac{1}{1-\delta^2}}.$$

From these formulas, we can see that the average fitnesses of the two strategies are non-linear functions of the fraction α of the c_0-players. As Fig. 10.5 shows, $v(c_0; \alpha\, c_0 + (1-\alpha)c_1)$ is a concave function of α, while $v(c_1; \alpha\, c_0 + (1-\alpha)c_1)$ is a convex function. (The proof is Problem 10.8.)

The intuition for the non-linearity of the average fitness function is that the meeting probabilities of various strategies do not coincide with the long-run shares of the strategies. In ordinary evolutionary games, which are essentially random matching games, if there are multiple strategies in the population, in each period the meeting

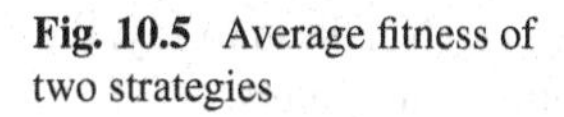
Fig. 10.5 Average fitness of two strategies

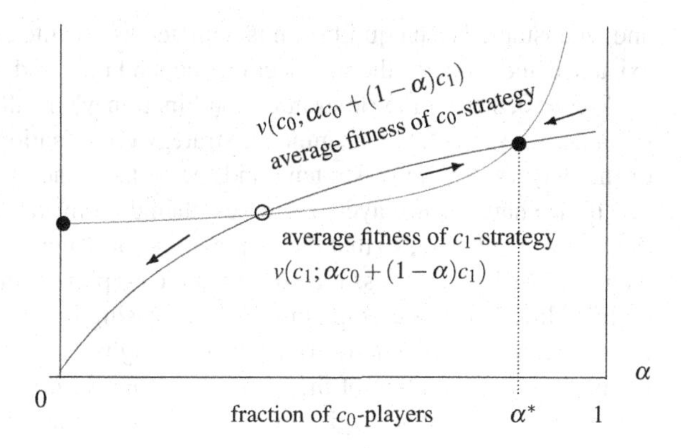

probability is the same as the share of the strategies by the Law of Large Numbers (see Duffie and Sun [2]). By contrast, in the VSRPD and with c_0- and c_1-strategies in the population, the same strategy pairs rarely dissolve (only by the death), and mostly the players who happened to be matched with a player of a different strategy return to the random matching process in the next period. Therefore, the long-run meeting probability of various strategies differs from the shares of the strategies, and instead depends on the initial strategy distribution and the player's own strategy.

For example, consider the case near $\alpha = 0$, where the population is predominantly occupied by c_1-players. A player endowed with the c_0-strategy would meet a c_1-player very often, and then (s)he must return to the random matching process in the next period, and again is likely to meet another c_1-player. Due to the advantage of a c_1-player against a c_0-player in a match, as α declines, not only the fitness of c_0-players decreases but also the decrease accelerates.

On the other hand, when the population is near $\alpha = 1$ so that most of the players are c_0-players, then a c_0-player is likely to meet another c_0-player, and (s)he can avoid going back to the random matching process to meet a c_1-player. As α declines, the decrease of c_0-players' fitness does not accelerate thanks to the long-term relationships among themselves. This is the intuition of the concavity for the average fitness of the c_0-strategy.

As for c_1-players, when α is near 0, they are likely to meet another c_1-player, and thus they cannot exploit c_0-players. Hence, their fitness does not increase very much as α increases. By contrast, when α is near one, they can meet c_0-players very often, which not only gives them a high one-shot payoff, but also a greater likelihood of meeting a "victim" in the next period. Thus, the fitness of c_1-players increases dramatically, as α increases.

The average fitness functions may intersect twice,[6] because of the concavity/convexity, as Fig. 10.5 illustrates. The intersection with the larger α is the stable one. In Fig. 10.5, two neutrally stable strategy combinations are shown as black circles,

[6]To be precise, a sufficient condition is needed. See Fujiwara-Greve and Okuno-Fujiwara [5].

and an unstable Nash equilibrium is depicted as a white circle. (Needless to say, the existence and value of the intersections depend on δ and other parameters.)

At $\alpha = 0$, the symmetric strategy combination where all players use the c_1-strategy is stable. At $\alpha = \alpha^*$, the asymmetric strategy combination is stable, under which α^* of the players use the c_0-strategy and the rest use the c_1-strategy. If α exceeds α^*, i.e., if the share of c_0-players increases, then their average fitness becomes less than that of the c_1-players, so that the c_1-players should increase their share to restore the balance. If α becomes less than α^*, then the c_0-players increase their share.[7]

In addition, as the average fitness is increasing in α, the asymmetric equilibrium consisting of the c_0- and c_1-strategies has a higher average payoff (for all players in the population) than that of the symmetric equilibrium of c_1-strategy. Therefore, if we restrict the set of strategies to c_0- and c_1-strategies, the asymmetric equilibrium is more efficient than the unique symmetric equilibrium of the c_1-strategy, which does not happen in the ordinary random matching games. The relationship between c_1-players and c_0-players is similar to the coordination game, because players want to be matched with a same-strategy player. However, as we have seen in Sect. 10.1, in a random matching game with the one-shot coordination game as the stage game, the (average) expected fitness of a player is linear in the fraction of its share, and a mixed-strategy equilibrium has less fitness than at least one pure-strategy equilibrium. Moreover, the mixed-strategy Nash equilibrium, i.e., co-existence of different strategies, is not stable.

By contrast, in the VSRPD model, the Prisoner's Dilemma part also influences the stability. The c_0-strategy has the advantage that, once matched with another c_0-player, they can establish a long-term cooperative relationship, while the c_1-strategy has the advantage that it can exploit c_0-players and can also establish a long-term cooperative relationship with another c_1-player after one period. Thus, the eventual long-term cooperative relationships can deter other strategies. The positive fraction of c_0-players means that they receive the cooperative payoffs from the onset of new partnerships, which contributes to higher social efficiency than that of the symmetric equilibrium of the c_1-strategy that makes everyone get a low payoff in the first period of all new partnerships.

The endogenous length of the stage game with the same partner thus affects the stability of strategy combinations significantly.

10.7 Replicator Dynamics

One of the most important models of dynamic evolutionary processes is the *replicator dynamics* model. For simplicity, assume that Σ consists of probability distributions over a set of finite pure strategies $S = \{1, 2, \ldots, K\}$, i.e., it is the $(K-1)$-dimensional unit simplex. We consider a long-run dynamic process for the share distribution of

[7]Rigorously speaking, we must prove that any mutant strategy cannot thrive when the population consists of α^* c_0-players and $(1-\alpha^*)$ c_1-players. This is done in [5].

the pure strategies, corresponding to a point in Σ. Time is continuous and a point in time is denoted as t.

At a point in time t, let $p_i(t)$ be the number of players with a pure strategy $i \in S$ and the total size of the population be $p(t) = \sum_{i \in S} p_i(t)$. Assume that $p(t) > 0$ at any t. The population size can change over time, and thus a *state* of the dynamic process is the share distribution of the K pure strategies:

$$\mathbf{x}(t) = (x_1(t), x_2(t), \ldots, x_K(t)) := \left(\frac{p_1(t)}{p(t)}, \frac{p_2(t)}{p(t)}, \ldots, \frac{p_K(t)}{p(t)}\right) \in \Sigma.$$

The dynamic process of the social state $\mathbf{x}(t)$ is defined indirectly by the size dynamics of pure strategies. The time derivative of a function is written such as $\dot{p}_i(t) = \frac{dp_i}{dt}$ and $\dot{x}_i(t) = \frac{dx_i}{dt}$.

When the social state at time t is $\mathbf{x}(t)$, the rate of change $\dot{p}_i(t) = \frac{dp_i}{dt}$ of the number of players with a pure strategy $i \in S$ is assumed to be proportional to the current number $p_i(t)$, as well as dependent on the birth rate $\beta(\geqq 0)$, the death rate[8] $\gamma(\geqq 0)$, and the fitness of the strategy i in the social state $\mathbf{x}(t)$, i.e., $u(i, \mathbf{x}(t))$. To formulate,

$$\dot{p}_i(t) = [\beta - \gamma + u(i, \mathbf{x}(t))]p_i(t). \tag{10.5}$$

The change of the total population size is

$$\dot{p}(t) = [\beta - \gamma + u(\mathbf{x}(t), \mathbf{x}(t))]p(t), \tag{10.6}$$

where $u(\mathbf{x}(t), \mathbf{x}(t)) = \sum_{i \in S} x_i(t)u(i, \mathbf{x}(t))$ is the *population average fitness* across all pure strategies.

By the definition,

$$p(t)\, x_i(t) = p_i(t).$$

Differentiating both sides and rearranging, we have

$$p(t)\, \dot{x}_i(t) = \dot{p}_i(t) - \dot{p}(t)\, x_i(t).$$

Substituting Eqs. (10.5) and (10.6) into the RHS, we get (for notational simplicity we omit t in the following):

$$\begin{aligned} p\, \dot{x}_i &= [\beta - \gamma + u(i, \mathbf{x})]p_i - [\beta - \gamma + u(\mathbf{x}, \mathbf{x})]p\, x_i \\ &= [\beta - \gamma + u(i, \mathbf{x})]p_i - [\beta - \gamma + u(\mathbf{x}, \mathbf{x})]p_i \\ &= [u(i, \mathbf{x}) - u(\mathbf{x}, \mathbf{x})]p_i. \end{aligned}$$

[8] The exogenous factors of birth rates and death rates can be omitted without affecting the resulting dynamic equation.

Dividing both sides by p, we have the *replicator equation* as follows:

$$\dot{x_i} = [u(i, \mathbf{x}) - u(\mathbf{x}, \mathbf{x})]x_i. \quad (10.7)$$

The replicator equation means that, for any pure strategy i, its share increases if and only if its share is positive and its fitness against the social state $u(i, \mathbf{x})$ is greater than the population's average fitness $u(\mathbf{x}, \mathbf{x})$. Since the sign of the time-derivative depends on the social state $\mathbf{x}$ at the time, it is possible that a pure strategy which increases its share at a state $\mathbf{x}$ decreases at the resulting (later) state $\mathbf{y}$.

Proposition 10.7 If a social state $\mathbf{x}$ is a Nash equilibrium, then it is a stationary point of the replicator dynamic (i.e., for any $i \in S$, $\dot{x_i} = 0$).

The proof of this Proposition is left to the reader as Problem 10.9. We show a few other important properties of the replicator dynamic.

Proposition 10.8 The replicator dynamic is invariant under positive Affine transformations of the fitness function.

Proof If the fitness function changes from u to $a \cdot u + b$ (where $a > 0$), the replicator equation becomes $\dot{x_i} = a\,[u(i, \mathbf{x}) - u(\mathbf{x}, \mathbf{x})]x_i$. This equation yields the same solution path as the one for (10.7). □

Under the assumption of a finite set of pure strategies, we can use a matrix representation of the fitness function. When the share distribution[9] is $\mathbf{x} = (x_1, x_2, \ldots, x_K)^T$, the fitness of a pure strategy i can be written as

$$u(i, \mathbf{x}) = \mathbf{e}_i^T U \mathbf{x},$$

where $\mathbf{e}_i$ is the indicator column vector in which the i-th coordinate is 1 and all other coordinates are 0, and U is the $K \times K$ payoff matrix such that the ij element is $u(i, j)$. The replicator equation can be written as

$$\dot{x_i} = [\mathbf{e}_i^T U \mathbf{x} - \mathbf{x}^T U \mathbf{x}]x_i.$$

This clarifies that the replicator equation is a polynomial equation of x_i. By the Picard-Lindelöf Theorem, for a given initial state, there is a unique solution orbit. Since the initial state belongs to the unit simplex, the solution orbit also stays within the unit simplex. (Proofs of these statements can be found in many mathematics books and are omitted.) By these properties, the replicator equation guarantees us a meaningful dynamic process of strategy share distributions. Clearly, if the process starts from an edge (where only one pure strategy exists in the population), then it remains there. In particular, if there are only two pure strategies, only the processes starting from an interior point are of interest.

[9]Consider the vector $\mathbf{x}$ as a row vector and $\mathbf{x}^T$ as its transpose, i.e., a column vector.

Proposition 10.9 For generic normalized 2×2 symmetric games, the replicator process starting from any interior point other than a Nash equilibrium state converges to a state corresponding to an ESS.

Proof Rename the strategies in Table 10.3 so that s_0 is strategy 1 and s_1 is strategy 2. The payoff matrix of a generic normalized 2×2 symmetric game is thus

$$U = \begin{pmatrix} a_1 & 0 \\ 0 & a_2 \end{pmatrix},$$

and the replicator equation for each pure strategy is (using $x_2 = 1 - x_1$)

$$\begin{aligned} \dot{x_1} &= [a_1 x_1 - (a_1 x_1^2 + a_2 x_2^2)]x_1 \\ &= [a_1 x_1 (1 - x_1) - a_2 x_2^2]x_1 \\ &= [a_1 x_1 - a_2 x_2]x_1 x_2, \end{aligned}$$

and $\dot{x_2} = -\dot{x_1}$. Note that a state can be identified by the share of strategy 1.
Case 1: When $a_1 \times a_2 < 0$.
Let $a_i > 0 > a_j$. Then the pure strategy i is the strict Nash equilibrium strategy, and thus the unique ESS. Because $a_i \, x_i - a_j \, x_j > 0$, we have $\dot{x_i} > 0$ and $\dot{x_j} < 0$. This implies that the replicator process converges to the distribution consisting only of the pure strategy i.
Case 2: When $a_1 > 0$ and $a_2 > 0$. (Coordination game.)
The rate of increase of the pure strategy 1 changes at $a_1 x_1 = a_2 x_2$, or $x_1 = a_2/(a_1 + a_2)$. If $x_1(0) < \frac{a_2}{a_1+a_2}$, then $\dot{x_1} < 0$ and, if $x_1(0) > \frac{a_2}{a_1+a_2}$, then $\dot{x_1} > 0$ holds. If the process starts at $x_1(0) = \frac{a_2}{a_1+a_2}$, then it stays at the Nash equilibrium state. Thus, if the replicator process starts from an interior point which is not a Nash equilibrium state, then it converges to one of the two pure-strategy states, and they are both ESSs.
Case 3: When $a_1 < 0$ and $a_2 < 0$. (Hawk-Dove game.)
This is similar to Case 2, but the direction is the opposite. If $x_1(0) < \frac{a_2}{a_1+a_2}$, then $\dot{x_1} > 0$, while if $x_1(0) > \frac{a_2}{a_1+a_2}$, then $\dot{x_1} < 0$. Hence from any interior point, the process converges to the mixed-strategy Nash equilibrium state $\frac{a_2}{a_1+a_2}$, and this is the unique ESS. □

Recall that as soon as we have three pure strategies, an ESS may not exist, and hence the replicator dynamic process may not behave well, either. For example, consider the rock-paper-scissors game in Weibull [18] with the following payoff matrix:

$$U = \begin{pmatrix} 1 & 2+a & 0 \\ 0 & 1 & 2+a \\ 2+a & 0 & 1 \end{pmatrix}.$$

Then, the system of replicator equations is

$$\dot{x}_1 = [x_1 + (2+a)x_2 - \mathbf{x}U\mathbf{x}^T]x_1$$
$$\dot{x}_2 = [x_2 + (2+a)x_3 - \mathbf{x}U\mathbf{x}^T]x_2$$
$$\dot{x}_3 = [x_3 + (2+a)x_1 - \mathbf{x}U\mathbf{x}^T]x_3.$$

We can simplify the analysis by looking at the movement of the product of the shares $x_1x_2x_3$ (or, the logarithm of it). Let $f(\mathbf{x}) = \log(x_1x_2x_3) = \log x_1 + \log x_2 + \log x_3$, then

$$\begin{aligned} \dot{f}(\mathbf{x}) &= \dot{x}_1/x_1 + \dot{x}_2/x_2 + \dot{x}_3/x_3 \\ &= (x_1 + x_2 + x_3) + (2+a)(x_1 + x_2 + x_3) - 3\mathbf{x}U\mathbf{x}^T \\ &= 3 + a - 3\mathbf{x}U\mathbf{x}^T. \end{aligned}$$

We can further simplify $\mathbf{x}U\mathbf{x}^T$. Note that $||\mathbf{x}||^2 = x_1^2 + x_2^2 + x_3^2$ implies that

$$1 = (x_1 + x_2 + x_3)^2 = ||\mathbf{x}||^2 + 2(x_1x_2 + x_2x_3 + x_1x_3).$$

By arrangements,

$$\begin{aligned} \mathbf{x}U\mathbf{x}^T &= x_1^2 + (2+a)x_1x_2 + x_2^2 + (2+a)x_2x_3 + x_3^2 + (2+a)x_1x_3 \\ &= 1 + a(x_1x_2 + x_2x_3 + x_1x_3), \end{aligned}$$

so that

$$\mathbf{x}U\mathbf{x}^T = 1 + \frac{a}{2}(1 - ||\mathbf{x}||^2).$$

Therefore, we have

$$\dot{f}(\mathbf{x}) = \frac{a}{2}(3||\mathbf{x}||^2 - 1).$$

Since the vector $\mathbf{x}$ is in the unit simplex, its squared norm $||\mathbf{x}||^2$ takes the maximal value of 1 (when the state is at an edge) and the minimal value of 1/3 (when the state is the center). Thus, except when the state is $s^* = (\frac{1}{3}, \frac{1}{3}, \frac{1}{3})$, $3||\mathbf{x}||^2 - 1 > 0$ holds. This means that the sign of the time-derivative of $x_1x_2x_3$ corresponds to the sign of a.

When $a = 0$, we have $\dot{f}(\mathbf{x}) = 0$, so that the solution orbit is a closed cycle such that $x_1(t)x_2(t)x_3(t) = x_1(0)x_2(0)x_3(0)$ for all $t \geqq 0$. Clearly, for any neighborhood V of $s^* = (\frac{1}{3}, \frac{1}{3}, \frac{1}{3})$, we can make the replicator process stay within V by making the closed cycle inside V. This implies that for the payoff matrix U with $a = 0$, $s^* = (\frac{1}{3}, \frac{1}{3}, \frac{1}{3})$ is *Lyapunov stable*.[10] Intuitively, the center state s^* is stable in the sense that with a small perturbation from it, the dynamic process stays near it forever.

[10] A state x^* is *Lyapunov stable* if, for any neighborhood V of x^*, there exists a neighborhood W of x^* such that, starting from any $x(0) \in W$, $x(t) \in V$ for any $t \geqq 0$.

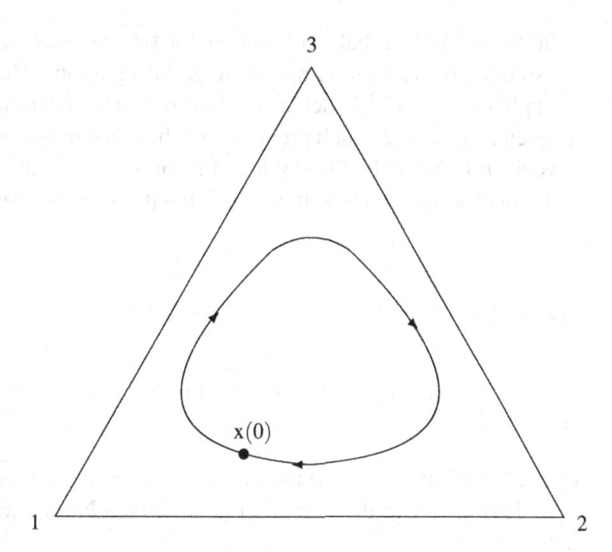

Fig. 10.6 Replicator dynamic process of R-P-S game with $a = 0$

When $a > 0$, the product of the shares keeps increasing, and thus the solution orbit moves inwards, towards the center s^*. In this case, with a small perturbation from s^*, the process comes back, which satisfies a stronger stability called the *asymptotic stability*.[11] By contrast, if $a < 0$, then the product of the shares decreases and the solution path moves outwards. Therefore, in this case s^* is unstable, which corresponds to the absence of a NSS. For this type of game, both the NSS concept and the replicator dynamic process imply that the center s^* is not stable.

To illustrate, we show a solution cycle for $a = 0$ in Fig. 10.6. Because $\mathbf{x}U\mathbf{x}^T = 1$ holds if, for example, the initial state has a relatively large share of strategy 1 such as $\mathbf{x}(0) = (0.6, 0.3, 0.1)$, then

$$\begin{aligned}\dot{x}_1 &= [x_1 + 2x_2 - 1]x_1 = 0.12\\ \dot{x}_2 &= [x_2 + 2x_3 - 1]x_2 = -0.15\\ \dot{x}_3 &= [x_3 + 2x_1 - 1]x_3 = 0.03\end{aligned}$$

so that strategy 2 decreases. Analogously, when strategy 3 has a relatively large share, then strategy 1 decreases. When strategy 2 has a large share, strategy 3 decreases. This makes the path cycle clock-wise.

An important property of the replicator dynamic process shown by Samuelson and Zang [14] is that for any strategy that is eliminated by an iterative elimination process of strictly dominated strategies, its share will converge to 0, starting from any interior state. Therefore, "irrational" strategies in terms of one-shot fitness will eventually disappear. However, strategies eliminated by an iterative elimination of weakly dominated strategies may not disappear.

[11] A state x^* is *asymptotically stable* if it is Lyapunov stable and there exists a neighborhood V such that for any $x(0) \in V$, $\lim_{t\to\infty} x(t) = x^*$.

In this way, the global movement of the strategy share distribution can be analyzed by a solution path for a system of differential equations. There are other dynamics than the replicator dynamic, such as the best response dynamic (see Sect. 9.3), and it has not been established which process is the best in what sense. Also, as Samuelson [13] surveys, unfortunately, the stable points of monotone dynamic processes (including the replicator dynamic) do not coincide with ESSs for general games.

Problems

10.1 Prove that the two definitions of ESS, Definition 10.1 and Definition 10.2, are equivalent.

10.2 Assume that $a_0 > 0$ and $a_1 > 0$ for the normalized symmetric 2×2 game in Table 10.3. Prove that the completely mixed Nash equilibrium strategy is not an ESS.

10.3 Consider a single population in which two players are randomly matched to play the following stage game. Each player is endowed with a role-dependent strategy, and the problability of becoming role 1 is 1/2.

role 1\role 2	L	D
L	0, 0	2, 1
D	1, 2	0, 0

(a) List all pure strategies.
(b) Find all ESSs including mixed strategy ones.

10.4 (Weibull [18]) Consider the Hawk-Dove Game in Table 10.9.

(a) What is the ESS of this game?
(b) Suppose that two mutants can emerge at the same time. Assume that the pure strategy Hawk and the pure strategy Dove emerge with the fraction $\epsilon/2$ each. That is, the post-entry distribution becomes $(1-\epsilon)x + \frac{\epsilon}{2} \cdot H + \frac{\epsilon}{2} \cdot D$, where x is the ESS in (a). Prove that x has lower post-entry fitness than that of at least one of the mutants.

Table 10.9 A Hawk-Dove game

	H	D
H	−1, −1	4, 0
D	0, 4	2, 2

10.5 In the Stag Hunt game with Cheap Talk (Sect. 10.3), prove that, for any message $m \in \{a, b\}$, the strategy mSS is a NSS.

10.6 For Selten's game in Fig. 10.3, prove that only $(R\ell, R)$ (among pure-strategy equilibrium strategies) is limit evolutionarily stable.

10.7 Consider the VSRPD game in Sect. 10.6. Prove that the d_0-strategy is not a NSS. (Hint: a positive fraction of c_1-players can invade the population of d_0-players.)

10.8 Consider the VSRPD game in Sect. 10.6. Prove that, when δ is sufficiently large, the long-run average fitness of the c_0-strategy

$$v(c_0, \alpha c_0 + (1-\alpha)c_1) = \frac{\alpha V(c_0, c_0) + (1-\alpha)V(c_0, c_1)}{\alpha \frac{1}{1-\delta^2} + (1-\alpha)}$$

is a concave function of α, and the average fitness of the c_1-strategy

$$v(c_1, \alpha c_0 + (1-\alpha)c_1) = \frac{\alpha V(c_1, c_0) + (1-\alpha)V(c_1, c_1)}{\alpha + (1-\alpha)\frac{1}{1-\delta^2}}$$

is a convex function of α. (Hint: Differentiate the functions and use $g > c > d > \ell$.)

10.9 Prove Proposition 10.7: If a social state $\mathbf{x}$ is a Nash equilibrium, then it is a stationary point of the replicator dynamic (i.e., for any $i \in S$, $\dot{x}_i = 0$).

10.10 Consider a symmetric 3×3 game, with the following payoff matrix for both players. Let the strategies be A, B, and C from the top. (This matrix is for the row player.)

$$U = \begin{pmatrix} 1 & 3 & 0 \\ 0 & 1 & 3 \\ 3 & 0 & 2 \end{pmatrix}.$$

(a) Find the (unique) symmetric Nash equilibrium.
(b) Prove that the Nash equilibrium strategy $\mathbf{x}^*$ is not an ESS.
(c) Prove (or explain as logically as possible) that the Nash equilibrium state $\mathbf{x}^*$ is asymptotically stable.

References

1. van Damme E (1987) Stability and perfection of Nash equilibria. Springer, New York
2. Duffie D, Sun Y (2012) The exact law of large numbers for independent random matching. J Econ Theory 147(3):1105–1139
3. Ellison G (1994) Cooperation in the Prisoner's Dilemma with anonymous random matching. Rev Econ Stud 61(3):567–588

4. Farrell J (1993) Meaning and credibility in cheap-talk games. Games Econ Behav 5(4):514–531
5. Fujiwara-Greve T, Okuno-Fujiwara M (2009) Voluntarily separable repeated Prisoner's Dilemma. Rev Econ Stud 76(3):993–1021
6. Gilboa I, Matsui A (1991) Social stability and equilibrium. Econometrica 59(3):859–867
7. Hamilton W (1967) Extraordinary sex ratios. Science 156(3774):477–488
8. Kandori M (1992) Social norms and community enforcement. Rev Econ Stud 59(1):63–80
9. Kohlberg E, Mertens J-F (1986) On the strategic stability of equilibria. Econometrica 54(5):1003–1037
10. Maynard-Smith J (1982) Evolution and the theory of games. Cambridge University Press, Cambridge
11. Maynard-Smith J, Price G (1973) The logic of animal conflict. Nature 246(2):15–18
12. Robson A (1990) Efficiency in evolutionary games: Darwin, Nash and the secret handshake. J Theor Biol 144(3):379–396
13. Samuelson L (1997) Evolutionary games and equilibrium selection. MIT Press, Cambridge
14. Samuelson L, Zang J (1992) Evolutionary stability in asymmetric games. J Econ Theory 57(2):363–391
15. Selten R (1983) Evolutionary stability in extensive two-person games. Math Soc Sci 5(3): 269–363
16. Swinkels J (1992) Evolutionary stability with equilibrium entrants. J Econ Theory 57(2): 306–332
17. Vickers G, Cannings C (1987) On the definition of an evolutionarily stable strategy. J Theor Biol 129(3):349–353
18. Weibull J (1995) Evolutionary game theory. MIT Press, Cambridge

Appendix
Mathematical Appendix

A.1 Basics of Topology on $\Re^n$

In this book we only use elementary finite-dimensional Euclidean space (real numbers) topology, and therefore the following supplementary materials are concentrated on this area of mathematics.

A.1.1 Definitions

Let $\Re$ be the set of real numbers, $\Re_+$ be the set of non-negative real numbers, and $\Re_{++}$ be the set of positive real numbers. For a finite natural number n, the n-dimensional Euclidean space is the n-times Cartesian product of $\Re$, written as $\Re^n$.

Definition A.1 For any two vectors $\mathbf{x} = (x_1, x_2, \ldots, x_n)$ and $\mathbf{y} = (y_1, y_2, \ldots, y_n)$ in $\Re^n$, *the distance* between them is written as $|\mathbf{x} - \mathbf{y}|$ and defined by

$$|\mathbf{x} - \mathbf{y}| = \sqrt{\sum_{i=1}^{n}(x_i - y_i)^2}.$$

To be precise, we can use any function d that maps two vectors into a real number which satisfies the metric axioms.[1] We use the above definition because it is the most intuitive one.

---

[1]Namely, for any $\mathbf{x}$, $\mathbf{y}$, and $\mathbf{z} \in \Re^n$, (1) $d(\mathbf{x}, \mathbf{y}) \geqq 0$, and the equality holds if and only if $\mathbf{x} = \mathbf{y}$, (2) $d(\mathbf{x}, \mathbf{y}) = d(\mathbf{y}, \mathbf{x})$, and (3) $d(\mathbf{x}, \mathbf{y}) + d(\mathbf{y}, \mathbf{z}) \geqq d(\mathbf{x}, \mathbf{z})$.

T. Fujiwara-Greve, *Non-Cooperative Game Theory*, Monographs in Mathematical Economics 1, DOI 10.1007/978-4-431-55645-9

Definition A.2 An *open ball* with the center $\mathbf{x} \in \Re^n$ and the radius $r \in \Re_{++}$ is written as $B_r(\mathbf{x})$ and is defined by

$$B_r(\mathbf{x}) = \{\mathbf{y} \in \Re^n \mid |\mathbf{x} - \mathbf{y}| < r\}.$$

Definition A.3 A set $X \subset \Re^n$ is *open* if, for any $\mathbf{x} \in X$, there exists $r \in \Re_{++}$ such that $B_r(\mathbf{x}) \subset X$.

Definition A.4 A set $F \subset \Re^n$ is *closed* if its complement $F^c = \{\mathbf{x} \in \Re^n \mid \mathbf{x} \notin F\}$ is open.

Definition A.5 A set $X \subset \Re^n$ is *bounded* if there exists $r \in \Re_{++}$ such that $X \subset B_r(\mathbf{0})$, where $\mathbf{0}$ is the origin.

Definition A.6 A set $X \subset \Re^n$ is *compact* if it is closed and bounded.

Definition A.7 A *sequence* in a set $X \subset \Re^n$ is a mapping from the set $\mathbf{N}$ of natural numbers to X.

We often write a sequence as $\mathbf{x}_1, \mathbf{x}_2, \ldots$ or $\{\mathbf{x}_k\}_{k=1}^{\infty}$. A sequence in the set X means that $x_k \in X$ holds for any $k = 1, 2, \ldots$.

When $\pi : \mathbf{N} \to X$ generates a sequence $\{\mathbf{x}_k\}_{k=1}^{\infty}$ in X and $\psi : \mathbf{N} \to \mathbf{N}$ is an increasing function from the set of natural numbers into itself, a sequence generated by $\pi \circ \psi : \mathbf{N} \to X$ is called a *subsequence* of $\{\mathbf{x}_k\}_{k=1}^{\infty}$ (because it picks up some of the vectors in the original sequence without going back the order).

Definition A.8 For any set $X \subset \Re^n$, a sequence $\{\mathbf{x}_k\}_{k=1}^{\infty}$ in X is said to *converge to* a vector $\mathbf{x} \in \Re^n$ if, for any open set $U \subset \Re^n$ containing $\mathbf{x}$, there exists $k_0 \in \mathbf{N}$ such that

$$\mathbf{x}_k \in U, \quad \forall k \geqq k_0.$$

This $\mathbf{x}$ is called the *limit* of the sequence.

Definition A.9 For any $X \subset \Re^n$ and $Y \subset \Re^m$, a function $f : X \to Y$ is *continuous at* $\mathbf{x} \in X$ if, for any sequence $\{\mathbf{x}_k\}_{k=1}^{\infty}$ in X which converges to $\mathbf{x}$, the sequence $\{f(\mathbf{x}_k)\}_{k=1}^{\infty}$ in Y converges to $f(\mathbf{x})$.

Alternatively, we can write that, for any sequence $\{\mathbf{x}_k\}_{k=1}^{\infty}$ in X and any $\epsilon > 0$, there exists $\delta > 0$ such that

$$|\mathbf{x}_k - \mathbf{x}| < \delta \Rightarrow |f(\mathbf{x}_k) - f(\mathbf{x})| < \epsilon.$$

Definition A.10 For any $X \subset \Re^n$ and $Y \subset \Re^m$, a function $f : X \to Y$ is *continuous on* X if it is continuous at any $\mathbf{x} \in X$.

Definition A.11 A set $X \subset \Re^n$ is *convex* if, for any $\mathbf{x}, \mathbf{y} \in X$ and any $\alpha \in [0, 1]$,

$$\alpha\mathbf{x} + (1 - \alpha)\mathbf{y} \in X.$$

Next, we give an important property of correspondences, or set-valued mappings. A *correspondence* F from X into Y, written as $F : X \to\to Y$, is a mapping such that for any $\mathbf{x} \in X$, $F(\mathbf{x}) \subset Y$.

Definition A.12 For any $X \subset \Re^n$ and $Y \subset \Re^m$, a correspondence $F : X \to\to Y$ is *upper hemi-continuous* at $\mathbf{x}_0 \in X$ if, for any open set $V \subset \Re^m$ which contains $F(\mathbf{x}_0)$, there exists an open set $U \subset \Re^n$ which contains $\mathbf{x}_0$ such that

$$F(x) \subset V, \quad \forall x \in U.$$

A correspondence $F : X \to\to Y$ is *upper hemi-continuous* if it is upper hemi-continuous at any $\mathbf{x}_0 \in X$.

Definition A.13 For any $X \subset \Re^n$ and $Y \subset \Re^m$, a correspondence $F : X \to\to Y$ is *non-empty-valued* (resp. *convex-valued, closed-valued*) if $F(\mathbf{x})$ is non-empty (resp. convex, closed) for any $\mathbf{x} \in X$.

Definition A.14 For any $X \subset \Re^n$ and $Y \subset \Re^m$, the *graph* of a correspondence $F : X \to\to Y$ is the set

$$gr(F) = \{(\mathbf{x}, \mathbf{y}) \in X \times Y \mid \mathbf{y} \in F(\mathbf{x})\}.$$

A.1.2 Useful Propositions and Kakutani's Fixed Point Theorem

Perhaps with the exception of Kakutani's Fixed Point Theorem, the proofs of the following famous and useful results are found in many books on topology.

Proposition A.1 *A set $F \subset \Re^n$ is closed if and only if, for any sequence $\{\mathbf{x}_k\}_{k=1}^{\infty}$ in F which converges to some limit $\mathbf{x} \in \Re^n$, the limit belongs to F, i.e., $\mathbf{x} \in F$.*

Proposition A.2 *A set $X \subset \Re^n$ is compact if and only if any sequence $\{\mathbf{x}_k\}_{k=1}^{\infty}$ in X has a convergent subsequence.*

Proposition A.3 *For any $X \subset \Re^n$, let $f : X \to \Re$ and $g : X \to \Re$ be continuous on X. Then the following functions are continuous on X:*

(a) $|f(x)|$,
(b) $af(x)$ *where* $a \in \Re$,
(c) $f(x) + g(x)$,
(d) $f(x)g(x)$,
(e) $1/f(x)$ *where* $f(x) \neq 0$,
(f) $\max\{f(x), g(x)\}$,
(g) $\min\{f(x), g(x)\}$.

Proposition A.4 (Bolzano-Weierstrass' Theorem) *Let $X \subset \Re^n$ be compact and $f : X \to \Re$ be continuous on X. Then f has a maximum and a minimum in X.*

Proposition A.5 *For any convex sets $X, Y \subset \Re^n$, $X \times Y$ is convex.*

Proposition A.6 *For any set $X \subset \Re^n$ and any compact set $Y \subset \Re^m$, a compact-valued correspondence $F : X \to\to Y$ is upper hemi-continuous if and only if its graph $gr(F)$ is closed.*

Theorem A.1 (Kakutani's Fixed Point Theorem) *Let $X \subset \Re^n$ be a non-empty, convex, and compact set. If a correspondence $F : X \to\to X$ is non-empty-valued, convex-valued, closed-valued, and upper hemi-continuous, then there exists $\mathbf{x}^* \in X$ such that*

$$\mathbf{x}^* \in F(\mathbf{x}^*).$$

A.2 Dynamic Programming

We briefly explain the necessary materials for dynamic programming to understand our analysis in this book. The original book by Bellman [1] is now cheaply available. Another classic is Ross [5]. For the fundamental result that a stationary discounted dynamic programming over the infinite horizon has a stationary solution, see Blackwell [2].

As a preparation, consider a simple two-period optimization problem of a decision maker. In each period, the decision maker receives some *reward* (a real number), which depends on her action choice as well as a "state", and the decision maker wants to maximize the sum of the rewards over the two periods. Let the set of feasible actions be A, and the set of states be S.

In the first period, a state $s_1 \in S$ is chosen exogenously. With or without the knowledge of the state, the decision maker chooses an action $a_1 \in A$. (The informational assumption depends on the actual problem at hand.) The reward is determined by a *reward function* $r(a_1, s_1)$. In the second period, a new state is generated by a transition function $s_2(a_1, s_1)$, depending on the history of the first period action and the state. The decision maker chooses an action $a_2 \in A$. Then a new reward $r(a_2, s_2)$ is given. (Notice that the sets of states and feasible actions are the same over time, and so is the reward function. This is a stationary problem. In general, the sets of states and actions may depend on histories and time periods. The transition of states can be stochastic as well.)

A special case of the above problem occurs when a player faces a twice-repeated game with perfect monitoring, and her opponents play pure strategies. For simplicity, consider a two-player stage game $G = (\{1, 2\}, A_1, A_2, u_1, u_2)$ with player i as the decision maker. The set of feasible actions is A_i for each period. The reward function is the stage game payoff function $u_i : A_i \times A_j \to \Re$ for each period. Let $s_j = (s_{j1}, s_{j2})$ $(j \neq i)$ be a strategy of player j, where $s_{j1} \in A_j$ is the first period action (based on no history) and $s_{j2} : A_i \times A_j \to A_j$ is the action plan for the second period by player j.

Assuming that A_i is finite, the optimization problem for player i is

$$\max_{a_{i1},\, a_{i2} \in A_i} u_i(a_{i1}, s_{j1}) + u_i\Big(a_{i2}, s_{j2}(a_{i1}, s_{j1})\Big).$$

This value is called the (optimal) value function, over the entire horizon of the problem, and can be written as $f(s_{j1})$ to show the dependence on the first period state. It is easy to prove that the above maximization can be decomposed as

$$\max_{a_{i1},\, a_{i2} \in A_i} u_i(a_{i1}, s_{j1}) + u_i\Big(a_{i2}, s_{j2}(a_{i1}, s_{j1})\Big)$$
$$= \max_{a_{i1} \in A_i} \Big\{u_i(a_{i1}, s_{j1}) + \max_{a_{i2} \in A_i} u_i\Big(a_{i2}, s_{j2}(a_{i1}, s_{j1})\Big)\Big\}.$$

In terms of the value function, the decomposition can be written as

$$f(s_{j1}) = \max_{a_{i1} \in A_i} \Big\{u_i(a_{i1}, s_{j1}) + f_2\Big(s_{j2}(a_{i1}, s_{j1})\Big)\Big\}, \tag{1}$$

with the value function from the second period on being

$$f_2\Big(s_{j2}(a_{i1}, s_{j1})\Big) = \max_{a_{i2} \in A_i} u_i\Big(a_{i2}, s_{j2}(a_{i1}, s_{j1})\Big).$$

The decomposition (1) of the value function is the idea behind backward induction and also the intuition for the infinite horizon dynamic programming. First, the decision maker solves the last period optimization, for each possible history (a_{i1}, s_{j1}). Then go back to the first period problem, taking $f_2\Big(s_{j2}(a_{i1}, s_{j1})\Big)$ as given.

A.2.1 Infinite Horizon Stationary Dynamic Programming with Discounting

Let us turn to an infinite horizon optimization problem. As before, let S be the set of states and A be the set of feasible actions for the decision maker, which are fixed over time. In the first period, an initial state $s_1 \in S$ is given, the decision maker chooses an action $a_1 \in A$, and a reward $u(a_1, s_1) \in \Re$ is obtained.

In periods $t = 2, 3, \ldots$, a state $s_t \in S$ is generated by a transition function $s_t : \{A \times S\}^{t-1} \to S$, and the decision maker chooses an action $a_t \in A$, which can depend on the history $h_{t-1} \in \{A \times S\}^{t-1}$. Let $\mathbf{a} = (a_1, a_2, \ldots)$ be a *strategy* of the decision maker such that $a_1 \in A$ is the first period action and $a_t : \{A \times S\}^{t-1} \to A$ is an action plan for period $t = 2, 3, \ldots$. Given a sequence of transition functions $\mathbf{s} = (s_1, s_2, \ldots)$ and a strategy $\mathbf{a}$, a history of actions and states is generated over the infinite horizon. The standard[2] discounted dynamic programming problem is to maximize the discounted sum of the rewards over the infinite horizon, given an initial state s_1. When the decision maker uses a strategy $\mathbf{a}$, the discounted sum of the rewards is written as[3]

$$V(\mathbf{a}; s_1) := \sum_{t=1}^{\infty} \delta^{t-1} \cdot u(a_t(h_t), s_t(h_t)),$$

where h_1 is degenerate and $h_t = (a_1, s_1, \ldots a_{t-1}(h_{t-1}), s_{t-1}(h_{t-1}))$ for $t \geq 2$. The (optimal) *value function* over the infinite horizon is defined[4] by

$$f(s_1) := \sup_{\mathbf{a}} V(\mathbf{a}; s_1).$$

A.2.2 Bellman Equation

Proposition A.7 *The value function satisfies the following equation, called the Bellman Equation:*

$$f(s_1) = \max_{a_1 \in A} \Big[u(a_1, s_1) + \delta f(s_2(a_1, s_1)) \Big].$$

Proof We first prove that the LHS is not more than the RHS.

For any strategy $\mathbf{a}$, the value function satisfies the following equation by the definition:

$$V(\mathbf{a}; s_1) = u(a_1, s_1) + \delta V(\mathbf{a}(a_1, s_1); s_2(a_1, s_1)),$$

where $\mathbf{a}(a_1, s_1) = (a_2(a_1, s_1), a_3, a_4, \ldots)$ is the *continuation strategy* (still over the infinite horizon) of the original strategy $\mathbf{a}$, specifying actions as if the initial state is $s_2(a_1, s_1)$. (Hence the "initial action" is $a_2(a_1, s_1) \in A$, and the action plans afterwards are the same as those of $\mathbf{a}$.) Clearly,

$$V(\mathbf{a}(a_1, s_1); s_2(a_1, s_1)) \leqq \sup_{\mathbf{x}} V(\mathbf{x}; s_2(a_1, s_1)) = f(s_2(a_1, s_1)).$$

---

[2]There are generalizations of the problem to allow other forms of evaluation for infinite sequences of rewards.

[3]For notational simplicity and following the tradition in this field, we suppress the dependence on the sequence of transition functions below.

[4]Even if the set of feasible actions is finite, there are infinitely many strategies, and thus the maximum may not exist, but the supremum exists. For a set $X \subset \Re$, its supremum (the least upper bound) $\sup X$ is the smallest y such that $y \geqq x$ for any $x \in X$.

Therefore,

$$V(\mathbf{a}; s_1) \leqq u(a_1, s_1) + \delta f(s_2(a_1, s_1))$$
$$\leqq \max_{a_1 \in A} \big[u(a_1, s_1) + \delta f(s_2(a_1, s_1))\big].$$

Since $\mathbf{a}$ was arbitrary,

$$f(s_1) = \sup_{\mathbf{a}} V(\mathbf{a}; s_1) \leqq \max_{a_1 \in A} \big[u(a_1, s_1) + \delta f(s_2(a_1, s_1))\big].$$

Next, we prove that the LHS is not less than the RHS. Let $a_1^* \in A$ be an action defined as follows:

$$u(a_1^*, s_1) + \delta f(s_2(a_1^*, s_1)) = \max_{a_1 \in A} \big[u(a_1, s_1) + \delta f(s_2(a_1, s_1))\big].$$

That is, a_1^* is an action that maximizes the sum of the first period reward and the (discounted) optimal value starting at the state $s_2(a_1, s_1)$. (The latter means that the choice of a_1^* only controls the "initial state" from the second period on.) By the definition of the supremum, for any s and any $\epsilon > 0$, there exists a strategy $\mathbf{a}'(s)$ such that $V(\mathbf{a}'(s); s) \geqq f(s) - \epsilon$.

Consider a strategy $\mathbf{a}$ which plays a_1^* in the first period and uses $\mathbf{a}'(s_2(a_1^*, s_1))$ from the second period on. Then

$$V(\mathbf{a}; s_1) = u(a_1^*, s_1) + \delta V\big(\mathbf{a}'(s_2(a_1^*, s_1)); s_2(a_1^*, s_1)\big),$$

so that

$$f(s_1) \geqq V(\mathbf{a}; s_1) \geqq u(a_1^*, s_1) + \delta f(s_2(a_1^*, s_1)) - \delta\epsilon.$$

By the definition of a_1^*, we have

$$f(s_1) \geqq \max_{a_1 \in A} \big[u(a_1, s_1) + \delta f(s_2(a_1, s_1))\big] - \delta\epsilon.$$

Taking $\epsilon \to 0$, the weak inequality continues to hold so that

$$f(s_1) \geqq \max_{a_1 \in A} \big[u(a_1, s_1) + \delta f(s_2(a_1, s_1))\big].$$

This completes the proof of the proposition. □

A.2.3 Unimprovable Strategies

Definition A.15 A strategy **a** is called *unimprovable in one step* if, for any initial state and any history until the tth period for each $t = 1, 2, \ldots$, no one-step deviation strategy, which chooses a different action in period t from **a** but follows **a** from the $t+1$th period on, gives a greater total discounted payoff than that of **a**.

By the Bellman equation, an optimal strategy is unimprovable in one step. (Interpret s_1 as an arbitrary history until period t.) In this subsection we show the other direction, that an unimprovable strategy in one step is an optimal strategy. (For a game theoretic exposition of the same result, see Fudenberg and Tirole [3], Theorem 4.2, and Mailath and Samuelson [4], Proposition 2.2.1.)

Lemma A.1 *If a strategy* **a** *is unimprovable in one step, then it is unimprovable in any finite steps.*

Proof In the first period, **a** is unimprovable in one step so that, for any s_1,

$$V(\mathbf{a}; s_1) = \max_{x_1 \in A}\Big[u(x_1, s_1) + \delta V(\mathbf{a}(x_1, s_1); s_2(x_1, s_1))\Big],$$

where $\mathbf{a}(x_1, s_1)$ is the continuation strategy of **a** after the history (a_1, s_1).

After the history $h_1 = (x_1, s_1)$, again **a** is unimprovable in one step so that

$$V(\mathbf{a}(h_1); s_2(h_1)) = \max_{x_2 \in A}\Big[\, u(x_2, s_2(h_1)) + \delta V\big(\mathbf{a}(x_2, s_2(h_1)); s_3(x_2, s_2(h_1))\big)\Big],$$

where $\mathbf{a}(x_2, s_2(h_1))$ is the continuation strategy of **a** after the history $(x_2, s_2(x_1, s_1))$. Combining the above equalities, we have

$$\begin{aligned} V(\mathbf{a}; s_1) &= \max_{x_1 \in A}\Big[u(x_1, s_1) + \delta \cdot \max_{x_2 \in A}\big\{u(x_2, s_2(x_1, s_1)) \\ &\quad + \delta V\big(\mathbf{a}(x_2, s_2(x_1, s_1)); s_3(x_2, s_2(x_1, s_1))\big)\big\}\Big] \\ &= \max_{x_1,\, x_2 \in A}\Big[u(x_1, s_1) + \delta u(x_2, s_2(x_1, s_1)) \\ &\quad + \delta V\big(\mathbf{a}(x_2, s_2(x_1, s_1)); s_3(x_2, s_2(x_1, s_1))\big)\Big]. \end{aligned}$$

That is, **a** is unimprovable in two steps. Repeating this argument for finitely many times, we have the conclusion. □

Proposition A.8 *Fix an initial state* s_1. *If a strategy* **a** *is unimprovable in one step, then* **a** *is an optimal strategy that attains* $f(s_1)$.

Proof Suppose that although **a** is unimprovable in one step, it is not optimal. Then there exist another strategy **x** and $\epsilon > 0$ such that

$$V(\mathbf{a}; s_1) + 2\epsilon \leqq V(\mathbf{x}; s_1).$$

By the discounting and given the $\epsilon > 0$ above, there exists a positive integer (a sufficiently distant future period) T such that, for any strategy **y** such that **y** and **x** coincide with the first T periods, their values are very close:

$$V(\mathbf{x}, s_1) - \epsilon \leqq V(\mathbf{y}, s_1).$$

In particular, we can take **y** such that it follows **a** after the Tth period. Combining the inequalities, we have

$$V(\mathbf{a}; s_1) + \epsilon \leqq V(\mathbf{y}, s_1).$$

However, **y** specified now is a strategy that differs from **a** in finite steps, and the last inequality contradicts the fact that **a** is unimprovable in finite steps by Lemma A.1. □

We can also generalize the above analysis for stochastic dynamic programming, where the states evolve stochastically based on the history. This problem corresponds to the case when opponents play mixed strategies in an infinitely repeated game.

References

1. Bellman R (2003) Dynamic programming. Reprint. Dover Publications, Mineola
2. Blackwell D (1965) Discounted dynamic programming. Ann Math Stat 26:226–235
3. Fudenberg D, Tirole J (1991) Game theory. MIT Press, Cambridge
4. Mailath G, Samuelson L (2006) Repeated games and reputations: long-run relationships. Oxford University Press, New York
5. Ross S (1983) Introduction to stochastic dynamic programming. Academic Press, San Diego

Index

T. Fujiwara-Greve, *Non-Cooperative Game Theory*, Monographs in Mathematical Economics 1, DOI 10.1007/978-4-431-55645-9

Printed in the United States
By Bookmasters